D1356116

14 Springer Series in Chemical Physics
Edited by Fritz Peter Schäfer

Springer Series in Chemical Physics

Editors: V. I. Goldanskii   R. Gomer   F. P. Schäfer   J. P. Toennies

Picosecond Phenomena II

Proceedings of the Second International Conference
on Picosecond Phenomena
Cape Cod, Massachusetts, USA
June 18–20, 1980

Editors
R. Hochstrasser   W. Kaiser   C.V. Shank

With 252 Figures

Springer-Verlag Berlin Heidelberg New York 1980

Series Editors

Professor Dr. Fritz Peter Schäfer

Professor Vitalii I. Goldanskii

Institute of Chemical Physics
Academy of Sciences
Vorobyevskoye Chaussee 2-b
Moscow V-334, USSR

Max-Planck-Institut für
Biophysikalische Chemie
D-3400 Göttingen-Nikolausberg
Fed. Rep. of Germany

Professor Robert Gomer

The James Franck Institute
The University of Chicago
5640 Ellis Avenue
Chicago, IL 60637, USA

Professor Dr. J. Peter Toennies

Max-Planck-Institut für Strömungsforschung
Böttingerstraße 6–8
D-3400 Göttingen
Fed. Rep. of Germany

Conference Chairman and Editor
Dr. **Charles V. Shank,** Bell Laboratories, Holmdel, NJ 07733, USA

Program Co-Chairman and Editors
Professor **Robin Hochstrasser,** University of Pennsylvania, Philadelphia, PA 19104, USA
Professor Dr. **Wolfgang Kaiser,** Technische Universität München,
D-8000 München, Fed. Rep. of Germany

Program Committee
A.J. Alcock   R.R. Alfano   D.J. Bradley   J. Ducuing   K.B. Eisenthal   S.E. Harris
E.P. Ippen   G.W. Robinson   S.L. Shapiro   S. Shionoya   D.A. Wiersma

Sponsored by
The Optical Society of America

Supported by Grants from
National Science Foundation   U.S. Army Research Office
Office of Naval Research   Air Force Office of Scientific Research

Industrial Support   Lasermetrics Inc.
Hamamatsu Corporation   Spectra Physics

ISBN 3-540-10403-8 Springer-Verlag Berlin Heidelberg New York
ISBN 0-387-10403-8 Springer-Verlag New York Heidelberg Berlin

Offset printing: Beltz Offsetdruck, Hemsbach/Bergstr. Bookbinding: J. Schäffer oHG, Grünstadt.
2153/3130-543210

Preface

The second international conference on the subject of Picosecond Phenomena was held June 18-20, 1980, in Cape Cod, Massachusetts. Scientists from a broad range of disciplines were brought together to discuss their common interest in ultrafast processes. This meeting was organized as a Topical Meeting of the Optical Society of America and was attended by 250 participants.

The conference reviewed the latest advances in the experimental and theoretical understanding of phenomena that occur on a picosecond timescale. New discoveries in electronics, chemical dynamics, solid state physics, and picosecond optics highlighted the interactions between chemists, physicists, biologists,and engineers who attended the conference. The enthusiasm generated by the rapid progress in the last two years and the pleasant Cape Cod weather resulted in a successful and enjoyable conference.

The conference owes a special thanks to Dr. Jarus Quinn, Joan Connon, and their colleagues at the Optical Society of America for doing a superb job in implementing the meeting arrangements and to the program committee for the selection and organization of the technical presentations. We gratefully acknowledge the financial support from the National Science Foundation, Office of Naval Research, the U.S. Army Research Office, and the AFOSR.

Philadelphia, Pennsylvania   *R. Hochstrasser*

Munich, Fed. Rep. of Germany   *W. Kaiser*

Holmdel, New Jersey   *C.V. Shank*
August 1980

Contents

Part II. *Advances in Optoelectronics*

Part III. *Picosecond Studies of Molecular Motion*

Part IV. *Picosecond Relaxation Phenomena*

Part I

**Advances in the Generation
of Picosecond Pulses**

Repetitive Mode-Locking in Thermally Compensated Phosphate Glasses at 5 Hz and Above

T. R. Royt
Naval Research Laboratory
Washington, D.C. 20375

The passively mode-locked Neodymium (Nd)-glass laser is an important source of high energy pulses for studies requiring temporal resolution below the 25 ps obtainable with Nd-YAG. On the other hand, the thermal conductivity of glass is over an order of magnitude smaller than that of YAG, and previous studies involving glass lasers have encountered thermal distortions and damage as repetition rates were increased beyond 0.2 Hz (1 to 10 pulse trains per minute) [1,2]. In addition, the low cross-section for stimulated emission and, also, the low dye concentrations required to avoid self-focusing and rod damage have caused irreproducible mode-locking with "misfirings" and multiple pulsing. As a result many experimentalists have turned from glass, its shorter pulses notwithstanding, and have elected to use YAG with its higher data rate, in order to reduce set up and alignment time, increase the size of statistical samples obtainable within realistic time limits, and improve detectivity through better signal-to-noise discrimination.

In this paper, a new Nd-glass laser is reported which is as reliable as Nd-YAG, produces pulses at 5Hz or better, and retains pulse widths that are an order of magnitude shorter than those of Nd-YAG. An intracavity Q-spoiler/dumper is triggered by the laser output so as to terminate oscillation above a predetermined intensity threshold. This allows the dye concentration to be increased by large amounts, thereby increasing the mode-locking probability and reducing the multiple pulsing. The tendency towards indiscriminant Q-switching is avoided by reducing the beam diameter in the rod with respect to the dye with an intracavity telescope. A Nd doped phosphate glass rod that is thermally compensated yields mode-locking at a high repetition rate without thermal distortions.

Experimental data have been taken for seven different varieties of Nd-doped phosphate and silicate glass laser rods (Hoya LHG-8, Owens-Illinois ED-2, EV-2, EVF-1, EV-4, Kiger Q-88, Q-100) having various stimulated emission cross-sections, thermal properties, and nonlinear indices. It has been found that 4 ps duration pulses with a TEM_{oo} transverse profile are generated at a rate at least as high as 5 Hz, with no thermal dependence, by using a new variety of phosphate glass (LHG-8, EV-4, or Q-100) which has a low thermal-optic coefficient. Pulses in the leading edge of the train are time-bandwidth limited. Spectral broadening, characteristic of self-phase modulation, was found to be present in the subsequent pulses of the train for all of the glasses, even for an n_2 as low as 0.69×10^{-13} esu (EVF-1), the lowest value currently reported for any laser glass. This is in contrast to a recent report that phosphate glass (LHG-5) could be used to generate unself-phasemodulated pulses throughout the train due to a low n_2 (1.13×10^{-13} esu) and a low dn/dT ($< 0.1 \times 10^{-6}/°C$) [3]. The LHG-5 glass is similar to the Q-88 used here. However, time resolved spectrograms show that under controlled cavity conditions and at high dye concentrations self-focusing destroys the cavity oscillations before self-phasemodulation has become strong. As a result, the pulses throughout most of the train have no self-phasemodulation, not because of a low n_2 or dn/dT, but because the n_2 effectively introduces a strong lens into the cavity that increases the cavity diffraction losses and reduces the net gain below threshold prematurely. Without Q-spoiling, pulse train envelopes at low dye concentration are regular and symmetric, while aberrations or self-termination occur only for high dye concentrations where self-phasemodulation and self-focusing strongly affect the losses and cavity stability. Streak camera data on the duration of

3

individual pulses are presented. The inter-train reproducibility is statistically determined, with pulse widths varying from 3 to 8 ps and mean deviations in peak pulse energy of 15%. The thresholded cavity Q-spoiler/dump, which uses a photodetector or sparkgap to trigger a $\lambda/4$ voltage to an intracavity Pockel's cell, precludes the unwanted higher intensity portion of the train that could cause damage and also rotates the polarization of the pulse by a total of $\lambda/2$ so that it is switched out of the cavity by a dielectric polarizer as shown in Fig. 1. As a result, both a pulse train and a single pulse are produced. (This may be useful for experiments that require pumping of a synchronously mode-locked dye laser [4].) A reproducible, high repetition rate source of high energy 4 ps pulses has thereby been achieved yielding trains and pulses like those shown in Fig. 2. Furthermore, absence of a thermal effect during repetitive operation of the laser at 5 Hz, suggests that one may be able to scale the laser to much higher rates and use amplifier rods with increased diameter for a given repetition rate.

CAVITY DUMPED MODE-LOCKED OSCILLATOR
a) Q-SPOILED/DUMPED CAVITY
b) HIGH DYE CONCENTRATION
c) TELESCOPE

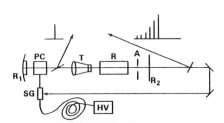

Fig. 1—Schematic diagram of high rep-rate cavity dumped oscillator. Pockel's cell (PC), sparkgap (SG), high voltage supply (HV), telescope (T), and thermally compensated rod (R) are shown. The single pulse is split out by a dielectric polarization selective mirror.

Fig. 2—Dual beam scope trace of train and single pulse output from the cavity dumped oscillator.

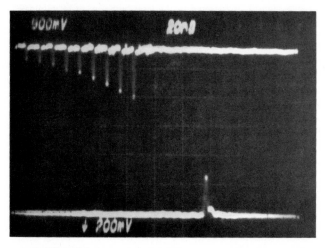

Stability of cavity gain and loss is essential for the proper operation of a glass laser. Mode-locking requires discrimination of a single fluctuation from the spontaneous emission contained in a random sample of duration 2L/c. While a finite sample will always have a ratio $I_1/I_2 > 1$ between the intensity of such a fluctuation and that of a competing fluctuation, the ratio is smaller for glass than for YAG, even after spectral narrowing, due to the larger bandwidth of glass. Discrimination of I_1 from I_2 (reduced multiple pulsing) is improved for a smaller net gain coefficient $g = a - l$, at a given dye concentration [5] (where a and $l = l_m + l_d$ are gain and loss coefficients for a cavity round trip, mirrors, and dye and where $a = l$ defines the free running threshold) according to (1).

4

$$\left.\frac{I_1}{I_2}\right|_{\text{final}} = \left.\frac{I_1}{I_2}\right|_{\text{initial}}^{\frac{l_d}{g}}. \tag{1}$$

For 10^3 fluctuations after spectral narrowing, there is a probability of 0.5 that I_1 and I_2 are related by (2). [6]

$$\left.\frac{I_1}{I_2}\right|_{\text{initial}} \geqslant 1.1 \tag{2}$$

In order for multiple pulsing to be minimized to the extent given by (3),

$$\left.I_2\right|_{\text{final}} < 0.1 \left.I_1\right|_{\text{final}} \tag{3}$$

for an unsaturated dye loss coefficient of $l_d = 0.25$, it is necessary that $g < 0.01$. Clearly, from a repetition rate standpoint, variations in g, caused by cavity changes, result in poor discrimination since the differential change in the final intensity ratio is $\sim \Delta\,g/g$. Also if $g < 0$, free-running oscillations will not occur and pulse selection cannot begin. From a reproducibility point of view (mode-locking statistics), it is evident from (1) that l_d should be made as large as possible.

Mode-locking also involves the probabilistic selection of a pulse that will saturate the dye for a given net undepleted gain. If the rate of increase in I_1 per cavity round trip at the onset of mode-locking is approximated by (4) where g_{n1} is a nonlinear gain given by (5),

$$\frac{dI_n}{dt} \cong g_u\,I_n + g_{n1}\,I_n \tag{4}$$

$$g_{n1} \cong \left(l_d - \frac{w_d^2}{w_a^2}\frac{l}{g_u}\frac{I_{d_{sat}}}{I_{a_{sat}}}\frac{I_{av}}{I_n}\right)\frac{I_n}{I_{d_{sat}}} \tag{5}$$

then, in order that $g_{n1} > 0$ for I_1 and $g_{n1} \leqslant 0$ for all other I_n, it is necessary that increases in l_d be compensated by decreases in the transverse mode size in the rod, w_a. Here $I_{d_{sat}}$ is the saturation intensity for the dye, $I_{a_{sat}}$ is the average depletion intensity for the rod, A is the amplification per cavity round-trip, w_d is the beam radius in the dye, and I_n and I_{av} are the intensities of the nth pulse and the cavity sample average. If l_d is too small, depletion of net gain for all fluctuations will occur before appreciable dye saturation or pulse discrimination has taken place, and mode-locking will not occur. A dye concentration that is too large yields large energy extraction with strong self-focusing and rod or mirror damage. For a conventional cavity we have found that it is necessary to maintain $0.15 < l_d < 0.3$, for which the probability for mode-locking does not exceed 30 to 50%. The probability for mode-locking was increased to $> 90\%$ with high dye concentrations ($l_d = 1.0$) when using the internal electronic Q-spoiler to limit the peak pulse intensity. This 90% probability is maintained and indiscriminant Q-switching is precluded at high dye concentrations by reducing w_a with the intracavity telescope.

The repetition rates for ED-2, EV-2, EVF-1, and Q-88 were limited by thermal distortion of the cavity. The silicate glass was poorest in this regard, extinguishing oscillation above 0.2 Hz at 40 J. The positive lens caused damage to the mirrors at higher pumping and repetition rates. This is in contrast to an early report that 5 Hz was possible [1]. However, the pumping cavity in that case was a double ellipse and may have provided more uniform thermal loading. The phosphate glasses EV-2, EVF-1, and Q-88 were much improved in several respects and it is expected that LHG-5 and LHG-7 would be similar. In fact, it has recently been shown that LHG-5 and LHG-7 exhibit strong thermal effects on mode-locking above 1 Hz. [7] In that work it was also found that LHG-8 yielded free-running oscillation at 30 Hz. This is reasonable in light of the mode-locking results on LHG-8 presented here. The higher cross-sections, σ_{se}, of the phosphate glass rods reduced the thermal loading at threshold. EV-2 has a negative lens and does not cause damage as the pumping is increased at higher distortions. The distortions prevented mode-locking above 0.5 Hz at 30 J. While EVF-1 has a good thermal-optic coefficient, σ_{se} is so low that it must be pumped hard. The thermally compensated phosphate glasses LHG-8, EV-4, and Q-100 all have low thermal-optic coefficients and high σ_{se}. EV-4 was mode-locked at 1 Hz with no apparent thermal limit and

LHG-8 was modelocked up to 5 Hz, the limit of the power supply, again without thermal effects and with no observed limit at 20J. These rates were all obtained with a gold plated specular single ellipse pumping head. Q-100 was operated at $l_d \approx 1.0$, which increases the pumping requirement by a factor of 2. It was also pumped with a 2 lamp diffusely reflecting head which increased the pumping requirement by a factor of 5. These conditions increased the thermal loading by a factor of 10. As a result Q-100 was mode-locked at only 1 Hz, although a specular ellipse would allow rates of 5 Hz.

The laser was designed to optimize the repetition rate and the emission quality. The pumping cavity used for the highest repetition rates was a gold coated specularly reflecting ellipse which provided efficient pumping of the red absorption bands of Nd^{3+} in the rod and produced a gain which was peaked near the rod axis in order to preferentially pump the TEM_{oo} mode and minimize the excess energy deposited in the rod. A diffusely reflecting pump cavity with 2 lamps has also been used up to 1 Hz successfully, although the requisite pumping energy is much higher. While mode controlling apertures at either end of the cavity are desirable in general, they are absolutely required with the latter head design. For the specular ellipse a single 4 mm bore xenon flashlamp was used with low level pumping to avoid inefficient pump emission in the blue that can result from driving a smaller bore lamp too strongly. The flashlamp discharge capacitor was accurately chargeable and resetable to the preflash energy ($\sim$ 20 J) with an error of $< 0.01\%$. The discharge was roughly matched to the fluorescent lifetime of 300 μsec for Nd^{3+} in glass in order to maximize efficiency. The dimensions of the Brewster faced rods were 1/4" $\times$ 3" (EV-4 had AR coated normal incidence faces). It should be noted the high humidity can etch phosphate rod faces and AR coating is advisable. The short rod length was selected to minimize the glass path-length per round-trip cavity transit in order to reduce the degree of effects involving n_2. The resonator was 1 m in length, with a $R = 100\%$ 10 m concave mirror and an $R = 70\%$ flat wedged output mirror. A contacted dye cell was placed on the 10 m mirror with a dye thickness of 0.6 mm. The dye had an unsaturated round trip transmittance of > 0.8 without and ~ 0.5 with the Q-spoiler. The pumping cavity was placed between 20 and 30 cm from the output mirror. The Q-spoiler was placed at the opposite end of the cavity. The resonator was constructed on a mechanically rigid and vibrationally isolated base with the finest quality mirror mounts and care was taken to support these mounts with strong thermally compatible materials. The environment was thermally controlled to $< 1°C$ and was protected from drafts. I would like to conclude by emphasizing the need for efficient pumping of any of these rods. For a 2" pumping length, 300 W of average pumping power should be considered a maximum limit, above which rod fracture will occur. This establishes a maximum repetition rate for mode-locking with 90% probability of ~3 Hz per inch of pumped rod length.

A convenient and reliable source of picosecond pulses has been obtained by addressing both the thermal and statistical aspects of picosecond pulse generation in passively mode-locked glass oscillators, complicated experiments requiring 5ps resolution that have been intractable in the past [4], due to long set up, alignment, and data accumulation time, will now be readily performed.

References

1. B. Fan, B. Leskovar, C.C. Lo, G.A. Morton, and T.K. Gustafson, IEEE J. of Quant. Electron., *10*, 9, 654, September 1974.

2. C. Langhoff, private communication.

3. J.R. Taylor, W. Sibbett, and A.J. Cormier, Appl. Phys. Lett., *31*, 3, 184, August 1977.

4. T.R. Royt and C.H. Lee, Appl. Phys. Lett. *30*, 332, April 1977.

5. B. Ya. Zeldovich and T.I. Kuznetsova, Sov. Phys. Uspekhi, *15*, 1, 25, July-August 1972.

6. G.H.C. New, Proc. IEEE, *67*, 3, 380, March 1979.

7. H. Kuroda, H. Masuko, S. Maekawa, T. Izumitani, J. Appl. Phys. *51*, 3, 1351, March 1980.

Fluorescence and Laser Action of "Old" (No. 9860) and New Mode-Locking Dyes

B. Kopainsky, A. Seilmeier, and W. Kaiser

Physik Department der Technischen Universität München
D-8000 München, Federal Republic of Germany

For more than a decade fast saturable absorbers have been used extensively to mode-lock and Q-switch Nd-lasers /1/. Especially, the commercially available dye Eastman No. 9860 has found wide application. In spite of the frequent use of this dye nothing is known about the fluorescence properties. It has been generally believed that this fast switching dye is not fluorescent.

In the course of a more detailed study of dyes in the near infrared we have collected detailed data on optical properties of dye No. 9860 and of new mode-locking dyes /2/. All dyes investigated exhibit strong absorption ($\varepsilon \approx 10^5$ cm^{-1} M^{-1}) around 1.06 µm. Integration over the absorption band allows an estimate of the radiative lifetime /2,3/, τ_2. If we assume that the previously measured absorption recovery times /4/, τ_1, are equal to the population lifetimes of the S_1 state, we can calculate the fluorescence quantum efficiency $\eta = \tau_1/\tau_2$ and obtain $\eta \approx 10^{-3}$ for No. 9860. On account of this estimate we have to expect a weak infrared fluorescence of these molecules.

Using appropriate excitation pulses and a sensitive detection system we were able to measure the fluorescence spectra. The molecules were excited by pulses of 150 mJ of a Q-switched Nd-laser and the fluorescence spectrum was recoreded using a calibrated spectrometer (f=25cm) with a PbS detector. The spectral resolution is 10 nm. As example we present results on dye No. 9860 and the new dye No. 15. In Fig. 1 the normalized fluorescence spectrum of dye No. 9860 is shown (full line) next to the absorption spectrum. We find a small Stokes shift of approximately 340 cm^{-1}. The structure due to the progression of skeletal modes

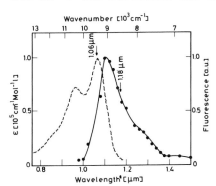

Fig.1 Fluorescence and absorption of dye No. 9860 in 1,2-dichloroethane

of the molecule is less pronounced in the fluorescence than in the absorption spectrum. The fluorescence quantum efficiency was measured to be $\eta \simeq 10^{-3}$. This value is in good agreement with the number estimated above.

The fluorescence spectrum of the new dye No. 15 /2/ is depicted in Fig. 2. The fluorescence is close to a mirror image of the absorption spectrum. The fluorescence spectrum extends to 1.7 μm. The Stokes shift is approximately 450 cm^{-1}. The quantum efficiency is found to be $\eta \simeq 10^{-3}$.

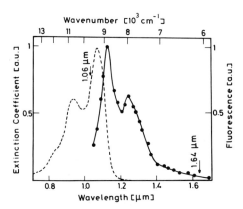

Fig.2 Fluorescence and absorption of dye No. 15

The small fluorescence quantum yield suggests a short fluorescence lifetime. To measure the fluorescence lifetime directly we chose a gain experiment /5/. A first ultrashort laser pulse at 1.06 μm excites the molecules and the amplification of a second delayed probe pulse at different wavelengths within the fluorescence spectrum is observed. The gain of the weak probe pulse monitors directly the momentary population of the excited electronic S_1 state. In our experiments, the tunable probe pulse is generated in a single-pass optical parametric system. For dye No. 9860 the observed gain of a probe pulse at a frequency of 1.18 μm (see r.h. arrow in Fig.1) is depicted in Fig. 3a. The data indicate a fluorescence lifetime of 7±1 ps.

It is interesting to compare this time constant with the result obtained from a bleaching experiment (see Fig.3b). In the latter experiment the transmission of the sample is first increased by an intense pump pulse at 1.06 μm. A weak probe pulse monitors the transmission recovery; i.e., the refilling of the S_0 state due to internal conversion $S_1 \rightarrow S_0$. From Fig. 3b an absorption recovery time of 7 ps is deduced. Within the experimental accuracy we find a lifetime of 7 ps for the lifetime of the S_1 state in two independent experiments.

The optical gain of dye No. 15 was investigated in a similar way. In Fig.4a the wavelength of light amplification is set at 1.64 μm which is at the tail of the fluorescence spectrum (see r.h. arrow in Fig.2). The experimental data suggest a fluorescence

8

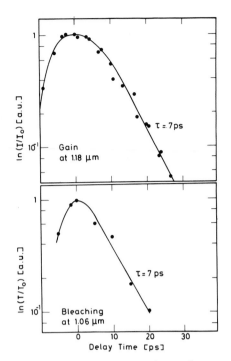

Fig.3 Transient gain and bleaching recovery in dye No. 9860

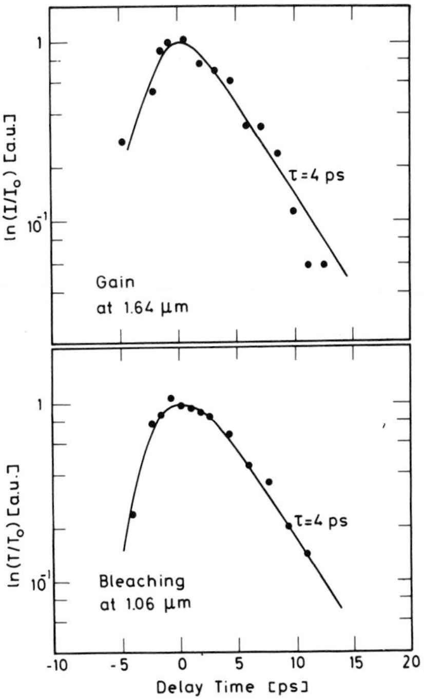

Fig.4 Transient gain and bleaching recovery in dye No. 15

lifetime of 4±1 ps, which is again in agreement with the time constant of the bleaching experiment depicted in Fig.4b.

Table 1 summarizes the experimental results for the investigated fast saturable absorbers. The new dyes (No.5 - 15) have a shorter relaxation time and a higher photochemical stability /2/ than dye No. 9860. When used as mode-locking dye in a Nd-glass laser the fast dye No. 5 provides ultrashort pulses of 1.7 ps duration compared to pulses of 3.2 ps obtained by No. 9860 in the same laser /6/ (Fig.5). Fig.5 demonstrates that the shot to shot variation of the pulse duration is small.

Dye	Measured τ [ps]	Absorption λ_{max} [μm]	Relative Stability
No. 9860	7	1.07	1
No. 15	4	1.07	40
No. 18	4	1.04	5
No. 5	2.7	1.09	40

Table 1

9

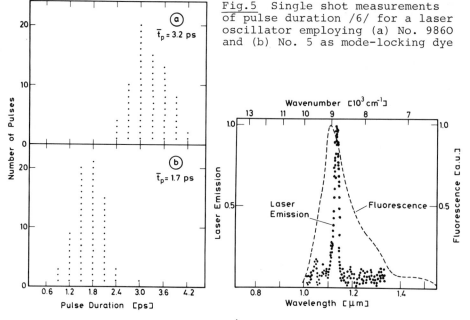

Fig.5 Single shot measurements of pulse duration /6/ for a laser oscillator employing (a) No. 9860 and (b) No. 5 as mode-locking dye

Fig. 6 Laser action in mode-locking dye Eastman 9860 in a 100 μm laser cavity. Note the reduction in spectral width of the laser emission compared to the spontaneous fluorescence Band

In the transient gain experiments light amplification is found during the early time of fluorescence. On account of this observation we predicted laser oscillation in fast mode-locking dyes in a resonator of sufficient feed-back and of short cavity lifetime. A laser cavity of 100 μm length was devised consisting of two mirrors with reflectivities of 100 % and 50 % between 1.0 μm and 1.3 μm. This optical resonator has a short cavity lifetime of 1.8 ps. The cavity was filled with dye No. 9860 at a concentration of 10^{-3} M. A pump pulse at 1.06 μm irradiated with approximately 10^{-17} photons for 5 ps an area of $1mm^2$ in the cavity. The dye laser output was spectrally analyzed by a spectrograph and a PbS vidicon.

The laser emission occurs near the peak of the fluorescence spectrum as seen from Fig.6, where the fluorescence is redrawn for comparison. The small red shift is due to reabsorption effects. The observed spectral width of the laser pulse is determined by the spectral resolution (300 cm^{-1}) of the spectrometer. According to numerical estimates, the duration of the dye laser pulse amounts to a few picoseconds only /7/. More detailed laser experiments with more efficient and highly stable infrared dyes will be discussed in a forthcoming publication /8/.

The authors are indebted to K.H. Drexhage for providing the new mode-locking dyes. They gratefully acknowledge the contributions of F.Wondrazek, J.K.Hallermeier, and C.Kolmeder.

10

References
/1/ A.J. DeMaria, D.A.Stetser, H.Heyman, Appl. Phys. Lett. 8
     (1966) 174
/2/ B. Kopainsky, W. Kaiser, K.H. Drexhage, Optics Commun. 32
     (1980) 451 ; G.A. Revnolds, K.H. Drexhage, J. Org. Chem. 42
     (1977) 885
/3/ J.B. Birks, in "Photophysics", Wiley, London (1970)
/4/ R.J. Scarlet, J.F. Figueira, H. Mahr, Appl. Phys. Lett. 13
     (1968) 7 ; D. v.d.Linde, K.F. Rodgers, IEEE QE-9 (1973) 960
/5/ D. Ricard, W.H. Lowdermilk, J. Ducuing, Chem. Phys. Lett. 16
     (1972) 617 ; C.V. Shank, E.P. Ippen, O. Teschke, Chem. Phys.
     Lett. 45 (1977) 29
/6/ C. Kolmeder, W. Zinth, W. Kaiser, Optics Commun. 30 (1979)
     453 ; C. Kolmeder et al., to be published
/7/ G.W. Scott, A.J. Cox, Proc. Soc. Photoopt. Instr. Eng. 113
     (1977) 25
/8/ W. Kranitzky et al., to be published

Stabilization of Solid-State Lasers for Picosecond Transient Spectroscopy

Kee-Ju Choi and M.R. Topp

Department of Chemistry, University of Pennsylvania,
Philadelphia, PA 19104, USA

Many recent developments in picosecond pulsed-laser technology have been motivated by the need for higher sensitivity, better measurement precision and, above all improved and reliable time-resolution in the study of primary photoprocesses. Although solid-state lasers are advantageous to us because of their intrinsic high power, they suffer from instability and the need to compromise between repetition-rate and pulse-duration.

The evolution of subnanosecond laser pulses in a passively mode-locked, pulsed solid-state laser has been analysed in detail in the literature [1-7]. The consensus favours a model in which, over a period of several tens of microseconds, mode-coupling via periodic and intensity-selective absorption saturation results in the evolution of short pulses close to the Fourier-transform limit. Subsequent amplification gives rise to the observed high-energy output, of which the constituent pulses are progressively broadened, both spectrally and in time.

A number of approaches have been used in the past to stabilise laser pulse-trains. Although the durations may be stabilised to some degree by cavity design, and Fourier-transform limited pulses may be generated from both YAG and glass lasers [8,9], (for example, by use of intracavity etalons) the typical minimum pulse durations remain ~8 ps for glass and >25 ps for YAG (frequently >50 ps in highly stabilised systems). The recent development of phosphate laser glasses has improved the efficiency and the thermal properties, principally allowing higher repetition rates and by virtually eliminating thermal lensing and self-focussed filament formation. Nevertheless, the output pulses are not significantly better than those available from an optimally stabilised silicate glass system.

A major problem in solid-state laser operation lies in the dual role of the passive Q-switching element. It principally functions as a modulator and, in order to operate on noise fluctuations in amplified fluorescence, the concentration is required to be very low. Its effect is to impose a ripple-modulation on the radiation field in the cavity, the dominant modulation frequency being that of the spacing between the cavity axial modes. Typically, the round-trip low-power transmission through the Q-switching element is maintained at 50%. However, once significant power is fed into the evolving short pulses, the amplitude modulation process saturates and the pulses can rapidly become distorted. One major problem is that, since the mode-locked pulse-train evolves from quantum-noise fluctuations over >10^{-5}s, a critical threshold condition obtains, and the form of the laser output is strongly dependent on experimental conditions, which usually cannot be reproduced from one shot to the next. The result is fluctuation in the pulse-train energy and shape as well as in the durations of the individual pulses.

The development of suitable laser devices for the laser fusion programme has resulted in the need for tailored pulses, principally based on stable mode-locked oscillators [3,4,10]. One approach, used with Nd^{3+}-YAG lasers, has involved active modulation for pulse-formation, combined with either active or passive Q-switching for output coupling and to reduce pulse-durations. The available pulses have the advantage that they can be accurately synchronised with external events, although the individual pulses are usually not as short as may be obtained by other means.

In order to produce shorter, high-power pulses a multistage system is used, in the first stage of which a pulse-train is generated in a routine manner. Subsequently, this output may be compressed in time, for example after the manner of Treacy [11], using dispersive elements. Alternatively, methods have been reported [12], for shortening picosecond pulses by repeated, but non-oscillatory passage through a saturable absorber-amplifier combination. The regenerative amplifier reported by Murray and Lowdermilk [10] is an oscillatory variant, but uses a cavity-dumping technique for single-pulse output. The development presented here uses an oscillatory system to generate whole trains of high-power pulses, with nearly 100% reproducibility [13].

Experimental

The central theme of our approach was to use a ring laser for short-pulse generation. The ring configuration allows ready separation of the output and input events, as well as giving a higher inversion-gain ratio for a cavity round-trip. The cavity functions as a regenerative amplifier (RA) for weak injected pulses, evolving into a stable oscillator at higher powers. Ultimately, the power is controlled by nonlinear loss processes, which compete effectively with the gain. It is important that the RA cavity operates below its self-oscillation threshold (controlled by the Q-switch concentration), so that the cavity oscillation can only be turned on by the injection of a moderately strong pulse.

Nd^{3+}-phosphate glass ring oscillator

The optimum experimental arrangement used pulses from a Nd^{3+}-phosphate glass ring laser oscillator to trigger the RA. The experimental arrangement is shown schematically in Fig. 1. A Pockels cell was used to direct a single pulse from the oscillator into the regenerative amplifier cavity, in the clockwise direction. The cavities were virtually identical, except for their length. The operation of the RA was effectively independent of the size and quality of the selected single pulse from the MO train. The two cavities do not have to be run synchronously - the adjustment of M_6 can be used for stroboscopic applications.

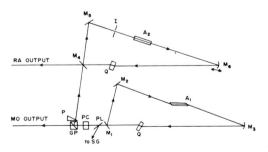

Fig. 1

The pumping level of the RA rod was adjusted to be significantly below the self-oscillation threshold, (typical Q-switch optical densities were 0.6-0.9). Otherwise, the configuration of the ring lasers was about the same as, for example, used by Eckardt et al [1]. Typically, both lasers were run at about the same pumping voltage of 4.2 kV, but the setting was only critical for the oscillator, as expected.

Output characteristics of the Nd^{3+}-glass RA

As Fig. 2 shows, the regenerative amplifier pulse-train does not evolve from a noise burst, but from a relatively organized, short, intense pulse. Because this pulse passes through the concentrated saturable absorber many times, and because the cavity is only above threshold for strong pulses, further pulse-shortening takes place. This is consistent with the observations of Murray and Lowdermilk [10] and of Penzkofer et al. [12]. The first pulse in Fig. 2 is the reflected part of the trigger pulse. The second and subsequent pulses are transmitted through M4. The initial single-pass gain is estimated to be ~5. For gain-saturated pulses, the limit is a gain of ~3.5 at a saturation energy per pulse of ~15 mJ inside the cavity.

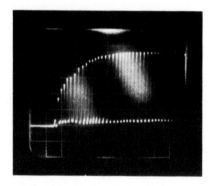

Fig. 2

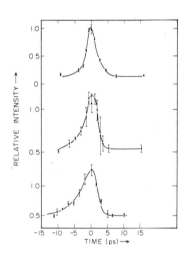

Fig. 3

To estimate the output pulse durations, the cross-correlation function between first and second-harmonic pulses was measured for the whole pulse-train. Both a single-shot two-photon fluorescence spot approach, and a split-beam multiple-shot photoelectric approach were used, with BBOT scintillator. The two-photon spots obtained with the RA were always much sharper and more reproducible than from the primary oscillator. Fig. 3 (upper) shows the multiple-shot profile. As far as could be determined the single-shot and multiple-shot methods gave identical profiles, confirming a high-degree of shot-to-shot stability. The cross-correlation profile had a width (FWHM) of ~3.5 ps indicating pulse-durations, averaged over the whole train, of <3 ps. Further, the pulse-profile of the <u>third</u> harmonic was measured by cross-correlation with second-harmonic pulses, using a split-beam photoelectric two-photon fluorescence arrangement, with naphthalene as the two-photon absorber.

14

The interesting results are shown in Fig. 3, where it is seen that the profiles are asymmetric, being appreciably broadened on the leading edge. Since
the fundamental plus second harmonic cross-correlation function was nearly
symmetric, the increased width could be attributed to group-dispersion in the
third-harmonic generation crystal (10 mm & 25 mm). This phenomenon could not
be observed with pulses from a normal oscillator.

Nd^{3+}-silicate-glass oscillator

An interesting case arises when the oscillator and regenerative amplifier
lasers are frequency-mismatched. For example, a Nd^{3+}-silicate glass oscillator was used to trigger the RA. From measurements of pulse-train profiles
at different wavelengths, it was observed that an induction time for the increase of laser pulse-train intensity is closely associated with the frequency spectrum: the pulses initially have a center wavelength near 1061 nm,
where the gain of the phosphate amplifier is relatively low. The initial
threshold condition for triggered oscillation is determined by the reduced
gain due to the frequency-bandwidth mismatch. Since the maximum gain lies to
shorter wavelengths, the spectrum is found to shift in that direction over the
first few pulses from the RA. Since the gain increases as the frequency
shifts, the threshold condition changes so that, even if the first pulses
suffer a net loss a delayed, but still triggered pulse-train can result. It
was determined that, apart from small differences in the time taken to reach
maximum intensity, the form of the output pulse-train was essentially independent of the intensity of the injected single pulse in the range of power levels used.

Nd^{3+}-YAG laser

The Nd^{3+}-YAG laser is potentially more useful in a general sense because, although pulse-duration is sacrificed, the repetition-rate is much higher than
for glass systems. The major difference between YAG and glass systems is in
the degree of inhomogeneous broadening in the laser transition. Thus, the
higher efficiency resulting from the reduced linewidth in a YAG crystal also
implies a lower gain-saturation threshold for a laser operated under the
usual low-gain conditions. The usual result for a mode-locked laser is that
the pulse-trains from Nd^{3+}-YAG tend to be shorter than those from Nd^{3+}-glass,
generally containing only a few percent of the energy.

Several important factors must be considered in the use of Nd^{3+}-YAG lasers
for transient spectroscopy. First, because of the available high repetition
rate this source is frequently used with signal-averaging techniques. The
actual experimental resolution is limited by both the absolute pulse duration
and the irreproducibility of the pulse from shot-to-shot. Also, since this
laser is a useful source for synchronously pumped dye laser applications the
duration, shape and reproducibility of the pulse-trains are directly reflected
in the quality of the dye-laser pulses.

Use of a ring regenerative amplifier configuration for Nd^{3+}-YAG readily
generates pulses ~12 ps in duration. Insertion of a non-linear loss into the
cavity [14] conveniently extends the pulse-trains to ~100 pulses by reducing
the oscillating pulse intensity and hence the drain on the amplifier. An
important consequence is the reduction of the pulse duration to ~7 ps, the
shortest yet recorded for a YAG laser. Pulse-train energies were 7-10 mJ,
with ~15% uncertainty. A major factor in achieving the high reproducibility
of this system was the use of a reliable pulse-extractor: a commercial
Krytron-triggered device was abandoned in favour of a home-made solid-state
unit. In the better unit, the jitter was <.5 ns, and reliable extraction of

15

single pulses was obtained, limited only by fluctuation in the quality of the oscillator pulse-train.

This work was supported in part by the Regional Laser Laboratory and by NSF grant CHE-78-25312.

References

1. R.C. Eckardt, C.H. Lee and J.N. Bradford, Appl. Phys. Lett. 19 420 (1971).

2. J.A. Fleck, Phys. Rev. B1 84 (1970).

3. I.V. Tomov, R. Fedosejevs and M.C. Richardson, Appl. Phys. Lett. 30 164 (1977); Rev. Sci. Instr. 50 9 (1979).

4. D.J. Kuizenga, Opt. Comm. 22 156 (1977).

5. V.S. Letokhov J.E.T.P. Lett. 7 25 (1968).

6. E.I. Moses, J.J. Turner and C.L. Tang, Appl. Phys. Lett. 28 258 (1976).

7. G.V. Krivoshehekov, L.A. Kulevskii, N.G. Nikulin, V.M. Semibalamut, V.A. Smirnov and V.V. Smirnov, Sov. Phys. J.E.T.P. 37 1007 (1973).

8. J.R. Taylor, W. Sibbett and A.J. Cormier, Appl. Phys. Lett. 31 184 (1977).

9. D.E. Cooper, R.W. Olson, R.D. Wieting and M.D. Fayer, Chem. Phys. Lett. 67 41 (1979).

10. J.E. Murray and W.H. Lowdermilk in Picosecond Phenomena (eds. Ippen, Shank and Shipiro) Springer Series in Chemical Physics, Volume 4 (Springer 1978) p. 281.

11. E.B. Treacy, Phys. Lett. 28A 34 (1968); J. Appl. Phys. 42 3848 (1971).

12. A.Penzkofer, D. von der Linde, A. Laubereau and W. Kaiser, Appl. Phys. Lett. 20 351 (1972).

13. K.J. Choi, M.R. Topp and L.A. Diverdi, J. Opt. Soc. Am. 70 607 (1980).

14. J.C. Comly, A. Yariv and E.M. Garmire, Appl. Phys. Lett. 15 148 (1969).

Excited Singlet State Absorption in Laser Dyes

S.T. Gaffney and D. Magde

Department of Chemistry, University of California at San Diego,
La Jolla, CA 92093, USA

1. Introduction

The variety of organic "dyes" available to cover the spectrum and the broad
tunability of each particular dye make the dye laser a versatile research
tool. The ultimate utility of these lasers will depend in large measure
upon the properties of the dyes available. Absorption by the relaxed
lowest excited singlet state, that is, by the upper level of the laser
transition, is one of the most important parameters determining the suita-
bility of a molecule as a gain medium [1,2].

We have examined excited singlet absorption (ESA) spectra for a number
of laser dyes and related molecules, including those listed in Table 1.
All show substantial, structured absorption in their ESA spectra which
correlates with their known behavior as laser media. In the spirit of a
poster presentation, we limit our discussion here to just an illustration
of the importance of ESA. We present our findings for two violet dyes,
POPOP and CA165. We will mention briefly our results for R6G in order to
promote this substance as a standard for ESA spectroscopy.

Table 1 Laser dyes for which ESA was measured

λ_{las}	Name	Abbreviation	Solvent
420	POPOP		toluene
425	7-dimethylamino-4-methylcarbostyril	CA165	ethanol
	(carbostyril 165)		
450	7-hydroxy-4-methylcoumarin	4MU	basic ethanol
	(4-methylumbelliferone; coumarin 4)		
480	7-diethylamino-4-trifluoromethylcoumarin	C481	dioxane
	(coumarin 35, coumarin 481F)		
510	coumarin 30	C515	ethanol
	(coumarin 515)		
590	rhodamine 6G	R6G	ethanol
	(rhodamine 590)		
620	rhodamine B	RB	ethanol
	(rhodamine 610)		

2. Experimental Procedure

Both POPOP and CA165 were obtained from Eastman Organic Chemicals. For ESA studies concentrations were in the range 0.05 to 0.5 mmol dm^{-3} and path lengths were 2 mm. Ground state absorption was measured on a Beckman Acta CIII spectrophotometer. Emission spectra were recorded with a "home built" apparatus and corrected manually. From the relative spontaneous emission spectra the relative cross sections for stimulated emission were calculated by multiplying by the fourth power of the wavelength [3]. To determine the absolute emission cross section, we simply assumed that its integral was equal to that of the lowest absorption band.

The ESA spectra were determined by picosecond flash photolysis and spectroscopy: a quasi-monochromatic photolysis or excitation pulse P prepared a fraction of the molecules in the desired excited state and a subsequent spectroscopic or probe flash of continuous wavelength distribution monitored the resulting changes in the absorption spectrum. Absolutely essential is the use of a double beam procedure. However, we now prefer a different method (adopted from Prof. Scott) for splitting the continuum into spectroscopic S and reference R portions than we used previously [4]. All pulses were derived from a TEM_{00} ruby laser passively mode-locked with DDI in methanol. A single picopulse was selected and amplified by four passes through a second ruby rod. Harmonic generation was followed by beam splitting with a dichroic mirror. The UV pulse at 347 nm was used directly for the P pulse. The energy incident on the sample ranged from 1/4 to 1 mJ in 1.5 mm^2. The fundamental beam, meanwhile, was focused with a 500 mm focal length lens into a 10 cm liquid cell. A mixture of H_2O/D_2O produced more uniform continua than did either liquid alone. The continuum was collimated and then divided into R and S using a partially reflective metallic mirror. The transmitted portion of the pulse was focused into the sample cell arriving 800 ps before P to become R. The reflected portion was delayed and then sent through the sample 200 ps after P to become S. The S and R beams were recollimated and then directed through a cylindrical lens onto different parts of the slit of a half-metre spectrograph. Detection was photographic. Spectral resolution was better than 2 nm.

Combining information from many laser shots requires one to monitor the concentration of the relaxed excited singlet S_1'. It is not sufficient to monitor just the energy of P, since the absorption is strongly nonlinear. We found that the best monitor for S_1' to be a measurement of the spontaneous fluorescence. On each laser shot the fluorescence signal was digitized and printed along with the numerical values for other monitors of laser performance. In order to calibrate the fluorescence monitor, the photolysis beam was expanded until it was incapable of inducing any nonlinear absorption. Then a knowledge of the low-level absorbance and of the P energy immediately determines the S_1' population. From such a calibration curve the population of S_1' may be determined for each shot, regardless of the nonlinear absorption of P which may occur under other conditions.

3. Analysis, Results, and Discussion

The molecular model assumed in our analysis is an augmented four state model. It considers transitions $S_0 \rightarrow S_1$ and $S_1' \rightarrow S_0'$ but adds higher states coupled to the relaxed excited singlet by absorption $S_1' \rightarrow S_n'$. Non relaxed states (sometimes called Franck-Condon states) S_1 and S_0' are assumed never to accumulate significant populations. Neither do the higher states S_n'. The 200 ps delay used in these experiments between the P and S flashes is sufficient to allow substantial, but by no means complete, rotational relaxation. Within the precision of low rep rate picosecond flash photolysis, it is adequate to assign the measured ESA to the rotationally averaged cross section. The other relevant relaxation processes, internal conversion from higher electronic states, vibrational thermalization, and solvent environmental relaxation are all much faster than rotation.

Given that the ground state population C_0 and the relaxed excited state concentration C_1 are known and not significantly perturbed by the spectroscopic flash S, the absorbance of S is given simply by Beer's law for the superposition of the two species present. Similarly, the absorbance of R is given simply by Beer's law for the total initial ground state population C_T. Of course, C_T (at R) equals $C_0 + C_1$ (at S). What we actually measure from the photographic spectra of R and S is the ratio of the transmittances or, equivalently, the difference of the absorbances:

$$\Delta A = A_S - A_R = \epsilon_{0 \rightarrow 1}(C_0 - C_T)\ell + (\epsilon_{1 \rightarrow 0} + \epsilon_{1 \rightarrow n})C_1\ell .$$

Here ℓ is the optical path length. The sign convention adopted requires that absorptivities associated with stimulated emission, namely $\epsilon_{1 \rightarrow 0}$, must be taken as negative. All parameters in this expression are known except the ESA cross section.

Before undertaking the study of the far more difficult blue-emitting dyes, we studied the rhodamines in order to test our procedures. When we began this work, there were serious discrepancies in results which had been reported for R6G. Now a consensus seems to be emerging which will make

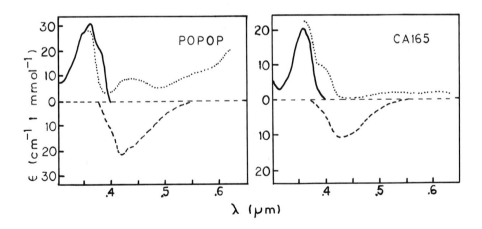

Fig.1 Spectra of two laser dyes. Solid line, ground state absorption; dotted line, excited singlet absorption; dashed line, stimulated emission.

available, for the first time, a standard reference substance which laboratories may use to test their procedures. Our ESA spectrum for R6G is qualitatively rather similar to the spectrum reported by DOLAN and GOLDSCHMIDT [5] and almost identical with the spectra obtained by ARISTOV and SHEVANDIN [6] and by HAMMOND [7].

Data for the two violet dyes POPOP and CA165 are incorporated into Fig.1. The most evident property of ESA in POPOP is that it is extensive and concentrated in the worst possible places. Significant ESA occurs at the lasing wavelengths around 420 nm, so that diminished gain and a high threshold would be anticipated. In addition, ESA is prominent near 350 nm, where one would like to pump. We could not quite extend our measurements down to 337 nm, where the nitrogen laser would be used as pump. However, a reported value [8] for the ESA cross section at 337 nm, $\sigma_{1 \rightarrow n} = 6.6 \times 10^{-17}$ cm^2, would lie exactly along the most plausible extrapolation of what we see as a peak falling off in that direction.

Carbostyril 165 exhibits much more promise than POPOP as a low threshold, efficient laser because of the low ESA over most of the luminescence band. However, we would expect limited efficiency in energy conversion for this dye when it is pumped with high power lasers operating in the near UV around 350 nm.

We close by reiterating our contention that ESA is an extremely important property in laser dyes under all excitation conditions, since it is a property of the upper level of the laser transition itself and may never be neglected under any excitation conditions.

Acknowledgement: This work was supported in part by the National Science Foundation and the Research Corporation.

References

1. O. Teschke, A. Dienes, and J. R. Whinnery: IEEE J. Quant. Electron. QE-12, 383 (1976).

2. E. Sahar and D. Treves: IEEE J. Quant. Electron. QE-13, 962 (1977).

3. O. G. Peterson, J. P. Webb, W. C. McColgin, and J. H. Eberly: J. Appl. Phys. 42, 1917 (1971).

4. D. Magde and M. W. Windsor: Chem. Phys. Lett. 27, 31 (1974).

5. G. Dolan and C. R. Goldschmidt: Chem. Phys. Lett. 39, 320 (1976).

6. A. V. Aristov and V. S. Shevandin: Opt. Spektrosk. 43, 228 (1977); trans. Opt. Spectrosc. (USSR) 43, 131 (1977).

7. P. R. Hammond: IEEE J. Quant. Electron. QE-15, 624 (1979). See also: P. Hammond: Laser Program Annual Report, Lawrence Livermore Laboratory 3, 10.73-10.76 (1978).

8. E. Sahar and I. Wieder: IEEE J. Quant. Electron., (corresp.) QE-10, 612 (1974).

Picosecond Pulse Generation with Diode Lasers

E.P. Ippen

Deparment of Electrical Engineering and Computer Science and Research Laboratory of Electronics, Massachusetts Institute of Technology, Cambridge, MA 02139, USA

D.J. Eilenberger
Bell Laboratories, Holmdel, NJ 07733, USA

R.W. Dixon
Bell Laboratories, Murray Hill, NJ 07974, USA

Interest in semiconductor picosecond devices has been stimulated by recent reports of ultrashort pulse generation by modelocking in GaAℓAs diode devices [1-3]. The transferability of this technique to longer wavelengths, where such pulses can be transmitted over long distances in optical fibers [4], has also been demonstrated [5]. Here we describe the generation of what we believe to be the shortest pulses to date with a semiconductor diode device. These results were achieved with a passively modelocked system consisting of a modified strip buried heterostructure (MSBH) GaAℓAs diode [6] with an external lens and mirror. It is especially important to note that this scheme allows us to generate pulses which are shorter in duration than a transit time of the diode laser itself.

Let us first contrast our passive modelocking with pulse generation by self-oscillation (pulsation), a commonly observed operating characteristic of many diode laser structures [7-9]. Although the physical mechanism may differ from case to case, self-pulsations may for our purposes be described by the Q-switching behavior illustrated in Fig. 1. Optical intensity within the diode begins to build up when the gain exceeds the loss, increases rapidly as the loss is saturated, and then collapses when the net gain is depleted. The process begins anew when the gain again exceeds the loss. It has been pointed out [10] that a necessary condition for such oscillation is that the "effective" absorber saturate with photon flux more easily than the gain.

Clearly, the pulses generated by self-pulsations must necessarily be at least several diode transit times in duration. This same limitation on pulse duration also applies when the gain is actively driven by r.f. current modulation [11] or short current pulses [12,13]. It is the addition of an external resonator that allows us to achieve pulses shorter than the diode itself. If the pulse energy is stored in an external resonator, the pulse can be shortened by many passages through the diode. The gain/loss dynamics in this situation can be as illustrated in Fig. 2. Such behavior is believed to produce subpicosecond pulses in the passively-mode-locked cw dye laser [14] and has been analyzed theoretically by New [15] and Haus [16]. The absorption must saturate more easily than the gain, and the round-trip time of the resonator must be long enough to allow some recovery but not so long that the gain begins to exceed the loss.

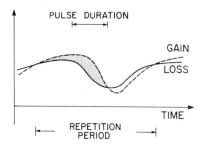

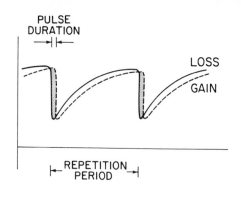

RELAXATION OSCILLATION

$$\sigma_A > \sigma_G$$
$$\tau_P > \tau_D$$

Fig.1 Dynamic behavior of gain and loss during self-pulsation of a diode laser.

$$\sigma_A > \sigma_G$$

$$\boxed{\tau_P < \tau_D}$$

Fig.2 Gain and loss dynamics under conditions of passive modelocking in an external resonator.

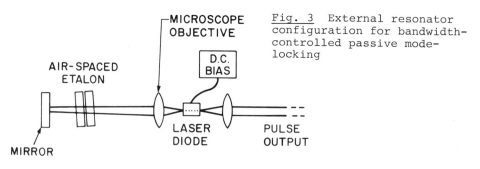

Fig. 3 External resonator configuration for bandwidth-controlled passive mode-locking

The experimental arrangement used to effect this mode of operation is shown in Fig. 3. The resonator consists of a 40X microscope objective, a variable-gap air-space etalon and a flat mirror. Almost every laser diode used was observed to produce pulses (passive mode-locking) in this resonator. The shortest pulses were obtained with modified buried strip heterostructure diodes [6]. These diodes have some advantage over other double heterostructure lasers in that reduced astigmatism of the emission pattern results in better coupling to the external resonator. A $\lambda/4$ ZrO antireflection coating on one of the diode faces was used to increase dependence on the external resonator and was crucial to ultrashort pulse generation.

The output pulse train was monitored with an avalanche photo-diode and a sampling oscilloscope. Actual pulse measurements were made by nonlinear autocorrelation using phase-matched SHG in $LiIO_3$. A sequence of such measurements that shows the effects of the intracavity etalon is shown in Fig. 4. The top trace,

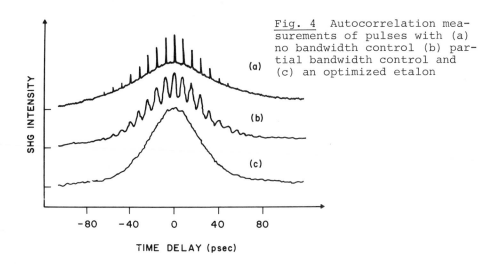

Fig. 4 Autocorrelation measurements of pulses with (a) no bandwidth control (b) partial bandwidth control and (c) an optimized etalon

made without any etalon, indicates pulses with considerable (random) sub-structure. Periodicity in the sub-structure is difficult to eliminate even with good AR coatings. In the frequency domain this corresponds to independent oscillation of several modelocked clusters centered at different longitudinal modes of the short diode. For the middle trace, an etalon was used to limit oscillation to only two of these clusters. Narrowing the etalon bandwidth somewhat more produced the transform-limited pulses measured in the lowest trace. An important feature of this sequence in Fig. 4 is that the introduction of an etalon into the resonator not only cleans up the substructure but actually makes the overall pulse envelope shorter. Measured pulse durations showed no significant variation as the external resonator length was changed over the range 7-20 cm. This is an indication of strong dependence on the external resonator and differs from behavior expected for relaxation oscillations weakly coupled to an external resonator [17].

Further pulse shortening could often be achieved after continuously operating the diode until an increase in threshold was observed. An autocorrelation trace of ultrashort pulse output after this aging process is shown in Fig. 5. We infer from this dramatic pulse shortening that the increase in threshold is associated with the introduction of an additional absorption loss, that this additional loss is saturable, and that it saturates more easily than the laser gain. Several proposed mechanisms [10,18] could satisfy these conditions.

In order to better understand the aging process, and to understand why some diodes produced shorter pulses than others, we monitored side emission from the active stripe by viewing through the top of the diodes with an infrared microscope. With current confinement by proton bombardment the top electrode metallization can be separated at the center to allow such viewing [19]. Concomitant with increases in threshold we observed the appearance and spreading of dark areas in the stripe emission [19]. Inter-

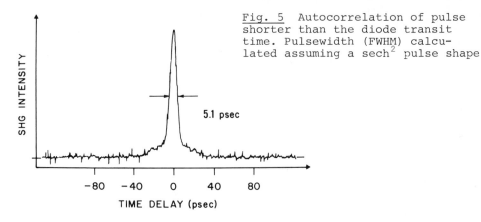

Fig. 5 Autocorrelation of pulse shorter than the diode transit time. Pulsewidth (FWHM) calculated assuming a sech^2 pulse shape

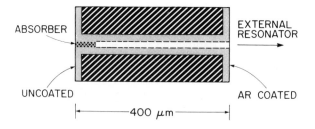

LASER DIODE
TOP VIEW (INFRARED)
MSBH

Fig. 6 Schematic showing experimental observation of induced (dark) absorption region near the end of the resonator

estingly, the shortest pulses (less than 6 psec) were always produced when the dark area formed near the face of the diode that served as the end of the resonator. Such an experimental observation is reproduced schematically in Fig. 6. A reason for such correlation may be improved saturability of an absorber under these conditions. An optical pulse shorter than the diode only produces a standing-wave field near the end face. Such a field results in a coherent coupling [20] between forward and backward wave that increases transmission.

We have demonstrated that the generation and propagation of ultrashort pulses in semiconductor waveguide devices is possible. (Note added in proof: In very recent experiments with similar diodes, pulses as short as 1.3 psec have been produced.) We look forward to experiments which will identify and characterize the physical mechanisms involved. We also expect the development of novel device structures that will incorporate and provide for control over stable saturable absorber elements.

We thank R. L. Hartman for providing the MSBH diodes and R. B. Lawry for the anti-reflection coatings. EPI gratefully acknowledges partial support by the Joint Services Electronics Program Contract DAAG-29-80-C-0104.

References

1. P.-T. Ho, L. A. Glasser, E. P. Ippen, and H. A. Haus, Appl. Phys. <u>33</u>, 241 (1978).

2. P.-T. Ho, Electron. Lett. <u>15</u>, 526 (1979).

3. J. P. van der Ziel and R. M. Mikulyak, J. Appl. Phys. (to be published).

4. D. M. Bloom, L. F. Mollenauer, Chinlon Lin, D. W. Taylor, and A. M. Del Gaudio, Opt. Lett. <u>4</u>, 297 (1979).

5. L. A. Glasser, Electron. Lett. <u>14</u>, 725 (1978).

6. R. L. Hartman, R. A. Logan, L. A. Koszi, and W. T. Tsang, J. Appl. Phys. <u>51</u>, 1909 (1980).

7. T. L. Paoli and J. E. Ripper, Appl. Phys. Lett. <u>15</u>, 105 (1969).

8. T. P. Lee and R. H. R. Roldan, IEEE J. Quant. Electron. <u>QE-6</u>, 339 (1970).

9. T. L. Paoli, IEEE J. Quant. Electron. <u>QE-13</u>, 351 (1977).

10. R. W. Dixon and W. B. Joyce, IEEE J. Quant. Electron. <u>QE-15</u>, 470 (1979).

11. H. Ito, H. Kokoyama, S. Murata, and H. Inaba, Electron. Lett. <u>15</u>, 738 (1979).

12. T. C. Damen and M. A. Duguay, Electron. Lett. <u>16</u>, 166 (1980).

13. T. Kobayashi, A. Yoshikawa, A. Morimoto, Y. Aoki, and T. Sueta, Digest of Tech. Papers, p. 667, XI IQEC Boston (1980).

14. C. V. Shank and E. P. Ippen, Appl. Phys. Lett. <u>24</u>, 373 (1974).

15. G. H. C. New, IEEE J. Quant. Electron. <u>QE-10</u>, 115 (1974).

16. H. A. Haus, IEEE J. Quant. Electron. <u>QE-11</u>, 736 (1975).

17. L. Figueroa, K. Lau, and A. Yariv, Appl. Phys. Lett. <u>36</u>, 248 (1980).

18. J. A. Copeland, Electron. Lett. <u>14</u>, 309 (1978).

19. R. L. Hartman, R. A. Logan, L. A. Koszi, and W. T. Tsang, J. Appl. Phys. <u>50</u>, 4616 (1979).

20. E.P. Ippen and C.V. Shank, *Ultrashort Light Pulses*, ed. by S.L. Shapiro, Topics in Applied Physics, Vol. 18 (Springer, Berlin, Heidelberg, New York 1977)

Bandwidth-Limited Picosecond Pulse Generation in an Actively Mode-Locked GaAlAs Diode Laser

M.B. Holbrook, W.E. Sleat, and D.J. Bradley

Blackett Laboratory, Imperial College, London, SW7 2BZ, England

While the first actively mode-locked CW diode lasers [1] produced 20 ps pulses, residual reflectivity at the cleaved facets interfered with the mode-locking process and the pulses contained temporal sub-structures. As is well known [2] from experience with other types of mode-locked lasers, elimination of spurious reflections inside the laser resonator is necessary to obtain reliable mode-locking with the generation of bandwidth-limited pulses. The complete suppression of internal mode-structure in a diode laser operating in an external cavity has recently [3] been obtained by anti-reflection coating a GaAs/GaAlAs DH laser with the stripe contact making an angle of 5° to the normal of the diode facets. In this manner a smooth lasing spectrum of bandwidth ∿1Å (FWHM) was obtained for operation at 10% above the lasing threshold. We now report the generation of the first bandwidth-limited pulses from an actively mode-locked CW diode laser. A streak camera driven continuously in synchronism [4,5] with the laser pulse train facilated optimization of the mode-locking process and gave direct measurement of the pulse shapes and durations. With second-harmonic intensity correlation measurements unambiguous pulse shapes are not obtained and it is not possible to detect low-intensity but long-lasting sections of the laser output containing a substantial proportion of the laser energy outside the ultra-short pulses themselves, as was found with the first synchronously pumped mode-locked dye lasers [4]. Excess background energy content within the mode-locked pulse train would be deleterious for applications in optical electronics.

The injection laser had a 18µm wide oxide insulated stripe, of length 0.5 mm. rotated on the chip by 5° from the conventional position normal to the cleaved

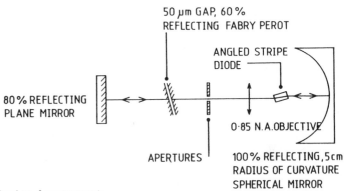

50 µm GAP, 60% REFLECTING FABRY PEROT

ANGLED STRIPE DIODE

80% REFLECTING PLANE MIRROR

0·85 N.A.OBJECTIVE

APERTURES

100% REFLECTING, 5 cm RADIUS OF CURVATURE SPHERICAL MIRROR

Fig.1   External Cavity Arrangement

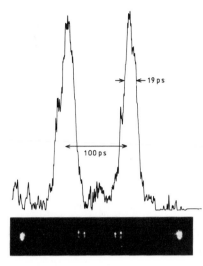

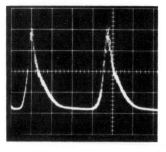

Fig. 2a <u>Top</u>. Microdensitometer trace of pair of pulses. (Ordinate linear density scale). <u>Bottom</u>. Streak photographs of pulses occur at the turning points of the sinusoidal deflection voltage

Fig. 2b Simultaneously recorded scanning Fabry-Perot spectrum of the pulses of Fig. 2 (Free spectral range 4.1 Å

facets. With anti-reflection coatings the lasing spectrum is free from chanelling due to sub-cavity resonances within the external cavity formed by the 100% reflectivity, 5 cm radius of curvature hemispherical mirror and the 80% reflectivity output plane mirror (Fig.1). A 0.85 NA microscope objective collimated the diode laser emission. Bandwidth control was obtained by a 50μm gap, wedged Fabry-Perot filter. An intracavity aperture prevented laser filimentation and multi-transverse mode operation. With the Fabry-Perot filter tuned to 860 nm the DC threshold was 142 mA.

Active mode-locking was produced when the DC bias current was reduced from its free-running level of 142 mA. to 120 mA. and R.F. modulation applied at the round-trip frequency of the resonator (375.5 MHz). Bandwidth limited pulse generation required that the D.C. bias current be kept between 120 and 122 mA. and that the R.F. modulation frequency be maintained on resonance to within 10 kHz. The streak camera had a Photochron II streak-tube [2] operating at a photocathode (S20) extraction field strength of 20,000 Vcm^{-1}. The tube was driven at the fourth sub-harmonic frequency of the diode modulation and the relative phases were arranged so that the pulses were incident during the linear portions of the sine wave deflecting voltage. Thus in Fig.2(a) the two pairs of pulses (generated in a Michelson interferometer optical delay line) are temporally dispersed in opposite directions. Calibration of the camera writing speed was easily obtained from the known separation of the pulse pairs. At the writing speed employed (10^9cm.s^{-1}) the camera resolution was ∿10 ps. This experimental arrangement with simultaneous real time observation of both the spectral and temporal profiles of the laser pulses, provided easy adjustments of the laser for optimum mode-locking performances. After deconvolving the camera resolution of 10 ps. a pulse of 16 ps. duration is obtained. The corresponding spectral profile of Fig. 2(b) gives a time-bandwidth product of $\Delta\nu\Delta t = 0.36$ to be compared with the theoretical values of 0.44 and 0.32 for Gaussian and Sech^2 shaped pulses respectively [2]. Under the conditions of operation, close to threshold, the average laser power was ∿1mW corresponding to a peak power of ∿1W in the

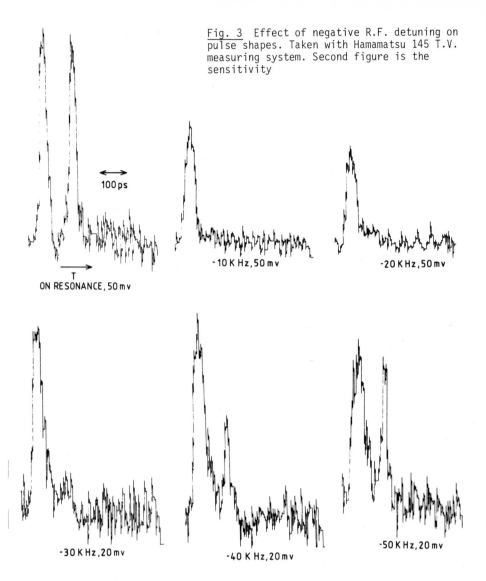

Fig. 3 Effect of negative R.F. detuning on pulse shapes. Taken with Hamamatsu 145 T.V. measuring system. Second figure is the sensitivity

100 ps

ON RESONANCE, 50 mv

-10 K Hz, 50 mv

-20 K Hz, 50 mv

-30 K Hz, 20 mv

-40 K Hz, 20 mv

-50 K Hz, 20 mv

pulses. The streak records also show that there is very little energy outside the bandwidth limited pulses.

The streak camera sensitivity to pulse form and low level signals has also enabled us to investigate the effect of R.F. frequency detuning. For this purpose a Hamamatsu 145 TV measuring system was employed to give an immediate visual record with rapid linear intensity profile recording. From Fig.3 it is clear that detuning seriously affects mode-locking. The qualitative characteristics of the traces obtained off-resonance are similar to those obtained from mode-locked gas lasers and synchronously pumped dye lasers [6]

Positive frequency changes result in a gradual pulse broadening while negative frequency detuning causes multi-pulsing. Formation of bandwidth-limited pulses was sensitive to the laser being biased just on threshold, since otherwise gain recovery leads to the formation of lower intensity satellite pulses. Using a sequence of short ($\sim$300 ps) pumping pulses (generated by a snap diode) instead of the sinusoidal R.F. modulation gave improved stability but resulted in a departure from bandwidth-limited operation ($\Delta\nu\Delta t = 0.9$). Power was increased by 30%.

The laser operated normally when the reflecting plane mirror was immersed in a dye cell ($\sim$0.2 mm path length) containing a 0.5×10^{-4}M solution of IR 140 dye. The dye solution was easily saturated and pulse restructuring occured. It is hoped to generate much shorter duration bandwidth-limited pulses in this way as already achieved with the "hybrid" mode-locked CW dye laser [6]. This result shows that mode-locked semi-conductor laser pulses can be effectively used for picosecond photochemistry, and also that intra-cavity studies of picosecond phenomena in semi-conductors could be easily carried out. Our results also confirm the power and convenience of the synchronously driven streak-camera as a diagnostic tool for picosecond optical electronics.

Acknowledgement

The authors wish to thank Dr. W. Sibbett, Dr. J.R. Taylor, Dr. J. Vukusic and Mr. D. Welford for helpful discussions and Mr. R.S. Morrison and Mr. J. Clarke for technical assistance. Financial support from the Science Research Council and the Wolfson Foundation is gladly acknowledged.

References

1. P.T. Ho, L.A. Glasser, E.P. Ippen and H.A. Hauss: Appl.Phys.Lett. 33, 241 (1978).

2. D.J. Bradley: In *Ultrashort Light Pulses* ed. by S.L. Shapiro, Topics in Applied Physics, Vol. 18 (Springer, Berlin, Heidelberg, New York 1977) and references therein.

3. M.B. Holbrook, D.J. Bradley and P.A. Kirkby: Appl. Phys. Lett. 36, 349 (1980)

4. D.J. Bradley: J.Phys. Chem. 82, 2259 (1978)

5. M.C. Adams, W. Sibbett and D.J. Bradley: Opt. Commun. 26, 273 (1978); Adv. Electron.Phys. 57, 265 (1979)

6. J.P. Ryan, L.S. Goldberg and D.J. Bradley: Opt. Commun. 27, 127 (1978)

A Simple Picosecond Pulse Generation Scheme for Injection Lasers

P.L. Liu, Chinlon Lin, T.C. Damen, and D.J. Eilenberger

Bell Laboratory, Holmdel, NJ 07733, USA

I. Introduction

Generation of short optical pulses from injection lasers is a topic of current interest. There is a great opportunity both in spectroscopy and high data rate communication for the maturing injection lasers.

So far, two approaches have been taken to generate picosecond optical pulses from injection lasers. In the first approach, conventional mode-locking is used [1,2]. Regularly spaced pulses as short as 5 psec or even shorter can be obtained [3]. However, in the active mode-locking scheme, the length of the external cavity must be carefully matched to the applied modulation signal, and/or the intrinsic self-pulsation rate. If passive mode-locking is used, the operation depends crucially on the saturable-absorption caused by dark line defects which usually occur at the later stage of the life span for injection lasers. In the second approach, fast gain switching is used. Due to the short internal cavity length, both build-up and decay of the light intensity may occur in picosecond time scale. By pumping the injection laser with either optical or electrical short pulse, single light pulse can be generated with each excitation pulse. Several demonstrations using this principle have been reported. For example, an optoelectronic regenerative pulser is used to generate 60 psec optical pulse [4]. Recently, a triggerable semiconductor laser which can produce 100 psec pulse is also reported [5]. There is an interesting demonstration in which a large sinusoidal signal is applied to the injection laser along with a dc bias to generate short optical pulse [6].

We report a simple scheme for generating short optical pulses from injection lasers. The principle used is still the fast gain switching. However, the distinguishing feature in our scheme is the use of an impulse train ("comb") generator for obtaining electrical pulses as short as 50 psec with an amplitude as large as 25 V across a 50 Ω load. By biasing the injection laser lower than threshold and driving it with this picosecond electrical pulse, we can generate light pulses as short as 40 psec.

II. Experiment

The setup is very simple. A sinusoidal signal in the 200 to
500 MHz range is used to drive a commercial comb generator
which produces short electrical pulses for pulsed pumping of
the injection laser. The circuit diagram is shown in Figure 1.

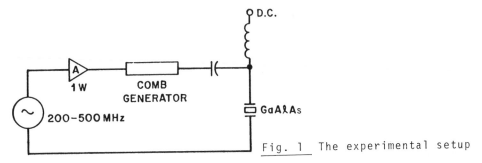

<u>Fig. 1</u> The experimental setup

The comb generator we use is a HP 33004A. Other comb genera-
tors or custom circuits built with similar characteristics can
also be used. Encapsulated in the comb generator are a step-
recovery diode and reactive filtering elements integrated into
a compact module. A short negative pulse is produced whenever
the step-recovery diode is changing from a forward biased condition to a reversely biased condition. The shortest pulse is
obtained when the r.f. driving frequency and power are properly adjusted. We operate the comb generator at 410 MHz and
drive it with 1W power. The electrical pulse observed on a
sampling oscilloscope with 25 psec rise time is shown in Fig.
2. The pulse has a full-width-at-half-maximum (FWHM) duration
of 50 psec and a voltage larger than 25V across a 50Ω terminal.

The short electrical pulse is combined with the dc bias through
a LC circuit or a broadband biasing network to drive the
injection laser.

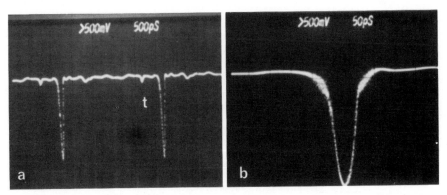

<u>Fig. 2</u> The electrical pulse generated by the comb generator

The injection lasers used in our experiment are GaAlAs DH lasers (~0.82 µm). The width of the laser is defined by shallow proton bombardment to 5 µm. The length of the laser is 380 µm. Both facets have Al_2O_3 high reflection coating to retard any dark line defect formation. The dc threshold is 104 mA.

When biased below threshold, light pulses are observed whenever the electrical pulses are applied. The magnitude of the optical pulse increases when the dc bias is raised. However, the pulses get broadened if the bias is too large. Sometimes, secondary pulses may be generated. The optimal bias for our laser is 90 mA. The light pulses are monitored by a fast Si photodiode (Spectra Physics 403B). The response of the detection system shows a FWHM of 80 psec when tested with a dye laser pulse of <5 psec duration. The light pulse from our injection laser has a FWHM duration of 88 psec as shown in Fig. 3.

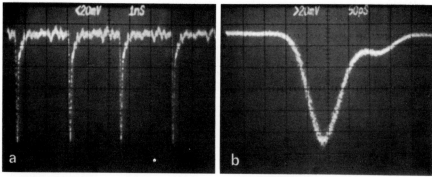

Fig. 3 Light pulse from the injection laser

By assuming a square-root-of-sum-of-squares rule, the deconvolved optical pulse width is 40 psec. An autocorrelation trace is obtained later. Assuming a Gaussian pulse, the FWHM is 42 psec, in good agreement with the direct detection result. Several other injection lasers are tested. The resultant pulse widths are also in the 40 to 50 psec range.

III. Discussion

Compared to other schemes, our setup is apparently much simpler. Neither optical nor electrical feedback is needed. Our key element is a compact, high performance commercial part, comb generator. The scheme does not depend upon the saturable-absorber and/or the self-pulsation phenomena associated with aged injection lasers although our pulses are not as short as that available by mode-locking. Compared to the case of sinusoidal r.f. drive, pulse drive definitely has better stability.

Preliminary study on the spectrum of our laser shows a shift toward shorter wavelength when operated in the pulsed mode.

The width of the spectrum is about 30 Å. Since the whole pulsing event occurs within few round trips, self-phase modulation due to fast change in population and chirping may occur. The shift in the center frequency of the spectrum reflects a lower temperature at the diode junction.

In conclusion, a very simple universal method for the generation of short electrical and optical pulses is demonstrated. Electrical pulses as short as 50 psec, 25 V in amplitude, and optical pulses of the order of 40 psec are obtained. Many applications of these fast electrical and light pulses are expected. For example, it has been pointed out that the synchronism between the two pulses can be used to monitor the response of the optical modulators [7]. The performance of actively mode-locked injection lasers may be improved by using the short electrical pulses. Pulse code modulation may also be achieved by phase shifting the electrical driving signal. The picosecond optical pulses available from various injection lasers at different wavelengths in the 0.8 to 1.7 μm region can also be used to study the modal and material dispersion properties in optical fibers with better resolution.

Acknowledgment

We would like to thank R. L. Hartman for providing us lasers. We also thank R. C. Alferness and M. A. Duguay for many stimulating discussions, and J. P. Heritage for the dye laser pulse used in testing the detection system response.

References

1. P.-T. Ho, L.A. Glasser, E.P. Ippen and H.A. Haus, Appl. Phys. Lett., *33*, 241 (1978)

2. L.A. Glasser, Electron, Lett., *14*, 725 (1978)

3. E.P. Ippen, D.J. Eilenberger and R.W. Dixon, *Picosecond Pulse Generation with Diode Lasers*, in this volume

4. T.C. Damen and M.A. Duguay, Electron. Lett., *16*, 166 (1980)

5. J.A. Copeland, S.M. Abbott and W.S. Holden, IEEE J. Quantum Electron., *QE-16*, 388 (1980)

6. H. Ito, H. Yokoyama, S. Murata and H. Inaba, Electron. Lett., *15*, 763 (1979)

7. N.P.Economou, R.C. Alferness and L.L. Buhl, *Novel Picosecond Optical Pulse Sampling Technique for Measuring the Response of High Speed Optical Modulators*, in this volume

Q-Switching of Semiconductor Lasers with Picosecond Light Pulses *

T.L. Koch, D.P. Wilt, and A. Yariv

Department of Applied Physics, California Institute of Technology, Pasadena, CA 91125, USA

Pulses tens of picoseconds in duration have been produced from semiconductor lasers both by injection of short current pulses [1] and by mode-locking [2, 3]. Below we describe a method for producing ultrashort pulses which employs a combination of electrical and optical pumping.

Double heterostructure GaAs-Al$_x$Ga$_{1-x}$As lasers were grown by liquid phase epitaxy with an active GaAs region thickness of $\sim$ 0.2 μm and a stripe width of $\sim$ 8 μm, as shown in Fig. 1. The junction was formed by a Zn diffusion up to the active region. Evaporated contacts of chrome-gold were etched off in a central region 50 μm long, while typical device lengths were 250 μm.

The essential feature of this device is that the unpumped central region offers considerable loss to the cavity, allowing the laser to be pumped to an initial gain, γ_i, in excess of what the threshold gain, γ_t, would be in the absence of the central region losses. A picosecond pulse from a CW mode-locked dye laser incident on the central region creates carriers which, by

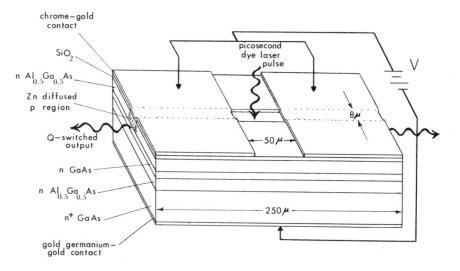

chrome-gold contact

SiO$_2$

n Al$_{0.5}$Ga$_{0.5}$As

Zn diffused p region

Q-switched output

n GaAs

n Al$_{0.5}$Ga$_{0.5}$As

n$^+$ GaAs

gold germanium-gold contact

picosecond dye laser pulse

V

8μ

50μ

250μ

Fig. 1 Geometry of Q-switched diode laser

*Research supported by the National Science Foundation

virtue of band filling , reduce the losses or actually provide additional gain. In either case, the cavity has effectively been "Q-switched" and is suddenly above threshold after illumination. Calculations indicate that the hot plasma created by the dye laser cools sufficiently in several pico- seconds, primarily by optical photon emission, to fill the states responsible for absorption at the diode wavelength when the optically excited plasma density is on the order of 10^{18} cm^{-3}. It is also essential for high speed response that the carriers be produced directly in the active region. To achieve this result, the aluminum content of the upper $Al_xGa_{1-x}As$ cladding layer was chosen such that the direct band gap was larger than the 2.03 eV energy of the dye laser photons, corresponding to $x \geq 0.5$ [4].

The lasers were operated with a 50 nanosecond pulsed current source syn- chronized to the cavity dumped output of the CW mode-locked dye laser at a repetition rate of 45 kHz; typical threshold currents were $\sim$ 100 mA. The pulse width of the dye laser was typically < 2 picoseconds as determined by SHG autocorrelation measurements in KDP. Fig. 2 shows the current input and light output of the diode laser at threshold. The spike in the light output coincides with the dye laser pulse illumination. The jitter in the current results from the current source being triggered 20 microseconds earlier by the preceding dumped dye laser pulse.

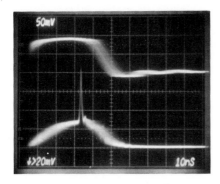

Fig. 2 Upper trace: current input to diode laser, 1 mV = 1 mA. Lower trace: light output of diode; spike in center is Q-switched output.

For the short term behavior governing the Q-switched spike, a simple analysis can be carried out by assuming that the dye laser pulse completely bleaches the losses of the unpumped central region and employing the familiar Q-switching equations.

$$\frac{d\phi}{d\tau} = \phi(\frac{\gamma(n)}{\gamma_t} - 1) \tag{1}$$

$$\frac{dn}{d\tau} = - \frac{\gamma(n)}{\gamma_t} \phi \tag{2}$$

where ϕ = photon density in the cavity, n is the carrier concentration in the pumped end regions, $\gamma(n)$ is the average gain of the cavity, γ_t is the threshold gain in the absence of the central region losses, and $\tau = t/t_c$, where t_c is the cavity lifetime in the absence of central region losses,

$$t_c = \frac{n_0}{c[\alpha_0 - \frac{1}{L} \ell n\ R]} \qquad (3)$$

For $\gamma(n)$, we adopt a semi-phenomenological gain coefficient given by

$$\gamma(n) \simeq an^2 - b \qquad (4)$$

with a and b fit to the numerical work of STERN [5].

As with all Q-switched lasers, the relevant time scale is t_c, which is typically several picoseconds for most diode lasers. The other important factor in determining the pulse width is the initial carrier density vs. the threshold carrier density. With pumped regions 100 μm in length each, and a central lossy region 50 μm in length, we estimate that $\gamma_i/\gamma_t \sim 2$ should be possible before the threshold of the unilluminated diode is reached. Eqs. (1), (2) and (4) are then easily integrated to yield a pulse FWHM τ_p of $\sim 5\ t_c$ or 10 picoseconds. The time delay τ_d between the dye laser illumination and the spike depends on the initial photon density. For $\phi_i/n_t \sim 10^{-5}$ to 10^{-8}, τ_d ranges from 20 to 35 picoseconds.

When biased below the threshold of the unilluminated state, relaxation oscillations are suppressed after the dye pulse because the diode returns to its low Q state as the unreplenished carriers in the central region recombine with a lifetime τ taken as 1 nanosecond. This behavior is shown in Fig. 3, measured with an avalanche photodiode and a sampling oscilloscope with a combined risetime of $\sim$ 100 picoseconds. The single spike is a detector limited response, indicating a diode laser pulse width τ_p < 200 picoseconds. Fig. 4 shows the same sample biased above the threshold of the unilluminated state, $I/I_t \simeq 1.1$. In this case we expect relaxation oscillations with a period [6]

$$T \simeq 2\pi \left(\frac{t_c\ \tau\ I_t}{I - I_t} \right)^{1/2} \qquad (5)$$

where t_c is now the cavity lifetime in the unilluminated state, taken as 1 picosecond with our device. Eq. (5) yields $T \simeq 630$ picoseconds, in rough agreement with Fig. 4

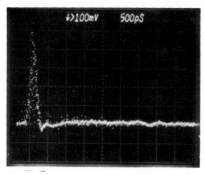 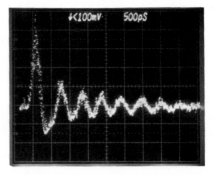

Fig. 3 Single Q-switched spike with diode biased below thresholf
Fig. 4 Same sample as Fig. 3 with diode biased at $I \simeq 1.1\ I_{th}$. Relaxation oscillations are clearly visible following Q-switched spike

Experiments are currently under way to measure the pulse width by a cross correlation of the semiconductor laser pulse and the picosecond laser pulse. This is done by phase matched parametric sum frequency mixing of the two pulses in a $LiIO_3$ crystal.

References

1. T. Kobayashi, A. Yoshikawa, A. Morimoto, Y. Aoki, T. Sueta, *Annual Meeting of Japan society of Applied Physics* 5a-P-4 (1978)

2. P.-T. Ho, L. A. Glasser, E. P. Ippen, H. A. Haus, App. Phys. Lett. 37 241 (1978)

3. L. A. Glasser, Elect. Lett. 14, 725 (1978)

4. B. Monemar, K. K. Shih, G. D. Pettit, J. App. Phys. 47, 2604 (1976)

5. H. C. Casey, Jr., M. B. Panish, *Heterostructure Lasers*, Pt. A (Academic Press, New York 1978), P. 164

6. A. Yariv, *Quantum Electronics* (John Wiley and Sons, Inc., New York 1975), P. 274

Picosecond Pulses from an Optically Pumped GaAs Laser

T.C. Damen, M.A. Duguay, and J.M. Wiesenfeld

Bell Telephone Laboratories, Holmdel, NJ 07733, USA

J. Stone and C.A. Burrus

Bell Telephone Laboratories, Crawford Hill Laboratory,
Holmdel, NJ 07733, USA

Gain switching techniques have been previously used to generate short
pulses on the nanosecond time scale. In order to extend gain switching to
the picosecond domain, one needs picosecond pumping pulses and an ultrashort
cavity. Our laser structure, shown in Figure 1, consists of a 1μ thick ribbon whisker of GaAs sandwiched between two dielectric mirrors and bonded to
them by means of transparent epoxy [1].

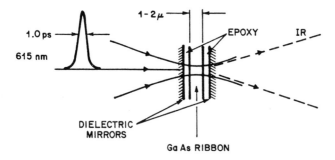

Fig. 1 Ultrashort cavity GaAs laser pumped by 615 nm dye laser pulse. The
GaAs ribbon has the thickness shown and a width of a few hundred microns.

The physical length of the cavity thus formed is 5 to 7μ, giving a round-trip
time of 0.08 ps. Picosecond pulses cavity-dumped from a mode-locked dye
laser are used to optically pump the ultrashort GaAs laser. The pump light
is focused on the sample by a microscope objective to a spot size of about
3 μm. The peak pump power entering the GaAs ribbon is about 1 kW, giving a
peak power density on the order of 10 GW/cm^2 at the focus.

The IR pulses from the GaAs laser are measured by an up-conversion sampling
technique as shown in Figure 2. Part of the 1-ps pulse from the dye laser
(615 nm wavelength) is used for pumping, while the other part is used for
probing the IR pulse. Probing (or sampling) is achieved by focusing both
pulses into a LiIO3 crystal set at the phase-matching angle for sum-frequency
generation into the UV. As the delay in the probe leg is varied, the UV
signal scans out the IR pulse time profile.

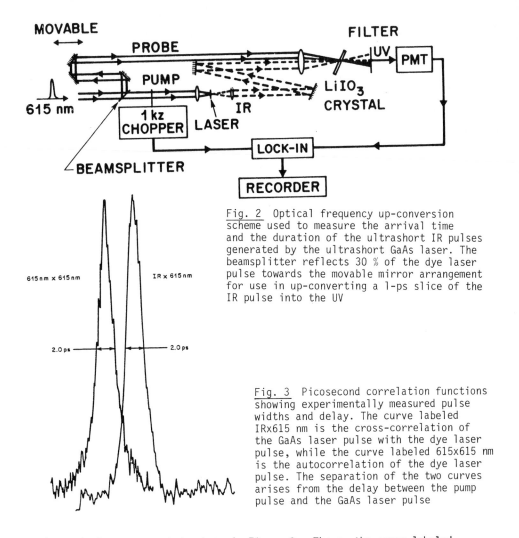

Fig. 2 Optical frequency up-conversion scheme used to measure the arrival time and the duration of the ultrashort IR pulses generated by the ultrashort GaAs laser. The beamsplitter reflects 30 % of the dye laser pulse towards the movable mirror arrangement for use in up-converting a 1-ps slice of the IR pulse into the UV

Fig. 3 Picosecond correlation functions showing experimentally measured pulse widths and delay. The curve labeled IRx615 nm is the cross-correlation of the GaAs laser pulse with the dye laser pulse, while the curve labeled 615x615 nm is the autocorrelation of the dye laser pulse. The separation of the two curves arises from the delay between the pump pulse and the GaAs laser pulse

A typical measurement is shown in Figure 3. There, the curve labeled IRx615 nm represents the IR pulse sampled by the 615 nm probe pulse. The curve labeled 615x615 nm represents the pump pulse sampled by the identically shaped probe pulse, i.e., it represents the autocorrelation function of the pump/probe pulse. Assuming a single-sided exponential shape for the dye laser pulse, the full width at half height of the 615 nm pulse is equal to one-half the autocorrelation curve width, viz. 1.0 ps [2]. Since the IRx615 nm curve is also 2.0 ps at half maximum, the IR laser pulse would also appear to be on the order of 1 ps. However, we believe for two reasons that the pulse duration is even shorter. First, there was jitter in the laser pulse curve caused by a ±15 percent amplitude fluctuations of the probe pulse: when the amplitude is varied intentionally, we have observed a variation in the time delay which is equivalent to 0.3 ±0.1 ps jitter. Correcting for this effect results in a pulse width of about 0.7 ps. Second, the spectral width of the laser line is 3 nm at 870 nm, corresponding to a lower bound on the pulse width of 0.4 ps.

Pulses longer than that shown in Figure 2 were also observed for different samples. Some pulses were as long as 20 ps, their exact shape depending upon pump pulse intensity. These effects are presently under investigation.

In conclusion, we have demonstrated a new limit, viz. 1 ps, to the pulse duration obtainable from GaAs lasers.

[1]J. Stone, C. A. Burrus and J. C. Campbell, J. Appl. Phys. (to be published).

[2]E. P. Ippen and C. V. Shank, *Ultrashort Light Pulses*, ed. by S. L. Shapiro, Topics in Applied Physics, Vol. 18 (Springer, Berlin, Heidelberg, New York, 1977)

Generation of Coherent Pulses of 60 Optical Cycles Through Synchronization of the Relaxation Oscillations of a Mode-Locked Dye Laser

J.-C. Diels, J. Menders, and H. Sallaba

Center for Laser Studies, University of Southern California, University Park, Los Angeles, CA 90007, USA

Coherent interaction experiments on single photon transitions have so far been restricted to dilute vapors [1] (collision times in the nanosecond range), or "frozen media" [2] (like ruby cooled down to liquid He temperature, for which the phase relaxation time of some transitions can exceed 100 ns[3]). There is a need for a stable, highly coherent source of pulses of about 0.1 ps to apply these diagnostic techniques to complex molecules at room temperature. We report the successful development of such a source, yielding bandwidth limited pulses of 0.12 ps duration with 300W peak power.

In the design of ultrashort mode locked dye lasers, an essential consideration is the absence of intracavity dispersion in the wavelength range of interest. To this end, the cavity is reduced to a bare minimum: no other intracavity element than the dye jet, which contains a solution of Rh6G, DODCI and malachite green in ethylene glycol is used [4]. The laser output mirror is a "third order reflector" with a flat reflectivity in the range 590 to 620 nm which defines the spectral range of operation. The dye mixture and mirror reflectance that give the shortest pulses are critical and interrelated. For instance, at a longer wavelength (using a mirror with a reflectance from 605 to 640 nm) the optimum dye mixture contains a higher concentration of DODCI.

When pumped close to threshold by a CW argon ion laser, the laser emits a train of mode-locked pulses. If the cavity does not exceed 60 cm, the pulses are spaced by the cavity roundtrip time, forming trains which contain pulses as short as 0.16 psec. The shortest pulses are emitted in bursts of a few μsec duration with an average periodicity of 20 μsec. These bursts, which are detected as peaks in the second harmonic signal, coincide with relaxation oscillations in the laser intensity, as shown in Fig.1. For larger pump powers or even at thresh-

Second Harmonic
Envelope

Pulse Envelope
(Fundamental)

←+←+ 10 μs

Fig.1 The laser intensity relaxation oscillations which coincide with bursts of short pulses

41

hold for cavity lengths exceeding 60 cm, a continuous train of
pulses is generated at twice the cavity round trip rate (two
pulses per cavity round trip time). Pulse repetition rates of
600 MHz or more can be observed. For the stable generation of
the shortest pulses, two stability problems have to be addressed:

   a)   long term stability: the laser has to operate a few
        mW above the threshold (1.5W) which evolves with dye
        temperature, age, and cavity losses.

   b)   short term stability: the randomness in the time of
        arrival of the short pulse bursts limits the use of this
        source, in particular, in combination with amplifier
        chains which have to be timed accurately.

Both long and short term stability problems have been corrected, the former by the use of active feedback, the latter
through the use of a pulsed modulation in the argon laser pump.

Long Term Stabilization

Unlike synchronously mode-locked systems, we measure that
optimum short pulse generation coincides with a maximum in the
average second harmonic of the dye laser output. Long term
stabilization is attained by adjusting the argon laser intensity, using a voltage responsive remote control, to maintain the second harmonic of the laser output at its peak.
This is done by making phase sensitive detection of a low
frequency modulation (10 Hz) introduced into the second harmonic through the argon laser remote control. The in-phase
component of that 10 Hz oscillation is measured with a lock-in
amplifier and fed back into the argon laser remote control.
The feedback loop seeks stable equilibrium at any local maximum of the second harmonic output. Hence the laser will
stabilize at the shortest pulse in the single pulse or the two
pulse per cavity roundtrip time mode, depending on the initial
pump power setting.

Short Term Stabilization

The ultrashort pulse operation of the dye laser appears to be
related to the relaxation oscillation observed at a frequency
of 50 kHz. By pulse modulating the argon laser intensity at
that same frequency we are able to synchronize these relaxation
oscillations, thus removing the randomness in the short pulse
production. A low voltage modulator is used to periodically
reduce the pump power intensity below threshold, for a few
microseconds, with an adjustable repetition rate. Fig.2
shows the evolution with time of the envelope of the synchronized pulse train, as measured by an avalanche photodiode.
It should be noted that the particular shape (in the microsecond time scale) of the pulse envelope as well as the duration of the pulses along the train depend strongly on the
exact frequency of the dips in the pump power.

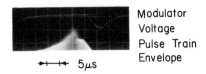

Modulator
Voltage
Pulse Train
Envelope

⊢┼┼⊣ 5μs

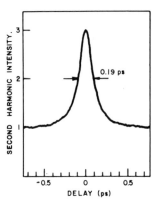

Fig.2 Evolution of the envelope of the synchronized pulse train

Fig.3 Conventional auto-correlation of a synchro-nized pulse train

Fig.2 shows the pulse envelope which occurs when the pulsed modulation has a frequency of 25 kHz. Boxcar integrated pulse diagnostics described below indicate that the train con-sists of bandwidth limited $sech^2$ shaped pulses of 0.12 psec duration for the first 15 μsec of the envelope. In the last few μsec before the modulator turns the laser off, the train consists of uncorrelated bursts of noise (inferred from an autocorrelation trace with a peak to background ratio of 2:1).

The ultrashort pulses were analyzed by boxcar integrated spectra, conventional autocorrelations with a peak to back-ground ratio of 3:1, and interferometric second-order auto-correlations with a peak to background ratio of 8:1. [4] The second-order autocorrelation shown in Fig.3 has a FWHM of 0.19 ps corresponding to a $sech^2$ shaped pulse of 0.12 ps. Spectral measurements confirm the symmetrical pulse shape and the absence of phase modulation.

Further proof of the absence of phase modulation is given by the interferometric second–order autocorrelation trace shown in Fig.4 . As the delay in the variable path arm of the inter-ferometer is decreased, the background is seen to remain constant until the onset of successive constructive/destructive interferences. The monotonic decrease of the envelope of the

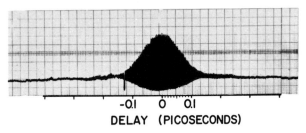

DELAY (PICOSECONDS)

Fig.4 Interferometric second-order autocorrelation of the synchronized pulse train (An imbalance in the beam splitting of the autocorrelator accounts for the departure from the 8:1 peak to background ratio.)

43

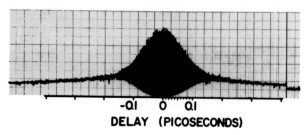

DELAY (PICOSECONDS)

Fig.5 Interferometric second order autocorrelation of the unsynchronized pulse train

destructive interferences (lower boundary of the plot) is strong evidence of absence of phase modulation or pulse to pulse frequency fluctuations.

By contrast, the unsynchronized pulse train (Fig.5) exhibits an autocorrelation trace that indicates either chirp or a periodic frequency drift along the pulse train.

Conclusion

We have demonstrated reliable and stable operation of a passively mode-locked dye laser which generates pulses of 0.12 psec duration and 300W peak power. The emission consists of 15 μsec long bursts of pulses at a rate of 250 MHz, with a repetition rate of 50 or 25 kHz. To understand this periodic structure of the synchronized pulse generation, the mechanism behind the relaxation oscillations needs to be explained. Proposals for models which account for this structure should predict several observed dependences of the pulsed operation on design parameters:

-For short pulse generation, a spectral bandwidth broader than the pulse bandwidth is required.
-Short pulse operation and the associated relaxation oscillations occur only for cavity lengths shorter than 60 cm, which would seem to test the usual assumption that the saturable absorber fully recovers over the interval between pulses.
-The degree of synchronization of the relaxation oscillations strongly depends on the frequency of the pump perturbation.

In particular, we have observed evidence that implicates a role for the dye in this behavior:

-Short pulse operation is limited by a dye lifetime of approximately one week.
-The laser operation is strongly dye temperature dependent.

This work was supported by the National Science Foundation under Grant No. ENG77-21435.

References

1 See for instance: B. Bölger and J.-C. Diels, Physics
  Letters, 28A, 401 (1968); R. E. Slusher and H. M. Gibbs,
  Physical Review A, 5, 1634 (1972)

2 S. L. McCall and E. L. Hahn, Physical Review, 183, 457
  (1969: J.-C. Diels and E. L. Hahn, Physical Review A, 10,
  2501 (1974)

3 I. D. Abella, N. A. Kurnit, and S. R. Hartmann, Physical
  Review, 141, 391 (1966)

4 J.C. Diels, E. VanStryland, D. Gold, *Picosecond Phenomena*, ed.
  by C.V. Shank, S.L. Shapiro, E.P. Ippen, Springer Series in
  Chemical Physics, Vol. 4 (Springer, Berlin, Heidelberg, New York
  1978)

Synchronization of an Active-Passive Mode-Locked Nd:YAG Laser and a Mode-Locked Argon Ion Laser

G.T. Harvey, G. Mourou, and C.W. Gabel

Institute of Optics and Laboratory for Laser Energetics,
University of Rochester, Rochester, NY 14627, USA

We report on the synchronization of pulses from an active-passive mode-locked Nd:YAG laser [1] to pulses from a mode-locked Ar ion laser. The jitter between the laser pulses was 18 ps (root mean square) with maximum deviations of ±30 ps over 18 shots. The addition of a synchronously pumped dye laser [2] to the YAG-Ar system would offer a picosecond tunable laser well synchronized to a high-power, high-energy laser. The low-energy dye pulses could also be amplified in dye cells pumped by the synchronous pulses from the mode-locked YAG laser. Applications include picosecond chemistry and laser fusion experiments.

The system used in the synchronization experiments is shown in Fig. 1. A 50 Mhz R.F. oscillator was employed to drive two acousto-optic modulators, one in each laser cavity.

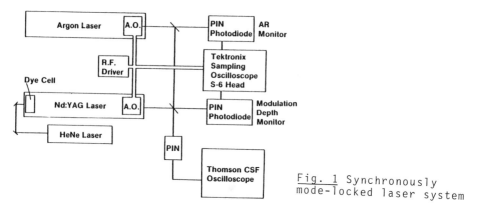

Fig. 1 Synchronously mode-locked laser system

The Nd:YAG laser also contained a flowing dye cell filled with a saturable absorber in contact with one of the laser end mirrors. The Nd:YAG laser produced 100 ns long trains of 45 ps pulses, the peak pulse having an energy of 150 uj. To monitor the modulation depth of the acousto-optic modulator in the YAG, we looked at the transmission of the HeNe alignment laser through the mode-locker with a PIN photodiode connected to a sampling oscilloscope. The depth of modulation at the HeNe laser wavelength was then used to calculate the proper modulation depth at the YAG wavelength. The sampling oscilloscope, triggered by the R.F. drive signal, was also utilized to monitor the argon laser cw pulse train. Jitter measurements were made with a PIN photodiode and a Thomson CSF 4.5 GHz. oscilloscope at a sweep speed of 500 ps./. div. The oscilloscope trace width at this speed corresponded to 20 ps. The jitter resolution for the RMS values was 12 ps.

Jitter in the YAG laser was highly dependent on the depth of modulation θ_m of the acousto-optic modulator (Fig.2).

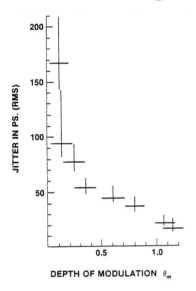

Fig. 2 Jitter as a function of modulation depth in the YAG laser.

θ_m is defined by the equation $T = \cos^2[\theta_m \sin\omega t]$ where T is the single pass transmission of the mode-locker and ω is the angular drive frequency [3]. The jitter is due to the insufficient buildup time to achieve a steady state pulse in the YAG cavity. Kuizenga [4] has shown that the number of round trips required to achieve steady state operation is

$$M = \frac{0.38}{g^{\frac{1}{2}}\theta_m}\left(\frac{\Delta f}{f_m}\right)$$

where g is the round trip amplitude gain, Δf is the linewidth and f_m is the modulation frequency. This equation leads to buildup times needed to achieve steady state operation on the order of 10 to 100 μs. In the active-passive YAG laser the buildup time is 4 μs. An increase in modulation depth results in a narrower time window and a decrease in the jitter of the system. This effect continues until the laser has sufficient buildup time to reach a steady state value or other causes of jitter such as mechanical or drive frequency instabilities predominate.

In addition to shot to shot jitter, long term drift is also a concern in the synchronization of mode-locked laser systems. Typical drift for the YAG-Ar system was 65 ps. over a 20 min. period. A major source of drift was due to small changes in the Argon laser intensity. We found that the phase of the Ar pulse was strongly correlated with the intracavity intensity. The phase shift was on the order of 10 ps. per milliwatt and was independent of whether the intensity change was due to a difference in cavity losses or a fluctuation in plasma tube current (Fig. 3).

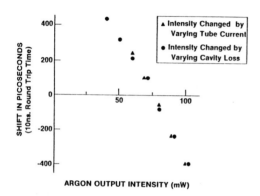

__Fig. 3__ Phase shift of the Argon ion pulse as a function of output power.

Another factor that affected the long term drift was a change in impedance of the acousto-optic modulator due to a temperature induced shift in the acoustic resonance. Cavity length changes resulted in only small phase shifts, a 1 μm increase leading to a 3 ps shift. A shift in drive frequency of 10 Hz produced a 1 ps shift.

1. B.C. Johnson and W.D. Fountain, International Meeting on Electronics Devices Tech. Digest, pp. 322-325 (1974).

    W. Seka and J. Bunkenburg, J. Appl. Phys. __49__, p. 2277 (1978).

2. C.K. Chan and S.O. Sari, Appl. Phys. Lett. __25__, pp. 403-406 (1974).

    J.P. Heritage and R.K. Jain, Appl. Phys. Lett. __32__, pp. 101-103 (1978).

3. V.N. Mahajan and J.D. Gaskill, Optica Acta __21__, pp. 893-902 (1974).

4. D.J. Kuizenga, D.W. Phillion, T. Lund and A.E. Siegman, Opt. Commun. __9__, pp. 221-226 (1973).

     This work was partially supported by the following
sponsors. Exxon Research and Engineering Company, General
Electric Company, Northeast Utilities Service Company,
New York State Energy Research and Development Authority,
the Standard Oil Company (Ohio), the University of Rochester,
and Empire State Electric Energy Research Corporation. Such
support does not imply endorsement of the content by any of
the above parties.

Pulse-Width Stabilization of a Synchronously Pumped Dye Laser

S.R. Rotman, C.B. Roxlo, D.Bebelaar, T.K. Yee, and M.M. Salour

Research Laboratory of Electronics, Massachusetts Institute of Technology, Cambridge, MA 02139, USA

The production of ultrashort pulses by the active mode-locking of dye lasers has been widely used for the last five years for the study of events which occur on picosecond time scales. However, experimental difficulties are posed by amplitude and pulsewidth fluctuations[1] caused by plasma insta-bilities in the Ar^+ laser, thermal drift in the cavity length of the dye laser, and electronic noise in the oscillator providing the signal for the acousto-optic mode-locking crystal. In this paper we will present the first attempts to stabilize the pulsewidth by using active feedback.

Our experimental setup (Fig. 1) includes an Ar^+ laser which is acousto-optically mode-locked at a repetition rate of 82 MHz; the 100 picosecond pulses from this laser pump a Rhodamine 6G dye laser. Pulses as short as 0.7 picoseconds (assuming a single sided exponential pulse shape) have been observed by removing all bandwidth limiting wedges and etalons from the cavity. The dye laser pulse width was measured on an intensity auto-correlator by observing second harmonic generation (SHG) in a KDP crystal. A DEC MINC computer controlled the dye laser cavity length with a transla-tion stage on which the dye laser output mirror was mounted.

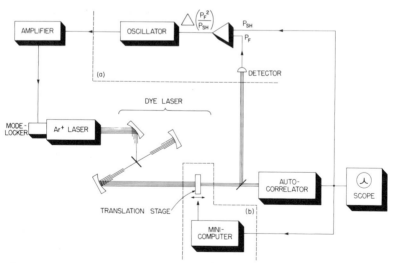

Fig. 1   Schematic of laser system: (a) fast analog feedback scheme, and
(b) digital scheme.

It is crucial that the cavity length of the dye laser be carefully matched to the repetition rate of the mode-locked Ar+ laser to assure the shortest pulse. [2] The second harmonic power P_{SH} is related to the fundamental power P_F and the pulsewidth t by $P_{SH} \sim (P_F)^2/t$.[3] The frequency dependence of the second harmonic power is dominated by pulsewidth variations. As the oscillator frequency of approximately 41 MHz is changed by 1 kHz, the second harmonic power varies by more than an order of magnitude while the visible power changes by less than 10%. At the frequency at which the shortest pulse occur, changes of as little as 10 Hz produce noticable changes in the pulse shape as observed on the autocorrelator. Similar results occur when the length of the dye cavity is changed; 10 Hz of the oscillating frequency corresponds to 0.5 microns of the 1.8 meter cavity.

The sensitivity of the pulsewidth to the cavity length and repetition frequency is the basis of our feedback scheme. From the above equation, one finds that, for small changes in P_F and P_{SH}, $\Delta t \sim 2 \Delta P_F/P_F - \Delta P_{SH}/P_{SH}$. The signal was fed back to the mode-locked oscillator as frequency modulation, thereby dynamically matching the Ar+ laser repetition rate to the dye laser cavity length. As seen in Fig. 2, the noise at frequencies less than 10 kHz was lowered by up to 10 dB at the expense of increased noise at higher frequencies. The bandwidth of the feedback system was limited to 10 kHz by the modulating crystal. The generation of the increased noise at higher frequencies must be due to nonlinear effects, presumably in the Ar+ laser system.

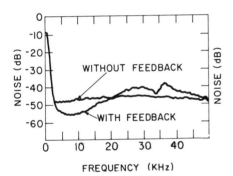

Fig. 2 Oscillograph taken from spectrum analyzer with 1 kHz resolution showing the result of feedback on the second harmonic output noise. The d c level of the second harmonic is -9 db (0.35 volts).

Noise spectra taken over a wider frequency range show that the second harmonic noise peaks around 150 kHz; in fact, most of the noise is concentrated in a 100 kHz band around this frequency. This is quite different from the spectra of the Ar+ output and the dye laser fundamental, which have flat profiles out to over 1 MHz. We believe that the high frequency band is due to temporal jitter in the Ar+ pulses which result in dye laser pulsewidth fluctuations. Further reductions in the pulsewidth will require a feedback system which has sufficient bandwidth to eliminate this noise.

Figure 1b shows the digital feedback scheme;low frequency noise and thermal drift were reduced by approximately 5 dB. This method compensated for thermal fluctuations in the length of the cavity to insure a perfect match to the Ar+ laser repetition rate (Fig. 3). The precise adjustment of the cavity length to better than 1 micron was made possible by this arrangement.

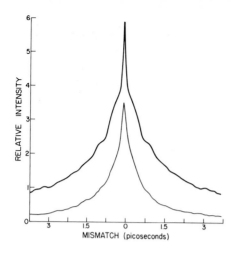

Fig. 3 The lower trace is the pulse resulting when feedback is applied for one hour; the pulsewidth is 0.7 picoseconds. The upper trace is the pulsewidth resulting from cavity mismatch after the system is allowed to drift for one hour without feedback; the pulse is 2.25 picoseconds. Note the correlation spike and the large amount of background signal with the mismatch pulse.

DYE AMPLIFIER SYSTEM

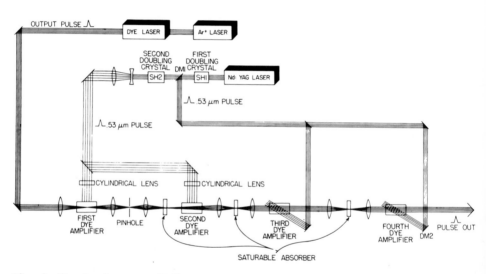

Fig. 4 The dye laser amplifier, pumped by a Nd:YAG laser at a ten Hertz repetition rate. A gain of 3×10^6 has been achieved.

A dye laser amplifier (Fig. 4) has been constructed to amplify the picosecond pulses generated by the synchronously-pumped dye laser system. Our amplifier, similar to that reported by R.L. FORK, C.V. SHANK, E.P. IPPEN, and A. MIGUS,[4] consists of a chain of four dye cells, pumped by an amplified Q-switched Nd:YAG laser which operates at a ten Hertz repetition rate and is synchronized electronically to the dye laser's picosecond pulses. A gain of 3×10^6 has been achieved with this system producing millijoule picosecond pulses with peak powers of a few Gigawatts.

In conclusion, we have reported the first demonstration of active feed-back applied to a synchronously pumped mode-locked dye laser to stabilize the pulses at the best possible shape and duration. Further application of this technique should allow the generation of shorter and more reproducible frequency tunable subpicosecond pulses.

References

1. R. H. Johnson, Proc. of the 1979 IEEE/OSA Conference on Laser Engineering and Applications, p. 84, 1978.

2. J. P. Heritage and R. K. Jain, Appl. Phys. Lett. 32, 101, 1978.

3. C. V. Shank and E. P. Ippen, App. Phys. Lett. 24, 373, 1974.

4. R. L. Fork, C. V. Shank, E. P. Ippen, and A. Migus, to be published.

Simultaneous Active AM Plus Passive FM with Q-Controlling Mode-Locked High Power Pulsed Laser

Yao Jian-quan

Tianjin University

China

After 1973 there were some new developments on the research of mode-locked solid-state lasers [1-10].

In order to obtain a stable and high-power ultrashort pulse a new mode-locked method is presented. We intend to combine active AM with passive FM and prelasing.

1. The Apparatus and Principle

The laser includes two resonators. The first one consists of laser medium (L), acousto-optical active AM mode-locked modulator (A), polarizer (P), Q-controller (C_1) and two mirrors (M_1, M_2). In the second resonator, except M_1, L and P, there are Q-controller C_2, optical Kerr effect modulator OKEM (F) and M_3.

The step voltage applied to electro-optical Q-controllers C_1 and C_2 were shown in Fig.2b,c. Before t_1 the two resonators were closed and the accumulation of population was provided. During t_1-t_2, the second resonator was still

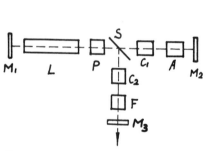

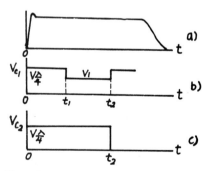

Fig. 1 The apparatus scheme (M_1, M_2: mirrors; M_3: output mirror; P: polarizer; L: laser medium; C_1, C_2: Q-controllers; S: beam-splitter; F: OKEM

Fig. 2 a) Optical pumping versus t; b) Voltage applied to C_1 versus t; c) Voltage applied to C_2 versus t

closed, but the first resonator was partially being opened. All the partial voltage V_1, the length of time that the V_1 is applied and the level of optical pumping can be adjusted. At the moment the resonator operates on prelasing which is just above threshold of the cavity. Because the mode-locked pulses repeatedly pass through AM-modulator, so pulsewidth can be compressed initially. After t_2, the first resonator was closed.

The OKEM consists of two standard quarter-wave plates and a CS_2 liquid Kerr cell. The transmission of pulse for a round trip through OKEM is a function of orientations and retardances of the wave plates and self-phase modulation (SPM) occurring in the cell. The parameters of OKEM are so selected as to guarantee that is has maximum transmission at the low power. Thus prelasering pulses pass through OKEM easily. A strong frequency chirp which is proportional to the power gain is produced. It is possible that the pulse is compressed to limiting duration.

2. The Prelasing Stage t_1-t_2

In this state by Siegman's theory [1] the steady-state mode-locked pulsewidth (τ_{p0}), the time (t_0) and numbers (M_0) of round trips which are necessary to reach steady state, and the transient pulsewidth of the M round trips are respectively given by

$$\tau_{p0} = \sqrt{2\ln 2}\ g^{\frac{1}{4}}/\pi(\theta_m f_m \Delta f_a)^{\frac{1}{2}} \tag{1}$$

$$t_0 = \Delta f_a/4\sqrt{2}\ g_0^{\frac{1}{2}}\theta_m f_m^2 \tag{2}$$

$$M_0 = \Delta f_a/2\sqrt{2}\ g_0^{\frac{1}{2}}\theta_m f_m \tag{3}$$

$$\tau_p(t) = \tau_{p0}/\left(\tanh \frac{M}{M_0}\right)^{\frac{1}{2}} \tag{4}$$

where θ_m is the depth of modulation, f_m is the frequency of modulation, Δf_a is the linewidth of laser medium, g_0 is the unsaturated round trip amplitude gain at prelasing stage.

Assuming that $f_m = 1 \times 10^8$, $\Delta f_a(\text{YAG}) = 1.2 \times 10^{11}$, $\Delta f_a(\text{Nd:glass}) = 3 \times 10^{12}$, we calculated different values for t_0, τ_{p0} and $\tau_p(t)$ for two laser mediums when g_0 and θ_m were taken (the results were neglected).

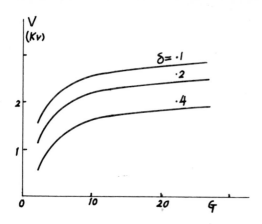

Fig. 3 The variation of V_1 versus G

Assuming that δ_0 is the other loss in the cavity besides C_1, G is total round-trip power gain at line center. C_1 is KD*P of longitudinal operation which is applied voltage V_1. By the threshold condition,

$$V_1 = \frac{\sin^{-1}\sqrt{1-\delta_0-1/G}}{2\pi n_0^3 \gamma_{63}/\lambda} .$$

(5)

The variation V_1 versus G is shown in Fig.3.

As mentioned above we get some results: 1) In order to obtain more narrow pulsewidth, we had better select θ_m as large as possible. 2) We must select suitable voltage V_1 by (5), and select a prelasing time as long as possible. 3) The optical pumping should adopt the power supply with multistage discharge circuit to make the pulsed laser as nearly quasi-CW, as shown in Fig.2a. 4) Any etalon effect must be avoided in two resonators.

3. The Stage of Compressing Pulsewidth

At the end of the prelasing stage, mode-locked pulses possess lower power and narrower pulsewidth. For example when $\theta_m = 0.6$ and $g_0 = 0.05$, for YAG:

(t=5 μs) = 75.8 ps , (t=10 μs) = 67.3 ps; for Nd:glass: (t=10 μs) = 46.9 ps, (t=100 μs) = 16.2 ps , (t=150 μs) = 14.5 ps.

At the moment t_2 the first resonator is being closed, but the second resonator is being opened simultaneously. In order to guarantee that the operation is reliable, it is necessary that both above-mentioned times should be synchronous. The parameters were so selected that the transmission of OKEM

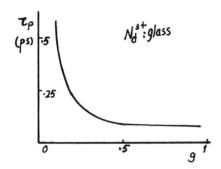

Fig. 4 The final pulsewidth versus g

at this moment must be near to one. Then the power of prelasing ultrashort
pulses will rapidly increase in the second resonator. When the power density
is high enough, the pulsewidth will be further compressed. Since the self-
phase modulation is produced by means of OKEM. This process is very fast.
The high power and narrow pulses are obtained.

The results indicate that the final pulsewidth approaches the limit of
transformation.

4. An Improved Scheme

In order to decrease the loss by beam-splitter, an improved device is presented (Fig.5).
The voltage applied to C_2 and the double $45°$ crystal C_1 are respectively
shown in Fig.6b,c. During $0-t_1$ laser is being closed. During t_1-t_2 the pulses
are prelasing in the resonator (M_1-M_2). After t_2 the prelasing pulses oscillate in the resonator (M_1-M_3). The process of three stages are shown in Fig.7.

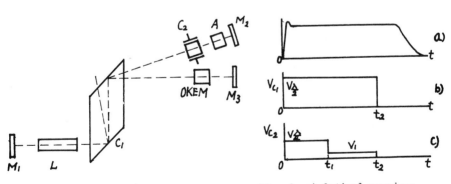

Fig. 5 An improved scheme

Fig. 6 a) Optical pumping;
         b) Voltage applied to C_1;
         c) Voltage applied to C_2

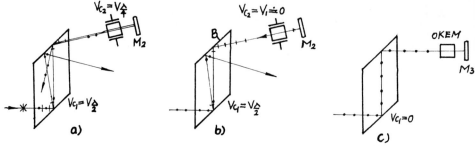

Fig. 7 a) closed stage; b) prelasering stage; c) stage of compressing pulse-width

In order to control the intensity of oscillation near threshold after the prelasing stage, we take $V_{c2} = V_1$. Thus the state of polarization of light which has passed through C_2 and turned back to surface B will rotate. In that time only its component oscillates in the cavity. In the moment t_2, the oscillation is transferred from the first resonator (M_1-M_2) to the second resonator (M_1-M_3). Because of mode competition, after t_2 only S component can oscillate. So it has not the loss which is present in the device shown in Fig.1.

5. Conclusion

This method cannot only maintain the features of higher stability and better reproducibility of active mode-locking and realize high-power output with prelasering, but also further compresses the pulsewidth by means of OKEM and obtains good synchronization. This apparatus is simple in structure and is suitable for various high-power pulsed lasers.

References

1 D.J. Kuizenga, D.W. Phillion, T. Lund, A.E. Siegman: Opt. Commun. 9, 221-226 (1973)
2 A.E. Siegman, D.J. Kuizenga: Optoelectronics 6, 43-46 (1974)
3 E.D. Jones, M.A. Palmer: Opt. Quantum Electron. 7, 520-523 (1975)
4 I.V. Tomov, R. Fedosejevs, M.C. Richardson: Appl. Phys. Lett. 29, 193-195 (1976)
5 I.V. Tomov, R. Fedosejevs, M.C. Richardson: Appl. Phys. Lett. 30, 164-166 (1977)
6 I.V. Tomov, R. Fedosejevs, M.C. Richardson: Opt. Commun. 21, 327-331 (1977)
7 I.V. Tomov, R. Fedosejevs, M.C. Richardson: Rev. Sci. Instrum. 50, 9-16 (1979)
8 D.J. Kuizenga: Opt. Commun. 22, 156-160 (1977)
9 A.J. Duerinckx, H.A. Vanherzeele, J.-L. van Eck, A.E. Siegman: IEEE QE-14, 983-992 (1978)
10 K. Sala, M.C. Richardson, N.R. Isenor: IEEE QE-13, 917-924 (1977)

A Model of Ultra-Short Pulse Amplification

A. Migus, J.L. Martin, R. Astier, and A. Orszag

Laboratoire d'Optique Appliquée, Ecole Polytechnique - ENSTA
F-91120 Palaiseau, France

1. Introduction

There has been recently important efforts devoted to the amplification of
subpicosecond pulses, with laser pumped dye amplifier, in view of numerous
applications in the domain of time—resolved spectroscopy. But still, little
has been done for the theoretical understanding of such experiments. We develop here a theoretical model of laser pumped dye amplifiers to be compared to
experimental arrangements which amplify subpicosecond pulses to gigawatt
peak power [1,2].

2. Energy stored in a dye cell :

A dye cell, concentration N, is uniformly excited on a region of width and
depth 2r, and length L (r << L). A simplified energy-level schema is considered with only the ground state S_0 and the first excited electronic state S_1
with lifetime τ. The pumping of a dye solution (Fig.1) with a pulse of intensity I_p and duration of a few τ results in the generation of an excited state
population N_1 and two counterpropagating waves of amplified spontaneous emission (ASE) ψ^+ and ψ^-. Assuming conditions of steady state, strong pumping
and important ASE, it can be shown [3,4] that the ASE flux can be considered
as monochromatic with a wavelength λ_m which can be deduced from the rate
equations. With these assumptions the equations take the simple form

$$\frac{d\psi^{\pm}}{dx} = (\sigma_{em} + \sigma_{am})\, \psi^{\pm} N_1 - \sigma_{am}\, N\, \psi^{\pm} \tag{1}$$

$$N_1 = N\, \frac{\sigma_p\, I_p + \sigma_{am}\, (\psi^+ + \psi^-)}{\sigma_p\, I_p + \tau^{-1} + (\sigma_{em} + \sigma_{am})\, (\psi^+ + \psi^-)} \tag{2}$$

where σ_p, σ_{am}, σ_{em} are, respectively, the absorption cross-section for the pump,
the absorption and emission cross sections at λ_m, and x the longitudinal coor-

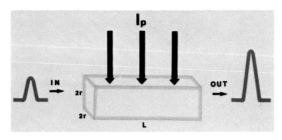

Fig.1 Geometry of pumping

dinate in the cell. This system gives exact analytical solutions. We present (Fig.2) the case of the first stage of our setup : a cell of Kiton Red ($N = 3.10^{17}$ cm^{-3}, $2r = 3.10^{-2}$ cm, $L = 3$ cm) pumped by a pulse of duration 12 ns, energy 30 mJ at wavelength 530 nm. More than the distribution of excited population, the important parameter is the density of excited molecules per transverse area unit

$$N_{st} = \int_0^L N_1 (x) \; dx \simeq \frac{\sigma_{am}}{\sigma_{em}} NL + \frac{1}{\sigma_{em}} \ln \frac{\psi^+(L)}{\psi^+(0)} \qquad (3)$$

because one excited molecule has the potential to deliver one photon at the wavelength of the amplified pulse. The total amount of stored energy in the cell is equal to

$$E_{st} = 4r^2 \; N_{st} \; h\nu_\ell$$

where ν_ℓ is the frequency of the amplified pulse.

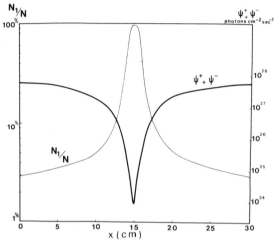

Fig.2 Flux of the A.S.E. and ratio of excited state population as a function of position in the cell (the parameters are in text).

For the above example ($\lambda_\ell = 613$ nm) we deduce $E_{st} \sim 20$ µJ which demonstrates that the permanent number of molecules reaching the state S_1 is much smaller than the number of photons in the pump pulse. This low efficiency can be explained as follows : the equilibrium is established within a time shorter than the expected time constant $\tau/1 + \sigma_\rho \; I_\eta \tau$ (and as a consequence shorter than the pump pulse duration) because of the depletion of S_1 due to the ASE ; then most of the pump photons coming after the steady state has been reached, do not increase the population N_1.

3. Amplification of ultrashort pulses

Once this steady state is reached a very short pulse of intensity I (x,t), t being its local time, travels in the excited medium. Writing the travelling wave equation along the longitudinal direction and the population rate equations [4] we get the pulse shape at the output x = L :

$$I(L,t) = \frac{\exp Z \exp E_0(t)}{1 + [\exp E_0(t) - 1] \exp Z} \quad I(0,t) \qquad (4)$$

and

$$G = \frac{\ln [1 + (\exp E_0(+\infty) - 1) \exp Z]}{E_0(+\infty)} \qquad (5)$$

the gain in energy (Fig.3) with Z a parameter of stored energy :

$$Z = (\sigma_\ell + \sigma_a) N_{st} - \sigma_a NL$$

σ_ℓ, σ_a being the emission and absorption cross sections at λ_ℓ and

$$E_0(t) = (\sigma_\ell + \sigma_a) \int_{-\infty}^{t} I(x,t') \, dt'$$

the normalized energy contained in the pulse from the far leading edge to the local time t.

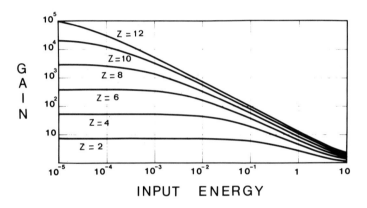

INPUT ENERGY

Fig.3 Gain of an amplifier stage as a function of the normalized input pulse energy , for different values of the stored energy parameter Z

We can conclude that a gain of 10^6 is impossible to reach with only one stage. An amplifier chain is needed, with increasing beam cross section. Nevertheless the amplifier stages are always driven in a near saturation regime, producing then distortion of the pulse shape. The analytical solution (4) takes into account the effects due to the pulse shape, and demonstrates the importance of a steep leading edge to avoid broadening. We present the shape of the amplified pulse for different input energy densities in two different cases : the input is gaussian (Fig. 4a) or is cosecant-like [5] (Fig. 4b).

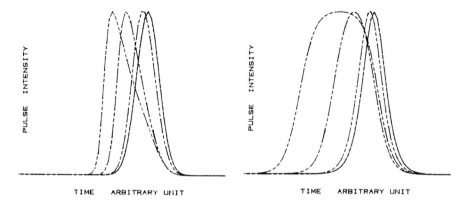

Fig.4   Normalized amplified pulse for a gaussian (Fig.4a on left) and
         cosecant like(Fig.4b on right) input pulse(continuous line)
         corresponding to three values of normalized input energy (1 ,
         0.01 and 0.0001 from left to right) for a stored energy
         parameter Z = 10 .

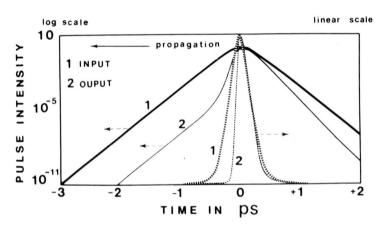

Fig.5 A cosecant like pulse through a malachite green jet. Curve 1
       (curve 2) is the input (output) and continuous (disconti -
       .nuous) line corresponds to a logarithmic (linear) scale .

4. Effect of saturable absorbers

To have not too much distortion we found that we had to avoid the case of
deep saturation of the gain, but also to steepen the leading edge before the
pulse was sent into an amplifier stage. For this last purpose a saturable
absorber stream separates one stage from the next.

   We solved numerically the equations of the saturable absorber in the case
of fast recovery, and for given input shapes. To realize the goal of producing
very steep leading edge while keeping a good transmission for the remaining
part of the pulse we found that very concentrate solutions had to be used, and
the beam tighly focused into the jet. As an example we present (Fig.5) the

pulse shaping of a cosecant like pulse, of energy 25 μJ, duration 0.5 psec, focused into a spot 200 μm diameter, through a Malachite green jet (small signal transmission 2.10^{-4} at 615 nm, recovery time : 2 ps) with an overall transmission close to 55 %.

5. Conclusion

The above theory gives results which are in reasonable agreement with the performance of different setups [1,2]. We could derive and check that a three stage dye amplifier pumped with a 30 MW pump pulse should amplify sub-picosecond pulses (from a passively mode-locked dye laser) to energy of the order of one to two millijoules, without significant broadening. In particular we demonstrated the need to control the overamplification of the leading edge by using saturable absorber and by avoiding very deep saturation of the gain.

Our setup has been used in numerous applications. One of these experiments [6] gives some informations about the shape of the pulses. One of the interpretations of the Fig. 3b in ref. [6] may reveal that the amplified pulses present an exponential leading edge with a subpicosecond time constant and a much steeper trailing edge, result which would be in good agreement with our analysis.

This work has been supported by the Direction des Recherches, Etudes et Techniques.

References

1. C.V. Shank, R.L. Fork, R.F. Leheny and J. Shah, Phys. Rev. Lett., 42, 112 (1979).

2. J.L. Martin, R. Astier, A. Antonetti, C. Minard and A. Orszag, C.R. Acad. Sc. Paris, B 289, 45 (1979).

3. U. Ganiel, A. Hardy, G. Neumann and D. Treves, IEEE J.Q.E. 11, 881 (1975).

4. A. Migus, C.V. Shank, E.P. Ippen and R.L. Fork, to be published.

5. H.A. Haus, C.V. Shank and E.P. Ippen, Opt. Comm., 15, 29 (1975).

6. J. Etchepare, G. Grillon, A. Antonetti and A. Orszag : this issue.

Picosecond Interferomatric Studies of CO_2 Laser Produced Plasmas

M.D.J. Burgess, G.D. Enright, R. Fedosejevs[1], and M.C. Richardson
National Research Council of Canada, Division of Physics,
Ottawa, K1A 0R6, Canada

1. Introduction

It has recently become apparent that the interaction of intense, short-pulse, CO_2 laser radiation with matter is dominated by collisionless absorption processes which couple energy primarily into fast electrons. This quickly leads to the formation of a two component plasma system with a relatively long lifetime, high density, thermal plasma adjacent to the solid target, surrounded by a much more transient collisionless superthermal plasma which rapidly propagates out of the interaction region. The diagnosis of these two regimes with high spatial and temporal resolution can provide considerable insight into the interaction physics relevant to current laser fusion studies.

2. The Thermal Plasma

2.1 The Electron Density Structure

The principal role, thus far, of picosecond time-resolved interferometry (PTRI) in the study of high-power, ns, CO_2 laser interaction with a plasma has been the determination of the thermal electron density profile for a range of incident laser energies between 1 and 30 J.[1] The experiments (see Fig. 1) utilized one beam of the COCO II laser system incident upon hollow glass microballoon targets together with a folded wavefront interferometer illuminated by synchronized 0.527 µm, 50-150 ps (FWHM) probe pulses generated by frequency doubling the output of an actively mode-locked neodymium phosphate glass laser system. Quantitative analysis of the interferograms revealed scale lengths and density steps through the critical density region respectively much less than a vacuum wavelength in extent and considerably higher than could be accounted for by a simple balancing of an isothermal plasma and the radiation pressure. In order to self-consistently explain these phenomena, a two-temperature description of the plasma distribution is required. Qualitatively, PTRI also revealed information about the overall thermal plasma morphology; its stability, its expansion velocity and the extent of any cratering, pinching or rippling of the critical density surface.[2]

2.2 Thermal Temperature Determination

In regions remote from the interaction zone, where pondermotive effects are assumed to be small, PTRI has also been used to obtain the temperature of this thermal plasma from the temporal variation of the electron density scalelength.[3] Knowledge of this temperature and the fractional laser energy absorption leads to an estimate for the amount of energy being transferred to the fast ions. For thin-walled (1 µm), empty, glass microballoon

---

[1] Present address: Projektgruppe für Laserforschung der Max-Planck-Gesellschaft, Garching, Federal Republic of Germany

targets of 70-150 μm diameter, irradiated from one side by 30 J laser pulses, at a power density of $\sim 10^{14}$ W cm^{-2}, more than 80% of the absorbed energy is decoupled from the target. It is, therefore, of great interest to determine some further properties of the superthermal plasma distribution.

3.   Effects of the Superthermal Plasma Expansion

3.1   Introduction

The superthermal plasma density is too low to be observed directly using optical interferometry. However, its interaction with, and subsequent ionization of, any neighbouring solid matter can indeed be seen. Evidence for the propagation of such an ionization front along the microballoon target support stalks has already been presented.[2] We here describe more recent work using different types of targets to elucidate the early time behaviour of the plasma.

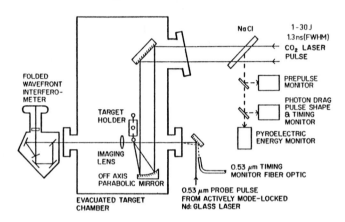

Fig. 1   Experimental Layout

3.2   Single Disc Experiments

Figs (2b-2d) show the result of irradiating 500 μm diameter, 12 μm thick aluminum discs (Fig. 2a) with about 4 J of CO_2 laser energy. After only 300 ps (Fig. 2b) a plasma has spread almost up to the edge of the disc in one direction and has travelled out of the field of view along the target support in the other. Some plasma has also formed on the rear surface. At a later time, but still during CO_2 irradiation (Fig. 2c), the rear surface is completely covered although the plasma in the centre appears to be relatively cold judging by its opacity and the small amount of fringe movement, resulting in the idea of the annular ring of plasma expanding inwards from the rim. Towards the end of the irradiating pulse (Fig. 2d), the toroidal nature of the plasma on the rear surface is even more apparent. From the expansion velocities off the front and rear surfaces (of Fig. 2c), the temperatures of the focal zone and central rear plasmas have been estimated at ~ 300 eV and ~ 30 eV respectively. Therefore, before the peak of the CO_2 laser pulse, both front and back of the discs are covered by plasma. However, significant heating of the plasma at the rear would seem to occur on a time-scale that is not inconsistent with classical thermal diffusivity, as the surface plasma that is formed can itself be a good conductor. It is also interesting to note that Figs 2c and 2d respectively

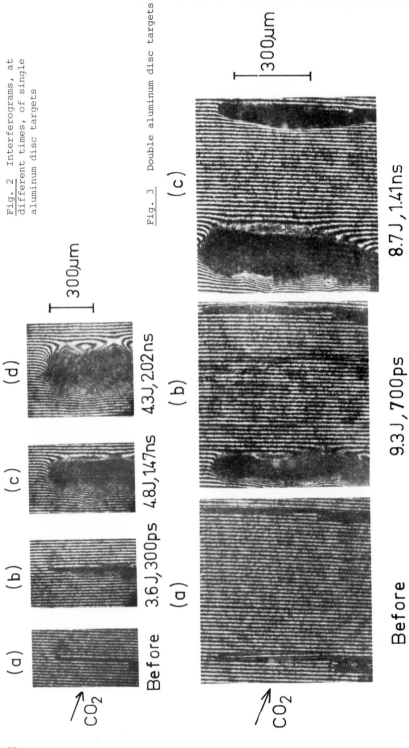

Fig. 2 Interferograms, at different times, of single aluminum disc targets

(a) Before
(b) 3.6J,300ps
(c) 4.8J,147ns
(d) 43J,202ns

300μm

Fig. 3 Double aluminum disc targets

(a) Before
(b) 9.3J,700ps
(c) 8.7J,1.41ns

300μm

CO_2

CO_2

66

show evidence of an overdense bump and cratering as found previously using microballoon targets.(2)

3.3 Double Disc Experiments

The effect of irradiating double-disc targets ($A\ell$, 12 μm thick, 500 μm dia-meter, 500 μm separation) with approximately 9 J of incident CO_2 laser en-ergy is shown in Fig. 3; at all times the irradiated disc behaves in a similar manner to the single targets described above. However, during ir-radiation (Fig. 3c), a substantial amount of energy has been transferred to the rear target even though the laser-induced thermal plasma has not expan-ded beyond 150 μm. Such interferograms, together with independent optical streak measurements and time-integrated x-ray photography are strong evi-dence for the transport of a significant fraction of the absorbed energy on a superthermal ion-acoustic time-scale by means of the super thermal coro-na.(4) Taking into consideration the delay in formation of the rear plasma, its temperature is also estimated to be ~30 eV. According to the present interpretation, before the superthermal corona reaches the second target it will already have been enveloped by the fastest, orbiting electrons which set up the ambipolar potential that accelerates the ions. If Fig. 3b is looked at closely, a small fringe shift at the front and back of the rear target can be detected, indicative of neutral vapour. At the time this interferogram was obtained, the leading edge of the superthermal corona should be within 100 μm of the rear foil showing that the integrated energy density carried by those electrons a few debye lengths ahead of the fast ion front is quite small. By placing several targets around an interaction region, the angular superthermal distribution could be mapped out.

3.4 Hollow cylinder experiments

In order to gain some insight into how the two temperature distribution affects the subsequent collapse of a microballoon, hollow glass cylinders were viewed end-on, to reveal the nature of their implosion. The glass cy-linders (Fig. 4a) were typically 300 μm long and 750 μm in diameter, with wall thickness in excess of 10 μm. At the earliest time (Fig. 4b) plasma already covers both the inner and outer surfaces. At later times (Figs 4c and 4d), the most interesting feature is the preferential expansion of the side opposite to the focal zone. However, it is not possible from the pre-sent observations to uniquely determine whether the cause is directly due to

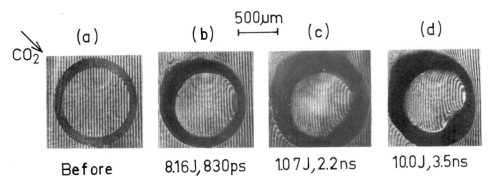

Fig. 4   Hollow cylindrical glass targets

hot electrons multiple passing or the superthermal corona created on the inside surface being accelerated across the intervening distance.

4.   Conclusions

The foregoing briefly illustrates the usefulness of current PTRI studies of laser fusion plasmas. The physical picture that emerges of the fastest electrons rapidly encircling the micro-targets generating their self-confining potential whilst also accelerating the superthermal corona is amply confirmed by the interferograms presented here. Coupled with other diagnostic techniques, it provides a powerful approach to understanding these highly transient and complex events.

Acknowledgements

The authors express their gratitude for the very able technical assistance of P. Burtyn and Y. Lupien, and G. Berry, K. McKee, W.J. Orr and R. Sancton.

References

1.   R. Fedosejevs, M.D.J. Burgess, G.D. Enright, M.C. Richardson:
      Phys. Rev. Lett. 43, 22, 1664 (1979)
2.   R. Fedosejevs, M.D.J. Burgess, G.D. Enright, M.C. Richardson
      (To be published)
3.   G.D. Enright, R. Fedosejevs, M.D.J. Burgess, D.M. Villeneuve,
      C. Joshi, N.H. Burnett, M.C. Richardson:
      Bull. Am. Phys. Soc. 24, 8, 1027 (1979)
4.   R.S. Marjoribanks, M.D.J. Burgess, C. Joshi, G.D. Enright
      M.C. Richardson:
      Bull. Am. Phys. Soc. 24, 8, 1027 (1979)

Part II

Advances in Optoelectronics

Recent Advances in Picosecond Optoelectronics

D.H. Auston, P.R. Smith, A.M. Johnson[1], W.M. Augustiniak,
J.C. Bean, and D.B. Fraser

Bell Laboratories, Murray Hill, NJ 07974, USA

The extremely short carrier lifetimes of amorphous and radiation-
damaged semiconductors have been utilized to make high-speed
photoconductive detectors and sampling gates with response times
of approximately 10 ps.

1. Introduction

The ability to make electronic measurements in the picosecond
time domain is important both for the study of the electronic
transport properties of materials and for the development of new
high-speed devices. One avenue towards the achievement of this
goal is the utilization of picosecond optical pulses to generate
and manipulate electronic signals in high-speed photoconducting
materials. Previous work[1] in this area has clearly demonstrated the feasibility of this approach. In this paper, we
report the results of recent experiments with some new materials
and circuit configurations which have produced a significantly
improved measurement capability.

2. Materials

Previous work on this topic utilized crystalline semiconductors
as the photoconducting materials. A distinct disadvantage of
these materials, however, is their relatively long carrier lifetimes. The most effective cause of short carrier lifetimes in
semiconductors are deep-level defect states or "traps". For
most applications, these states are undesirable and much effort
has been expended to grow relatively "defect-free" crystals. Our
requirements here, however, run counter to the traditional
approach and suggest that materials with high defect densities
may have unique applications for picosecond optoelectronic
devices. Specifically, we have found that amorphous semiconductors and radiation-damaged semiconductors, both of which contain
very high densities of structural defects, are extremely valuable
photoconducting materials for picosecond application.

In the case of amorphous semiconductors, we utilize the dis-
tinction between the different modes of conduction in extended
and localized states. [2] With suitable optical excitation above
the mobility gap, we can produce a short current impulse due to
extended state conduction which rapidly terminates by relaxation
to low-mobility localized states. The density of localized
states and hence the carrier lifetime depend strongly on the
method of preparation of the amorphous films and can vary from
3 ps to 300 ps. Typical extended state mobilities are of the

---

[1] Physics Department, City College, CUNY, New York, NY 10023, USA

order of 1 cm^2/Vs. The extremely high resistivities of these
films (typically 10^3 to 10^7 Ωcm) offer the additional advantage
of low dark current. The details of the transport properties of
these films are the subject of a companion paper [3] in this
volume.

In the case of radiation-damaged semiconductors, we have
utilized ion-implantation of varying doses to continuously reduce
the carrier lifetime by introducing a specific density of defect
states. For example, we have found that the implantation of
0.8 MeV O$^+_2$ into silicon-on-sapphire (SOS) films can reduce the
lifetime from approximately 1 ns to 8 ps with doses ranging from
3×10^{11} to 3×10^{15} ions/cm^2. Due to the residual short-range order
of the crystalline lattice, the mobilities are higher by at least
an order of magnitude than comparable samples of amorphous films.

3. Detectors

We recently reported [4] a high-speed photoconductive detector
which utilized a thin film of amorphous silicon grown by chemical
vapor deposition. Its response to optical pulses from a mode-
locked Rh6G dye laser was 40 ps (full width at half-maximum) when
measured with a sampling scope having a 25 ps aperture. More
recently, we have found that radiation-damaged silicon-on-sapphire
films can also be used to make high-speed detectors suitable for
use with picosecond pulses. They have the advantage that the
carrier lifetime can vary over a wide range, depending on the
radiation dose, and also they have much higher mobilities than
completely amorphous films and are consequently much more sensitive. Fig. 1 is an example of a detector made from a silicon-on-
sapphire film which was irradiated with a dose of 9×10^{14} O$^+_2$
ions/cm^2 at 0.8 MeV. On the sampling scope, the response was
measured to be 55 ps FWHM. The sensitivity of this detector is
approximately 20 times greater than a comparable detector using
a completely amorphous silicon film grown by chemical vapor
deposition.

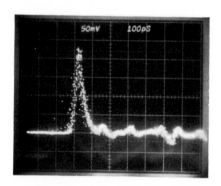

Fig. 1 Sampling scope response
of radiation-damaged silicon-on-
sapphire detector. The incident
optical energy per pulse was
approximately 0.3 nJ and the
bias was 65 V.

4. Electronic Autocorrelations

When the material response times become faster than approximately
50 ps, sampling oscilloscopes are no longer adequate. To make
accurate electronic measurements in the time range below 50 ps,

we have developed a technique [5] which correlates the response
of two photoconductors. One photoconductor acts as a detector,
and the output signal from it is used to bias the second photoconductor which acts as a sampling gate. If the total charge
from the second photoconductor is then measured, one finds that
the response is proportional to the second-order correlation
function:

$$Q(\tau) \propto \int_{-\infty}^{+\infty} V(t) \ V(t+\tau) \ dt$$

where τ is the relative time delay between the optical pulses
incident on the two photoconductors, and $V(t)$ is the response
of a single photoconductor. As example of the measurement
scheme is shown in Fig. 2. In this case, the material was an
amorphous silicon film prepared by evaporation in ultra-high
vacuum. The autocorrelation measured 11 ps FWHM. An approximate deconvolution of this result indicates that the material
response was pproximately 4 ps, and the circuit approximately
4 ps. (The optical pulses were 3.5 ps.)

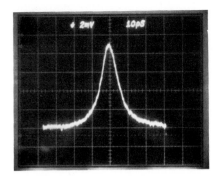

Fig. 2 Electronic auto-
correlation of an amorphous
silicon photoconductive film
prepared by evaporation.

5. Sampling Gate

A straightforward extension of the electronic autocorrelation
measurement technique is to use two photoconductors which are not
identical, one of which has a known very fast response. The
known, fast material then acts as a true sampling gate to measure
directly the response of the other material (or device). An
example of this scheme is illustrated in Fig. 3, in which a
sputtered amorphous silicon film is used as a sampling gate to
measure the response of a silicon-on-sapphire film. Since the
response is virtually jitter-free, the signal can be averaged
to give good signal-to-noise. Although the rise-time is not as
fast in this case due to the somewhat slower response of the
sputtered film, the result clearly illustrates the utility of
this technique. Notice also the absence of reflections, a
problem that usually complicates conventional sampling scope
measurements.

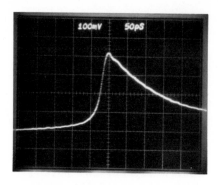

Fig. 3 Opto-electronic sampling gate using a sputtered amorphous silicon film correlated with a slower silicon-on-sapphire sample.

6. Conclusions

The materials and measurement techniques we have outlined provide the basis for an electronics measurement capability in the range of approximately 10 ps. With further refinements, we expect to be able to improve the resolution to approximately 3 ps. We are currently investigating numerous applications in the area of transport measurements and device evaluations. A companion paper [3] in this digest deals with the application of these techniques to the study of electronic transport in amorphous silicon.

References

1. For a review, see: D. H. Auston, "Picosecond Opto-Electronics" to be published in Progress in Quantum Electronics.

2. See, for example, N. F. Mott and E. A. Davis, "Electronic Processes in Non-Crystalline Solids", Clarendon Press, Oxford (1979).

3. A. M. Johnson, D. H. Auston, P. R. Smith, J. C. Bean, J. P. Harbison, and D. Kaplan, "Picosecond Photoconductivity in Amorphous Silicon", see proceedings of this conference.

4. D. H. Auston, P. Lavallard, N. Sol, and D. Kaplan, Appl. Phys. Lett. $\underline{36}$, 66 (1980).

5. D. H. Auston, A. M. Johnson, P. R. Smith, and J. C. Bean, to be published in Appl. Phys. Lett..

Optoelectronic Switching in the Picosecond Time Domain and Its Applications

G. Mourou and W. Knox

Laboratory for Laser Energetics, and

M. Stavola

Department of Chemistry, University of Rochester, Rochester, NY 14627, USA

The laser-activated photoconductive switch [1-4] allows generation of picosecond risetime electrical pulses that are in picosecond synchronism with optical pulses. The availability of such a driver has led to a number of applications that include ultrafast active pulse shaping [5], jitter-free streak camera operation [6], active pre-pulse suppression in a laser driven fusion system [7] and more recently the generation of powerful microwave bursts [8].

Electrooptic devices working in the picosecond domain often require picosecond switching synchronism and excellent voltage amplitude stability. Signal averaging with a streak camera is one example where amplitude fluctuations of less than 1% are required. This constraint precludes the use of pulse bias techniques in many cases because of the shot to shot voltage variation and the timing fluctuation between the laser and the high voltage bias pulse. Here, the photoconductive switching of a kilovolt DC bias by Au-doped Si held at liquid nitrogen temperature is demonstrated [9]. The amplitude and temporal stability of the electrical pulse obtained with this technique is illustrated in a streak camera application where a picosecond jitter is reported. Using this picosecond accuracy a picosecond optical transient lifetime has been measured. In addition to the complete alleviation of bias timing and amplitude stability problems DC photoconductive switching at cryogenic temperature should lead to KHz-repetition rates. It should be noted that while highly Cr-doped GaAs at room temperature [3,4], exhibits similar voltage hold off capability as the Au-doped Si used here at cryogenic temperature, the inability to obtain pulse durations of more than a few hundred pico-seconds restricts the room temperature use of such material severely in electrooptic applications where square pulses that are several nano-seconds in duration are required. Also, Si, with its band gap at 1.12 eV as opposed to that of GaAs at 1.42 eV, can absorb a near infrared excitation laser more efficiently. Furthermore, the absence of the thermal runaway allows one to apply DC fields well into the impact ionization field range leading to

 a) optimized trigger sensitivity in photoconductive switching

 b) avalanche multiplication switching.

Au doped Si switch description [9]

Nearly intrinsic Si (3×10^4 Ωcm) as is used in pulsed bias, room temperature switching devices, was found to be an unacceptable switching element at liquid nitrogen temperature. At low voltage a ten-fold increase in resistivity was observed followed by a premature bulk

dielectric breakdown that was monitored on a nanosecond time scale for
an electric field strength of ~1000 V/cm. An interpretation of these
results that is well known from the early work on dielectric breakdown
in insulating crystals [10] is that at low temperature, the freezing out
of phonon motions increases the mean free path of charge carriers so
that their kinetic energy can exceed the impact ionization energy even
for small electric field strengths leading to a collapse of the dielectric
breakdown field strength. To enhance the dielectric breakdown field
strength to the 100 kV/cm range at liquid nitrogen temperature Au
impurities were purposely added to decrease the carrier mean free path.

To maintain a high electrical bandwidth the switching element is
integrated into a microstrip transmission line on a sapphire substrate.
The block of Au-doped Si is 1mm x .5mm x 2mm. The 1mm thick sapphire
substrate is fastened to a brass plate that is in contact with a liquid
nitrogen cold trap. The apparatus is enclosed in a chamber at a pressure
of 1 millitorr to prevent condensation. The stripline is biased up to
3kV. Upon irradiation by a laser pulse the Au-doped Si becomes conducting
by photoconductivity, thereby discharging the biased stripline. The
fast pulse produced is shown in Figure 1. A 1.5 nsec charge line biased
to 2.4 kV was used for this oscillogram. The laser energy of 150 µJ
± 10% that activates the Si-switch is about 20 times the value required
to provide 50% switching efficiency. Laser energy and optical pulse
shape fluctuations are believed to be responsible for the residual
jitter of a few picoseconds.

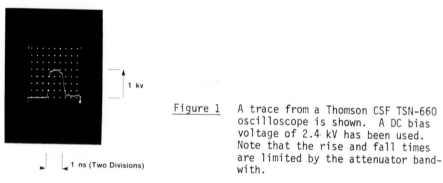

1 kv

1 ns (Two Divisions)

Figure 1   A trace from a Thomson CSF TSN-660
oscilloscope is shown. A DC bias
voltage of 2.4 kV has been used.
Note that the rise and fall times
are limited by the attenuator band-
with.

Picosecond high power switching applied to a streak camera [6,11]

Among all of the applications requiring picosecond high power
switching we will describe the one which from our point of view is the
most demanding: the picosecond jitter streak camera. In order to
eliminate shot-to-shot fluctuations, picosecond switching accuracy and
pulse amplitude fluctuations less than one precent have to be achieved.
We recall that the switching time fluctuations give rise to zero time
fluctuations whereas voltage fluctuations translate into sweep speed
fluctuations. DC-biased switching is used to provide the ramping voltage
for the streak camera. An extremely stable short and long term operation
has been observed making possible signal averaging techniques and signal
multiplexing. In addition, the low timing jitter allows a local sweep
speed calibration to be made that does not involve the generation of a
train of pulses with a bounce etalon.

76

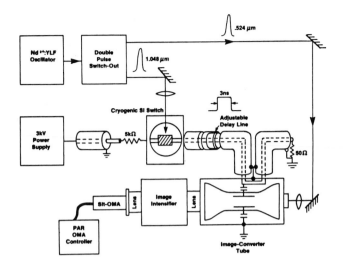

Figure 2 Experimental configuration. A single 3P-ps pulse from a mode-locked Nd3:YAG laser is frequency doubled. A silicon block which interrupts the continuity of a 50-Ω coaxial line becomes conducting under the action of the IR laser pulse and an electrical pulse is generated of length corresponding to the charge-line length with rise time limited by the optical pulse width. This pulse is applied to the deflection-plate of an image converter tube and the swept image is recorded by an intensified OMA

Figure 2 shows the experimental arrangement. Two pulses are selected from the train of a mode-locked Nd^{3+}:YLF oscillator, one by a double Pockels cell switch out system (contrast > 10^5) and one by a single Pockels cell switch out system (contrast > 10^3), and are directed along separate beam lines. The higher contrast pulse is used to irradiate a 2mm gap cryogenic Si-switch. The electrical pulse generated provides the deflection voltage for a Photochron II streak tube (S20 photocathode). The lower contrast pulse is frequency doubled and directed towards a sample cell at the entrance of the streak camera. The resolution of the system was focusing limited to about 5 or 6 psec, less than the width of any temporal structure to be resolved. An OMA II was used to collect and store large numbers of streaked traces that could then be averaged and background corrected. In Figure 3 are presented streaked traces of the average of 30 laser shots, separated by a delay time corresponding to the addition of 2.1mm of glass to the beam line. From the index of refraction of the glass we estimate the time delay should be 3.9 psec in agreement with the observed average shift of 3.9 ± .5 ps. Also shown in Figure 3 is the difference of the two traces. The amplitude of the signal difference is proportional to the introduced time delay. The signal to noise ratio of the signal difference indicates that, by averaging only over few tens of shots this technique can give rise to subpicosecond precision.

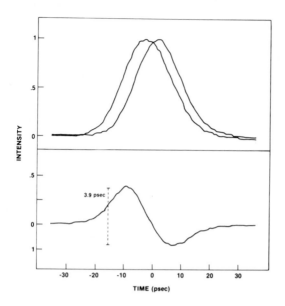

Figure 3
Two series of 30 accumu-
lated shots of the
optical excitation. The
optical delay path has
been offset by 3.9 ps.
In the bottom, the
difference between the
two series showing the
sub-picosecond timing
precision. Optical
pulse width: 18 ps.

This ability to measure small time delays has been used to measure
the fluorescence time of the purple membrane (PM) [12] which is a
fraction of the 20 psec laser excitation pulse width. The streaked
signal, averaged over 280 shots is shown in Figure 4. It has previously
been shown that if the transient response does not exceed 1/3 of the
laser pulse width, the time shift between the peak of the laser excitation pulse and the fluorescence signal is equal to the response time of
the system at the 1/e decay point. However, the energy of the photo-
electrons emitted from the photocathode of the streak tube in response
to the infrared P.M. fluorescence is less than those emitted in response
to the green stimulus light. This leads to different streak characteristics and precludes use of the green stimulus as the reference. To
alleviate this problem, the zero-time reference was provided by the 1/2
rise time of the long lifetime (150 psec) dye crystal violet in glycerol.
The crystal violet peak emission wavelength matches with the purple
membrane emission peak. A small correction of 0.9 psec is introduced
to take into account the finite crystal violet solution lifetime.

ACKNOWLEDGEMENT

This work was partially supported by the following sponsors: Exxon
Research and Engineering Company, General Electric Company, Northeast
Utilities Service Company, New York State Energy Research and Development Authority, The Standard Oil Company (Ohio), The University of
Rochester, and Empire State Electric Energy Research Corporation. Such
support does not imply endorsement of the content by any of the above
parties.

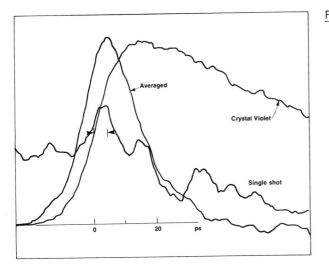

Figure 4 Fluorescence signals at $\geq$ 700 nm of the reference long-lifetime dye crystal violet, single shot P.M. fluorescence, and P.M. fluorescence integrated over 280 shots are displayed. Single-shot data is inadequate for lifetime determination because of the 2.4×10^{-5} quantum efficiency of the P.M.[11]. The integrated fluorescence signal yielded a 4.0 ± 0.3 psec fluorescence decay of the P.M.

References

1. D. Auston, Appl. Phys. Lett. 26, 101 (1975).

2. P. Lefur and D. Auston, Appl. Phys. Lett. 28, 21 (1976).

3. C. Lee, Appl. Phys. Lett. 20, 84 (1977).

4. G. Mourou and W. Knox, Appl. Phys. Lett. 35, 492 (1979).

5. J. Agostinelli, G. Mourou and C. Gable, Appl. Phys. Lett. 35, 731 (1979).

6. G. Mourou and W. Knox, Appl. Phys. Lett. 36, 492 (1980).

7. G. Mourou, J. Bunkenburg and W. Seka, Optics Commun. To be published.

8. G. Mourou, D. Blumenthal, C. Stancampiano and C. Gable. Submitted to Appl. Phys. Lett.

9. M. Stavola, M. Sceats and G. Mourou, Optics Commun. To be published.

10. N. F. Mott and R. W. Gurney, Electronic Processes in Ionic Crystals (Dover, New York, 1964) p197.

11. M. Stavola, G. Mourou and W. Knox, Optics Commun. To be published.

12. R. Frankel, G. Mourou, W. Knox and M. Stavola. Submitted to Nature.

Novel Picosecond Optical Pulse Sampling Technique for Measuring the Response of High-Speed Optical Modulators

N.P. Economou[1], R.C. Alferness, and L.L. Buhl
Bell Laboratories, Holmdel, NJ 07733, USA

Efficient high-speed optical waveguide switches and amplitude modulators will be important components for utilizing the large available bandwidth of single-mode lightwave systems. In addition to amplitude modulation for digital coding, fast optical switches may prove very useful for high-speed time division multiplexing. Presently, several guided-wave optical switch/modulators have been reported with demonstrated modulation bandwidth of approximately 1 GHz [1,2]. For high-speed operation, accurate switching speed measurements under electrical pulse operation are hampered by the finite speed of the pulsing circuit and the photodetector. Generally, it is necessary to estimate the actual device switching time by deconvolving the response time of the pulser and detector from the measured output pulse [1]. Undesirable optical damage effects generally preclude the high optical power levels required for fast photodiodes or for nonlinear pulse correlation techniques [3]. We have developed a new measurement technique utilizing picosecond optical sampling pulses to overcome the limitation of finite detector speed.

The experimental arrangement is shown schematically in Fig. 1. A mode-locked argon laser synchronously pumps a Rhodamine 6G dye laser to produce a train of optical pulses at $\lambda \simeq 0.6$ μm with pulse width $\sim$5 psec [4]. These pulses are coupled into the waveguide modulator via a lens through an end-polished face. The RF mode-locker oscillator signal, frequency doubled to equal the argon laser pulse repetition frequency, also drives an electrical comb generator. The comb generator, by differentiating the output of a fast step-recovery diode, produces pulses with $\sim$60 psec rise and fall times and peak voltage of $\sim$15 volts at the repetition frequency of the RF driving signal. These pulses, synchronous with the optical sampling pulses, drive the modulator. The temporal shift between the electrical and optical pulses at the modulator is controlled with the electrical phase shifter (Fig. 1) or alternatively with an optical delay line. The modulator output, measured with the photomultiplier, varies with the temporal overlap of the optical sampling pulses and the electrical drive

---

[1] Present address: Lincoln Laboratories, Lexington, MA, USA

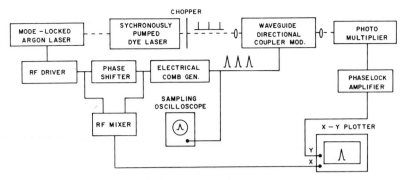

Fig. 1 Experimental Arrangement

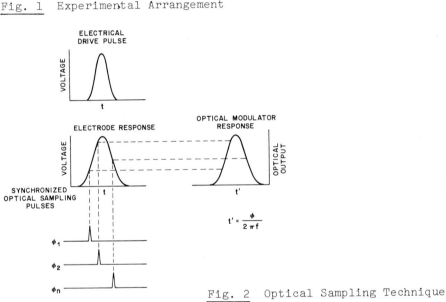

$$t' = \frac{\phi}{2\pi f}$$

Fig. 2 Optical Sampling Technique

pulses as broadened by the modulator electrodes lumped electrical characteristics. The technique is shown schematically in Fig. 2. Thus, the modulator time response is mapped out by measuring the averaged output optical power as a function of the time shift between the electrical and optical pulse trains. This time shift is given by the RF mixer difference signal (Fig. 1). Because the technique sums over many pulses, a fast detector is not required and the average optical power can be kept low.

This measurement technique has been employed to measure the speed of newly designed fast electrooptically switched waveguide directional couplers made with Ti-diffused lithium niobate waveguides. We have achieved significantly reduced device length compared to previous switches without a corresponding increase in the required drive voltage by maximizing the electrical-optical interaction in the crystal. Because the switching time scales approximately with length [1], reduced device length results in a proportional reduction in switching time.

The directional coupler was fabricated by photolithographically defining, on a Z cut Y propagating lithium niobate crystal, two titanium metal strips (∿650 Å thick) 1 μm separation over an interaction length of 750 μm. The metal was diffused into the crystal at 980°C for four hours resulting in single-mode waveguides for both the TE and TM polarizations [5]. After diffusion, electrodes were carefully aligned over the waveguides using a two-step metalization process to buildup a chrome/aluminum electrode thickness of ∿2000 Å. A thick layer is necessary to minimize electrode resistance. The uniform mismatch electrodes are each 30 μm wide with a 1 μm separation. The crystal ends were cut and polished to allow endfire coupling. The device was evaluated with TM polarized light allowing utilization of the strong r_{33} coefficient.

The compact lateral geometry of this device - the narrow waveguides and small interwaveguide and interelectrode gaps - results in a large applied electric field in the waveguide region for given electrode voltage. As a result, in spite of the short interaction length the modulator output can be modulated with ∿ -8 dB extinction with a 6 volt modulation voltage and 6 volt dc bias.

The measured device response together with the electrical drive signal from the comb generator is shown in Fig. 3. The measured 10 percent to 90 percent risetime of the optical modulator output pulse is ∿125 psec. Deconvolving the corresponding drive pulse risetime of ∿60 psec yields an effective switching time of ∿110 psec. In a nonreturn-to-zero digital format this corresponds to a data rate of ∿9 Gbs or an analog modulation rate of ∿4 GHz. This result is in reasonable agreement with the expected RC bandwidth ignoring parasitics and assuming R = 50Ω and 0.7 pf, calculated from the electrode geometry [1], of Δf ∿ 9 GHz.

ELECTRICAL DRIVE
SIGNAL

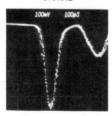

MODULATOR RESPONSE
VIA SAMPLING

200 psec

MODULATOR 10% TO 90% RISETIME

τ_m ~ 110 psec

Fig. 3 Measured
Modulator Response

In summary we report a novel technique for measuring the switching time of high-speed optical switch/modulators without requiring a fast photodetector. With this technique switching times as short as ~60 psec (limited by the electrical pulse width) can be measured with a resolution limited only by the mode-locked dye laser pulse width.

References

1. P. S. Cross and R. V. Schmidt, IEEE Journ. of Quant. Electron., to be published; see also, R. V. Schmidt and R. C. Alferness, IEEE Journal of Circuits and Systems, CAS-26, p. 1099 (1979).
2. O. Mikami, J. Noda and M. Fukuma, Trans. of the IECE of Japan, Vol. E 61, p. 144 (1978).
3. M. Maier, W. Kaiser and J. A. Giordmaine, Phys. Rev. Letters, 17, p. 1275 (1966).
4. See for example, R. K. Jain and J. P. Heritage, Appl. Phys. Lett., 32, p. 44 (1978).
5. R. V. Schmidt and I. P. Kaminow, Appl. Phys. Lett., 25, p. 458 (1974).

Ultrafast Magnetophotoconductivity of Semiinsulating Gallium Arsenide

R. Moyer, P. Agmon[1], T.L. Koch, and A. Yariv

Department of Applied Physics, California Institute of Technology, Pasadena, CA 91125, USA

Fast photoconductive switches have been investigated by a number of experimenters [1-3]. The primary objectives in these efforts have been the production of ultrafast current pulses (for Pockels-cell triggering, etc.), the demonstration of these switches as fast photodetectors, and the determination of the photocarrier recombination properties of the materials used.

The high absorption coefficient of Cr:G--s for visible light results in a critical dependence of photocarrier lifetimes on the surface recombination velocity. Excess carriers experience both surface recombination and bulk recombination; the former effect is due to recombination states created at the surface lattice termination or by impurities adsorbed onto the surface from the outside, and the latter is due to chromium recombination centers introduced during crystal growth. Single wavelength photoconductivity data alone are insufficient to establish reliable values of the surface recombination velocity and bulk recombination lifetime. The two phenomena act in concert, and must be decoupled in some way. One method of accomplishing this is to prepare the Cr:GaAs samples with various etchants known to passivate the surface to varying degrees. The bulk recombination effects would remain constant, while the surface recombination velocity would be prescribed by the surface treatment. Another method of separating the bulk from surface effects is by the application of a magnetic field.

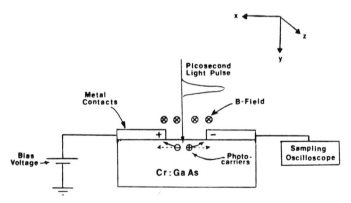

Fig. 1 Geometry for magnetophotoconductivity measurements

---

[1] Present address: Hewlett Packard Company, Palo Alto, CA 94304, USA

The field impels the carriers toward or away from the surface according to the direction of the field. Thus, one effect can be made more or less significant, facilitating a reliable calculation of the two recombination parameters.

Photoconductivity measurements are taken using the geometry shown in Fig. 1. Metallic contacts on a Cr:GaAs chip are separated by a gap. An electric field E is applied, but no current flows across the highly resistive gap. A picosecond light pulse is directed onto the gap, and the resulting electron-hole pairs allow current to flow for the duration of their existence. If a magnetic field B is applied as shown, the carriers (bearing opposite charges and traveling in opposite directions) are both drawn toward or away from the surface. Assuming no carrier density dependence in the x or z directions, we write a diffusion equation for the carrier density n

$$\frac{\partial n}{\partial t} = D\frac{\partial^2 n}{\partial y^2} - \frac{n}{\tau} - \mu^2 \, EB \, \frac{\partial n}{\partial y} \tag{1}$$

subject to the boundary conditions

$$\frac{D}{n}\frac{\partial n}{\partial y}\Big|_{y=0} = S - \mu^2 EB, \quad n(t=0) = n_o e^{-\alpha y} \tag{2}$$

where α is the absorption coefficient, S is the surface recombination velocity, D is the ambipolar diffusion coefficient, τ is the bulk recombination rate, and μ is the carrier mobility.

The solution is

$$n(y,t) = \frac{n_o}{2}\exp(-\frac{t}{\tau} - \frac{y^2}{4Dt} - k(y + Dkt))\Big\{W(\beta\sqrt{Dt} - \frac{y}{2\sqrt{Dt}}) + W(\beta\sqrt{Dt} + \frac{y}{2\sqrt{Dt}})$$

$$- \frac{2\gamma}{\gamma - \beta}[W(\beta\sqrt{Dt} + \frac{y}{2\sqrt{Dt}}) - W(\gamma\sqrt{Dt} + \frac{y}{2\sqrt{Dt}})]\Big\} \tag{3}$$

where $k = \frac{\mu^2 EB}{2D}$, $\gamma = \frac{S}{D} - k$, $\beta = \alpha - k$ and $W(x) = \exp(x^2)\,\mathrm{erfc}(x)$.

The normalized current flowing at any time is proportional to the y-integral of the charge density:

$$i(t) = \frac{\exp(-\frac{t}{\tau} - Dk^2 t)}{(k-\beta)(k-\gamma)(\gamma-\beta)}\Big\{k(\beta^2-\gamma^2)W(k\sqrt{Dt}) + \beta(\gamma^2-k^2)W(\beta\sqrt{Dt})$$

$$+ \gamma(k^2-\beta^2)W(\gamma\sqrt{Dt})\Big\} \tag{4}$$

Figure 2 shows actual magnetophotoconductivity data obtained from a fast (25 psec rise time) sampling oscilloscope. The left trace is a current pulse obtained under the application of a magnetic field to drive the charges away from the surface; the right trace was obtained with the opposite direction of the magnetic field. The difference in pulse width and height demonstrates the effect of the magnetic field on the overall photocarrier lifetime. The surface of the gallium arsenide was prepared with a passivating citric acid etch [5]. Figure 3 shows the data obtained from a gallium arsenide surface which was activated through mechanical polishing. Note the treater effect of the magnetic field on the current pulse.

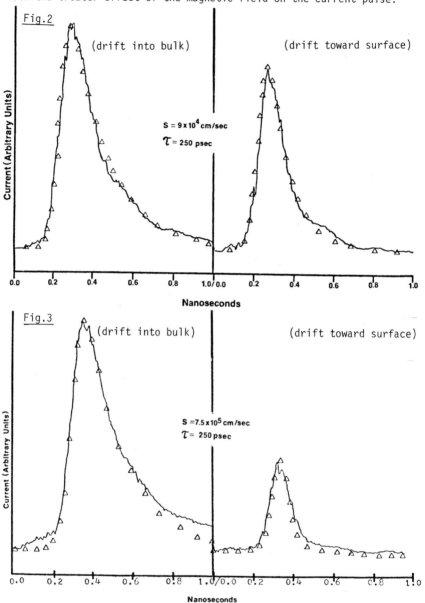

Fig. 2 Magnetophotoconductivity of a passivated Cr:GaAs surface (B = ± 15 kgauss)

Fig. 3 Magnetophotoconductivity of an activated Cr:GaAs surface (B = ± 15 kgauss)

Superimposed on the traces is a curve of (4) passed through a 25 psec Gaussian filter, with the parameters of surface and bulk recombination determined through a nonlinear least-squares fitting routine. The symbols represent the theoretical curve, and the solid line is the experimental curve. The bulk recombination lifetime was determined to be 250 picoseconds, while the surface recombination velocities were found to be 9×10^4 and 7.5×10^5 cm/sec in Figures 2 and 3 respectively. The bulk value was in good agreement with that obtained from high-intensity CW illumination experiments [4]. The surface recombination velocities also show good agreement with data obtained from luminescence experiments on samples with identical surface preparations. These results conclusively demonstrate the significant contribution of surface effects on the speed of Cr:GaAs opto-electronic switches. This measurement technique offers a means of simultaneously determining the surface recombination velocity and bulk recombination rate of transiently generated photocarriers.

References

1. R. A. Lawton and A. Scavannec, Elect. Lett. 11, 74 (1975).

2. C. H. Lee, A. Antonetti, and G. Mourou, Opt. Commun. 21, 158 (1977).

3. D. H. Auston, Appl. Phys. Lett. 26, 101 (1975).

4. S. S. Li and C. I. Huang, J. Appl. Phys. 43, 1757 (1972).

5. G. P. Peka and L. G. Shepel, Sov. Phys. Sol. St. 14, 2025 (1973).

Optoelectronic Modulation of Millimeter-Waves by Picosecond Pulses

Chi H. Lee and P.S. Mak

Department of Electrical Engineering, University of Maryland,
College Park, MD 20742, USA

A.P. DeFonzo

Naval Research Laboratory, Washington, D.C. 20375, USA

1. Introduction

We report here a new class of device - optoelectronic millimeter-wave modulator. The interaction of millimeter-wave with optically induced electron-
hole plasma in a semiconductor facilitates the control of the phase and/or
amplitude of the wave as it propagates through the waveguiding structure.
Conversely, millimeter-waves serve as excellent probe on the transient behavior of the charge carriers induced by picosecond optical pulses. The
plasma density at which the value of the plasma frequency equals the frequency of the millimeter-wave is in the range of 10^{13} to 10^{14}/cm^3. This
shows that millimeter-waves are sensitive to the optically induced plasma
layer in the waveguide.

The basic principle of optical control of millimeter-waves is illustrated
schematically in Fig.1. The propagation constant, k_z, in an interval ΔL of a
rectangular semiconductor waveguide is changed to k_z' by illuminating the
broadwall with optical radiation. The energy of the optical radiation is
chosen to be in excess of the semiconductor band gap. The absorbed light
generates electron-hole pairs resulting in a change of the complex index of
refraction of the semiconductor thereby altering the boundary conditions of
the waveguide and changing the propagation constant. A millimeter-wave
launched into the waveguide from a tapered transition experiences amplitude
and/or phase modulation while propagating through the illuminated interval,
ΔL. The ratio of amplitude to phase modulation depends on the density and
geometry of the plasma. For example, if the net number, N, of electron-hole

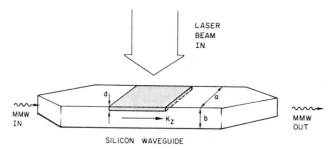

Fig.1 Schematic diagram of an optically controlled phase shifter and loss
modulator. K_z is the propagation vector in the guide, d is the injected
depth of the electron hole plasma layer, ΔL is the interval of guide illuminated, a is the width of the guide and b is the height of the guide.
Guides used in experiments were fabricated from high resisting silicon.

pairs within a depth, d, of the broadwall surface is such that density, $n = N/a \times d \times \Delta L$, yields a skin depth $\delta < d \ll b$, at the millimeter wave frequency, the effect of the plasma is much like that of a metal in contact with the guide. This yields a nearly pure phase shift, ϕ, given by the relationship,

$$\phi = (K_z - K'_z)\Delta L$$

where K'_z is approximately equal to the propagation constant of an image guide in the interval ΔL. The principle of phase shifting in this case is similar to that found in p.i.n. diode waveguide phase shifters [2]. With respect to the latter mechanism, the present approach has advantages of simplicity, high isolation, low insertion loss and better control of the plasma millimeter-wave interaction volume.

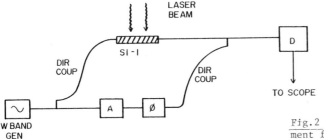

Fig.2 Experimental arrangement for phase shift measurements.

Experimental Setup and Results

To demonstrate the efficiency of the approach and to verify the model, we performed the following experiment. A semiconductor waveguide was mounted in a conventional millimeter wave bridge operating at 94 GHz. As shown in Fig.2. The oversized dielectric waveguide of cross-sectional dimensions a = 2.38mm and b = 1mm was fabricated from a single crystal silicon slab polished and oriented with the (111) plane parallel to the broadwall. Two adiabatic transitions were provided at either ends for coupling from and to conventional W-band metal waveguide. The bridge was nulled with the phase

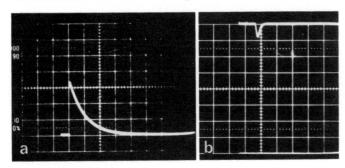

Phase shift 200 µs/div   Attenuation 200 µs/div

Fig.3. Oscilloscope traces. (a) Phase shift. The trace shows the deviation from null of the output of a millimeter-wave bridge. (b) Dynamic loss. The dip is transmitted power occurs sometime after the onset of laser excitation. The sweep rate is 200 µs/div in both cases.

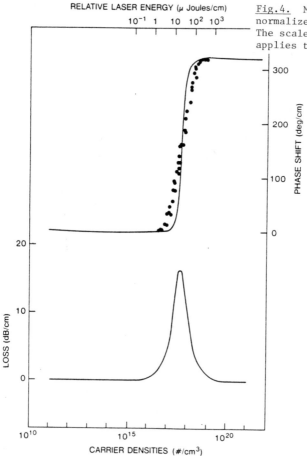

RELATIVE LASER ENERGY (μ Joules/cm)

Fig.4. Measured phase shift normalized to units of deg/cm. The scale for the upper abscissa applies to the reduced data points.

CARRIER DENSITIES (#/cm³)

shifter and insertion loss of the unilluminated silicon waveguide in one arm balanced against a variable attenuator and phase shifter in the other. The static insertion loss of the silicon guide was less than 1 dB. A 30 pico-second pulse of .53μ light from a mode-locked Nd:YAG laser and KDP doubling crystal was used to illuminate a 2.38 x 1.6mm² broadwall area. This created an electron hole plasma within .5μ of the silicon surface. The resulting transient modulation of the phase and amplitude of the millimeter wave prop-agating in the silicon guide was monitored with a high speed Schottky barrier detector and displayed on a Tektronix 7904 oscilliscope. Typical traces are shown in Fig.3.

The dependance of phase shift on initial carrier density was determined by varying the light intensity. The data points for phase shift per unit length of illuminated guide as a function of plasma density are shown in the upper portion of Fig.4 superimposed on a theoretical curve. A theoretical loss curve shown in the lower portion of the figure illustrates the dis-placement in density of maximum loss with respect to maximum phase shift. This displacement is consistent with the delay of 10 to 100 nanoseconds ob-served in the maximum attenuation shown in Fig.3b with respect to the maxi-mum phase shift shown in Fig.3a.

Comparison of Theory and Experiment

The mechanism for phase shift and attenuation was satisfactorily described in terms of a model based on the Marcatili approximation [3] which reduces the determination of the propagation coefficient to the solution of a four layer boundary value problem in which the layer containing the optically injected carriers was characterized by a complex index of refraction derived from semiclassical Drude theory. Details of the analysis are presented elsewhere [4]. The computed theoretical results for optically induced millimeter wave phase shift is shown in Fig.5. The curves were calculated for a guide identical to that used in the experiments and the results are given for a variety of injection depths ranging from 1μ to 500μ. The plasma density at which the value of the plasma frequency equals the frequency of the millimeter-wave is $4 \times 10^{13}/cm^3$. We expect the plasma layer to have no effect on the propagating millimeter-wave when the plasma density is less than this value. This is evident in Fig.5 even for the case when the plasma layer thickness is $500\mu m$. As the density increases, more interactions between the plasma and the wave are taking place. However, the millimeter-wave is not totally screened from the occupied volume until the skin depth is comparable to the layer thickness. In this density regime, the millimeter-wave penetrates into the plasma layer causing loss. One expects the maximum loss to occur in this density range. As the density increases further, the skin depth decreases. When the skin depth is equal to the thickness of

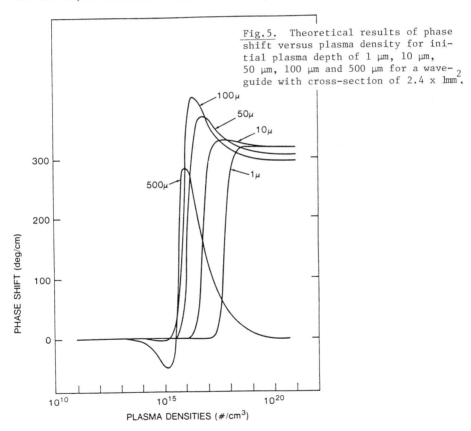

Fig.5. Theoretical results of phase shift versus plasma density for initial plasma depth of 1 μm, 10 μm, 50 μm, 100 μm and 500 μm for a waveguide with cross-section of $2.4 \times 1mm^2$.

the layer, the plasma region can be regarded as metallic. The dielectric waveguide becomes an image line. The phase shift of the millimeter-wave is due to the doubling of the dimension of the waveguide. At extremely high densities, calculated values of the phase shift agree with those obtained for an image line. The curves in Fig.5 reveal that a thicker plasma layer can introduce the same phase shift as a thinner layer does but at a lower density. The 1μ phase shift curve is also shown superimposed on the data point in Fig.4. The agreement with experiment is good. The transient behavior in Fig.3, is accurately accounted for by a reduction in carrier density in time resulting from ambipolar diffusion. The high insertion losses, obtained by deep optical injection, were used to create a high speed loss modulator readily capable of switching a gating millimeter wave on time scales less than one nanosecond [5].

Additional model calculations were performed for silicon waveguides with $1 \times 0.5 \text{mm}^2$ cross-sections. The results are qualitatively similar to those shown in Fig.5 but with a factor of 50 increase in magnitudes, yielding phase shifts as large as 1° per nanojoule of absorbed optical energy. Additional improvements in performance could be obtained utilizing other semiconductor materials such as GaAs.

Conclusions

Optical control of milimeter waves is a viable alternative to both p.i.n. diodes and ferrites. Our present analysis indicates that with the proper choice of materials, dimensions and operating temperature, as few as 1×10^9 absorbed photons should produce phase shifts of 1° or attenuations of 1 dB.

References

1. I.P. Kaminow, J.R. Carruthers, E.H. Turner, L.W. Stultz, Appl. Phys. Lett. 22, 540 (1973).
2. H. Jacobs, M.M. Crepta, IEEE Trans. MTT-22, 411 (1978).
3. E.A.J. Marcatili, Bell Syst. Tech. J. 48, 2074 (1969).
4. C.H. Lee, P.S. Mak, A.P. DeFonzo, IEEE Journal of Quantum Electronics, QE-16, 277 (1980).
5. C.H. Lee, S. Mak, A.P. DeFonzo, Electron. Lett., 14, 733 (1978).

Measurements of Ultrashort Light Pulses with a Multichannel Two-Photon Photoconducitvity Detector

Chi H. Lee and P.S. Mak

Department of Electrical Engineering, University of Maryland,
College Park, MD 20742, USA

1. Introduction

Since the first successful generation of picosecond optical pulses [1], there
has been a lot of progress made in the measurement technique for such ultrashort pulses [2-8]. So far a number of instruments are capable of displaying
a picosecond optical signal in a laser shot with picosecond resolution, but
each one suffers some degree of shortcoming. On of the promising techniques
for the measurement of sub-nanosecond optical pulses involves using the multiphoton conductivity effect in semiconductors as proposed by PATEL [9]. LEE
and JAYARAMAN [10,11], after observing two-photon and three-photon conductivity effects in semiconductors using nanosecond or picosecond laser pulse
excitation, used this technique to measure the pulse width of picosecond
pulses from a mode-locked Nd:glass laser. The multiphoton conductivity technique is analogous to the two-photon fluorescence (TPF) technique. The major
difference between these two techniques is the method of detection. In the
TPF method the fluorescent signal from a dye is measured, while in the two-
photon conductivity (TPC) or three-photon conductivity methods the photoconductive signal from a biased semiconductor is measured. This difference
is naturally reflected in the construction of the detector. We report here
the operation of a seven-channel prototype TPC detector. It is a second
order laser intensity autocorrelator which can measure the pulse width of
mode-locked Nd:glass laser pulses in a single laser firing. A simple peak
detector was used to relay the signals from the two-photon detector to a
displaying system. The successful demonstration of the TPC detector has led
us to believe that, due to the simplicity of the device, a refined version
of it which may consist of a layered thin film semiconductor and/or charge
coupled device with multichannel data processing capability can be built
with relatively low cost.

2. Two-photon Conductivity Technique

The operation of the detector is based on the TPC technique illustrated in
Fig.1a. This technique is completely analogous to that used in the thin cell
TPF method by SHAPIRO and DUGUAY [12] except for the replacement of the dye
cell by a stack of semiconductors as shown in Fig.1b and 1c. For two-photon
electron-hole pair generation, we require a semiconductor with band-gap energy
E_g such that $h\nu < E_g < 2h\nu$, where ν is the frequency of the laser pulse. For
the measurement of mode-locked Nd:glass laser pulses with photon energies of
1.17eV, $CdS_{0.5}Se_{0.5}$ is a suitable material, having a band gap of 2.0eV. For
the measurement, the ultrashort light pulses are split and then recombined
to have the pulses propagating in opposite directions. In the path of the

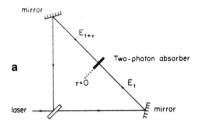

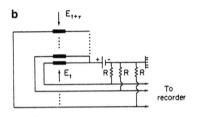

a

mirror

$E_{t+\tau}$

Two-photon absorber

$\tau=0$ E_t

laser   mirror

b $E_{t+\tau}$

E_t R R R

To
recorder

Fig.1 (a) Schematic experimental
arrangement for the measurement
of second-order autocorrelation
curve of an ultrashort laser pulse.
(b) Multi-channel detector. The
two-photon absorber is a spatially
separated stack of $CdS_{0.5}Se_{0.5}$
crystals. (c) Arrangement of
$CdS_{0.5}Se_{0.5}$ crystals.

c

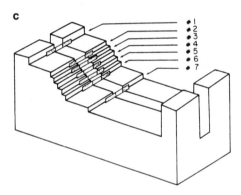

colliding pulses is a stack of seven pieces of thin (0.25mm) $CdS_{0.5}Se_{0.5}$
crystal. When positioned properly, each semiconductor piece registers a
photoconductive signal which represents a particular point on the second
order correlation curve. Thus, if the photoconductivity changes in the seven
crystals are monitored simultaneously, it is possible to measure the intensity correlation curve, or the pulse width, of the picosecond laser pulses
in a single shot. To accurately measure the correlation function of the
ultrashort pulses, we must first establish the fact that the photoconductivity
signal depends quadratically on the laser intensity. For this purpose, the
TPC of the samples must first be measured.

3. Two-photon Conductivity in $CdS_{0.5}Se_{0.5}$

Under steady-state excitation BASOV et al [13] and YEE [14] have derived an
expression for the conductivity change ΔG due to two-photon absorption

$$\Delta G = \frac{c}{a}|e|(\mu_e + \mu_h)A_0\tau[I_0^2 L/(1 + KI_0 L)] \, , \qquad (1)$$

where c/a is a geometric factor of the sample, μ_e and μ_h are the electron and hole mobilities, respectively, A_0 is a constant related to the band parameters, τ is the lifetime of the charge carrier, I_0 is the incident laser intensity, L is the thickness of the sample and $K=2h\nu A_0$ is the two-photon absorption coefficient in $cmMW^{-1}$. If the width of the exciting pulse is less than the lifetime of the charge carrier, the effect is transient and the change of conductivity will be reduced. It is appropriate in the transient regime to replace the lifetime τ in (1) by Δt_p, the duration of the exciting light pulse.

The change in conductivity ΔG in a sample due to the laser pulses can be computed from the voltage change V across the series resistor R as depicted in Fig.1b. The photoconductive signal V varies as a function of the laser intensity I_0 according to the expression [15].

$$V = \frac{V_0 Rc I_0^2 |e| (\mu_e + \mu_h) A_0 \tau L}{a(1+KI_0L) + Rc|e|(\mu_e+\mu_h)A_0\tau I_0^2 L} \quad . \tag{2}$$

At high laser intensities, V saturates at the bias voltage V_0. At low laser intensities, V becomes a quadratic function of I_0. For a sensitive two-photon detector, we want to operate the photoconductors well below the saturation region and inside the slope-two region. With an appropriate choice of values for the parameters R, c, V_0, and a, we control the laser intensity so that

PHOTOCONDUCTIVE SIGNAL

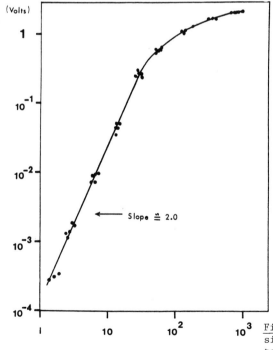

Fig.2 Two-photon photoconductive signal verses relative laser intensity

RELATIVE LASER INTENSITY

the optimum two-photon conductivity signals are measured. It has been determined that the combination of R=150 ohms, V_0=20 volts, c=1mm, and a=1.5mm applies best to our experiment. A typical response of the high resistivity ($\rho \approx 10^7$ ohm-cm) $CdS_{0.5}Se_{0.5}$ samples to laser illumination is shown in Fig.2. The slope of the curve prior to saturation was found to be two, indicative of a two-photon process. The response of the samples enabled us to use them to construct a TPC detector.

4. Experimental Results and Discussion

The TPC detector is a two-stage device: the $CdS_{0.5}Se_{0.5}$ samples constitute the first stage and the second stage is a seven-channel peak detector which is used to relay the conductivity changes in the samples to a recorder. Each channel detects the peak of the photoconductive signal from the corresponding sample and stores it. Stored signals are then retrieved one by one from the peak detector using a seven-way switch and read on an oscilloscope or digital voltmeter. In a more advanced version of this device we envisage clocked readout of the peak detector stored voltages and continuous oscilloscope display of the autocorrelation function.

Before making pulse width measurements, the crystals constituting the detector were calibrated in a position where the laser pulses could not overlap temporally. These calibration signals provided a baseline for the measured signals when the pulses overlapped. The seven-channel detector was then set at a position where the laser pulses would overlap [16]. After a single laser shot, the ratios of the signals from the detector when laser pulse overlap occurred to the calibration signals were obtained and plotted as a function of the position of the crystals in the detector. The result of a typical measurement is shown in Fig.3. The pulse width is estimated to be 10 ps. This agrees well with previous TPF and TPC measurements [10]. The contrast ratio here is 2.8. Comparison with the results obtained by LEE and JAYARAMAN

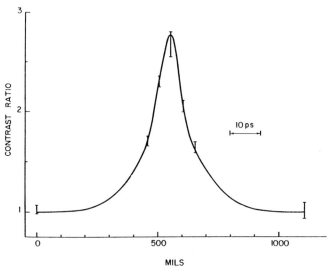

Fig.3 Measured correlation curve with a contrast ratio 2.8:1. The measured pulse width is $\approx$ 10 ps.

[10] demonstrates that the present detector gives a higher contrast ratio. This improvement in the contrast ratio is due mainly to the vast reduction in experimental time and number of laser shots, and consequently laser fluctuations, involved in taking the data. The thickness of the $CdS_{0.5}Se_{0.5}$ crystals used limited the resolution of the detector to 2 ps. Thinner crystals and single ps pulses should give better resolution and even higher contrast ratio. The error bars were due to d.c. drifts of the channels in the peak detector.

These data show that a 7-channel two-photon detector is sufficient to resolve the pulse width of a picosecond pulse. This TPC detector, though simple in design and construction, proved to be a convenient device to use for these measurements. Our device would be ideal for the measurement of picosecond pulses from Nd:YAG lasers which have beam quality superior to the glass laser used in the present work.

In conclusion, we have demonstrated measurements of the pulse width of picosecond laser pulses with a $CdS_{0.5}Se_{0.5}$ two-photon detector used in conjunction with a seven-channel peak detector. Improvements in the two-photon detector structure and the interface electronics between detector and data processor would better reveal the capabilities of this technique. We believe that a refined model of the device would be very competitive with existing techniques for the characterization of ultrashort light pulses. This research was supported by NSF, grant number ENG-78-06862.

References

1. A. J. DeMaria, D. A. Stetser, and H. Heyman, Appl. Phys. Lett., 8, 174 (1966).
2. J. A. Giordmaine, P. M. Rentzepis, S. L. Shapiro, and K. W. Wecht, Appl. Phys. Lett., 11, 216 (1967).
3. J. A. Armstrong, ibid 10, 16 (1967).
4. A. J. DeMaria, W. H. Glenn, J. J. Brienza, and M. D. Mack, Proc. IEEE, 57, '2 (1969).
5. R. C. Eckardt and Chi H. Lee, Appl. Phys. Lett., 15, 425 (1969).
6. M. C. Richardson, IEEE J. Quantum Electron., QE-9, 768 (1973).
7. T. Netzel, P. M. Rentzepis, and J. Leigh, Science, 182, 238 (1973).
8. G. C. Vogel, A. Savage, and M. A. Duguay, IEEE J. Quantum Electron., QE-10, 642 (1974).
9. C. K. N. Patel, Appl. Phys. Lett., 18, 25 (1971).
10. Chi H. Lee and S. Jayaraman, Opto-electronics, 6, 115 (1974).
11. S. Jayaraman, Ph. D. Thesis, University of Maryland (1972).
12. S. L. Shapiro and M. A. Duguay, Phys. Lett., 28A, 698 (1969).
13. N. G. Basov, A. Z. Grasyuk, I. G. Zubarev, V. A. Katulin, and O. N. Krokhin, Sov. Phys. JETP, 23, 366 (1966).
14. J. H. Yee, Appl. Phys. Lett., 14, 231 (1969); 15, 431 (1969); Phys. Rev. 86, 778 (1969).
15. P. S. Mak, Ph.D. Thesis, University of Maryland (1979).
16. M. Maier, W. Kaiser, and J. A. Giodmaine, Phys. Rev. Lett., 17, 1275 (1966).

Part III

**Picosecond Studies
of Molecular Motion**

Anisotropic Absorption Studies of Orientational Motion

G.S. Beddard, G.R. Fleming[1], D.B. McDonald[1], G. Porter,
D. Waldeck[1], and M. Westby

Davy Faraday Research Laboratory, The Royal Institution,
21 Albemarke Street, London WIX 4BS, England

[1] Department of Chemistry and James Franck Institute,
The University of Chicago, Chicago IL, USA

The rotational correlation function of medium–sized molecules in solution has
been measured directly both by absorption [1,2], and emission techniques [3,
4]. SHANK and IPPEN [5] have described a technique based on induced dichroism
which provides an attractive alternative to the above approaches. In their
anisotropic absorption technique a strong, polarized, pump pulse creates a
hole in the ground state distribution of orientations. The return of the
orientational distribution to equilibrium is probed by a weak pulse polarized
at 45° to the exciting pulse. Differential absorption of the components of
the probe polarized parallel with, or perpendicular to, the excitation polarization induces a time-delay dependent transmission of the probe through an
analyzer polarizer.

Experimental

A synchronously pumped dye laser (CR590/03)/Ar^{+} laser (CR12) combination provided the picosecond pulses. We have studied pulse to pulse reproducibility
and the influence of coherence on autocorrelation measurements by using a
zero background correlation set-up [6] to measure cross correlations of the
form $\int I_n(t) I_{n-m}(t+\tau)dt$, when n labels an individual pulse and m the number
of round trips separating the pair [7]. The results for $m = 6$ are shown in
Fig. 1. When the dye laser cavity is set for optimum pulse length with no
discernable structure or satelite pulses in the autocorrelation trace, the
cross correlation (n,n-m) function (d) is indistinguishable from the autocorrelation (n,n) function (e). Fig.1a-c show the results where the dye
laser cavity length is incorrectly set and partial mode locking results.
Autocorrelation traces such as the curve in Fig.1c are characteristic of a
noise burst [8]. For bandwidth limit noise structure the autocorrelation
has the form

$$G(\tau) = G_p(\tau) \quad [1 + G_N(\tau)] , \tag{1}$$

where $G_p(\tau)$ represents the autocorrelation of the pulse envelope and $G_N(\tau)$
is a gaussian function resulting from the noise bandwidth. For pulse envelopes significantly longer than the noise bandwidth limit (1) can be approximated by

$$G(\tau) = G_p(\tau) + G_N(\tau). \tag{2}$$

The contributions from G_N and G_p can be easily observed in Fig.1a-c where
there is a large cavity mismatch. In cases where there is little or zero
mismatch, however, the contributions are much more difficult to distinguish
(Fig.1d,e). Applying (2) to the pulse in Fig.1e we find $\Delta t_p = 5ps$ (for

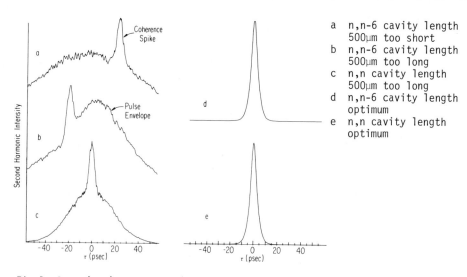

a n,n-6 cavity length
   500μm too short
b n,n-6 cavity length
   500μm too long
c n,n cavity length
   500μm too long
d n,n-6 cavity length
   optimum
e n,n cavity length
   optimum

<u>Fig.1</u> Auto (n,n) and cross (n,n-m) correlation function measurements

$\Delta\tau$ = 5.7ps) whereas the conventional method has been to divide $\Delta\tau$ by between 2 and 1·41 giving Δt_p of 2·9 - 4ps.

The noise spike in Fig.1a,b marches to one side or the other of the broad base, depending whether the dye cavity is too long or too short [7]. Consider the case where the dye laser cavity is too long. The leading portion of the pulse is amplified more than the trailing portion because of gain saturation, and the position of the peak of the pulse envelope is advanced. The result is that dye pulses circulate with the same period as the pumping pulses, even if the cavity length is mismatched. The substructure on the pulse, however, is reproduced during the amplification process, but becomes shifted relative to the pulse envelope. In other words, the substructure accumulates a phase shift on each round trip. Thus, in the n,n- m cross correlation the noise spikes add in phase for a value of τ different from zero. This argument predicts a linear dependence of spike displacement on both m and cavity mismatch (in μm). Both linear dependences are observed experimentally [7]. Only when the cavity periods of the pump and pumped lasers are equal will the coherence spike be centered on the envelope portion of the correlation trace. The position where the n,n- 6 cross correlation trace becomes symmetric provides the first experimental determination of the dye laser length for perfect synchrony. In our experiments it corresponds to the optimum pulse duration, as might be expected on intuitive grounds, but in contrast to theoretical predictions requiring slightly shorter [9], or slightly longer [10] cavities than resonant length for optimal pulse duration.

Anisotropic absorption

The vertically polarized output of the dye laser is split into two beams. The strong exciting beam is focused by a 5 cm focal length lens into the sample. The probe beam traverses a variable delay and is then focused into the sample by the same lens. The probe beam polarization is rotated by a half wave plate such that it is 45° to the polarization of the pump. The

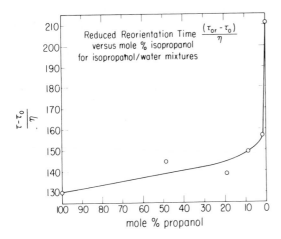

Fig.2 Reduced reorientation time of oxazine versus mole per cent for isopropanol water mixtures

probe beam intensity through a crossed analyzer polarizer is measured as a function of time delay. The excited state lifetimes were measured by the time correlated single photon counting technique. The measured instrument response function has a duration of ~290 ps and the estimated time resolution is ~50 ps [11]. For our experimental geometry [5,12] the observed time dependent intensity of the probe beam ($T(t)$) is (for small induced dichroism):

$$T(t) = \text{const } [r(t) \, K(t)]^2 \qquad (3)$$

where $r(t)$ is the rotational correlation function and $K(t)$ the excited state decay function [3]. In the most straightforward case where both $r(t)$ and $K(t)$ are single exponentials with characteristic time τ_{or} and τ_{ex} respectively, $T(t)$ is a single exponential with decay time

$$\tau_M^{-1} = 2(\tau_{or}^{-1} + \tau_{ex}^{-1}). \qquad (4)$$

Results

The observed anisotropic absorption signal (after the coherence spike [5, 13]) could be well fit by a single exponential decay in all the solvents we studied. The fluorescence decays were also single exponential. In a series of alcohols, deuterated alcohols, water, heavy water, and acetone the orientation time showed linear behavior with viscosity and was in accord with stick hydrodynamics assuming a prolate shape for oxazine for all the solvents except H_2O and D_2O, where the rotation was significantly slower than expected [12]. In this paper we discuss our results in mixed propanol water solutions.

The reduced rotation time $(\tau_{or} - \tau_{or}(o))/\eta$ vs mole per cent isopropanol is plotted in Fig.2. ($\tau_{or}(o)$ is the zero viscosity intercept the plot of τ_{or} vs η plot for all solvents). Over the range 100% - 20% alcohol the reduced reorientation time changes very slowly. Between 10% and 0% alcohol and reduced reorientation time changes extremely rapidly, indicating a marked deviation from simple hydrodynamic behavior with a stick BC. Fig.3 shows the excited state lifetime as a function of mole per cent propanol. There is a

103

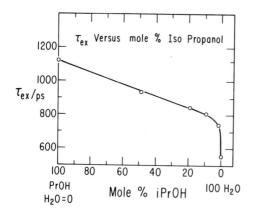

Fig. 3 Oxazine excited state
lifetime versus mole per cent
for isopropanol water mixtures

rapid decrease in lifetime between 1% propanol and pure water corresponding
to the rapid increase in rotation time over this range in Fig.2.

Discussion

The results in Fig.3 suggest a possible explanation of the striking behavior observed in Fig.2. The measured rotational correlation function contains only ground state motion if the excited state is long lived compared
to the reorientation time. However, if the two lifetimes are comparable the
measured decay contains a contribution from molecules that have rotated in
the excited state and subsequently returned to the ground state. In H_2O the
excited state has decayed by 40% during the 1/e time of the rotational motion.
Thus, our results require a very large change in τ_{or} in the excited state.
Such a large change in τ_{or} seems unlikely based on the results of EICHLER et
al. [14], additionally in the 19% (mole %) isopropanol water solution the
excited state has decayed by 50% by τ_{or} and yet no anomalous rotation results.
We are thus confident that the data in Fig.2 represent a significant increase
in the rotational diffusion coefficient in pure water as compared with alcohol.

One possibility is to seek an explanation of Fig.2 in terms of the structure of the solvent. It is possible that a very rapid change in the water
structure occurs in the range 0-10% propanol possibly from the full three
dimensional random network of pure water to the linear chain like structures
of pure alcohols [15,16]. However, we must still explain why the particular
structure of water inhibits the rotational motion of oxazine significantly
more than is implied by the viscosity (assuming stick BC). A possible alternative interpretation is that there is a specific interaction of the oxazine
with propanol. More experiments are required to clarify these interesting
results, and hopefully shed light on the influence of solvent structure and
specific interactions on molecular rotation.

Acknowledgements

The reorientation measurements were carried out at the Royal Institution
through the support of the Science Research Council. The pulse structure
studies were carried out at Chicago and supported by the Louis Block Fund
and the Camille and Henry Dreyfus Foundation.

References

1. T.J. Chaung and K.E. Eisenthal, Chem. Phys. Letters $\underline{11}$, 368 (1976).
2. A. Von Jena and H.E. Lessing, Chem. Phys. $\underline{40}$, 245 (1979).
3. G.R. Fleming, J.M. Morris and G.W. Robinson, Chem. Phys. $\underline{17}$, 91 (1976).
4. K.G. Spears and L.E. Cramer, Chem. Phys. $\underline{30}$, 1 (1978).
5. C.V. Shank and E.P. Ippen, Appl. Phys. Lett. $\underline{26}$, 62 (1975).
6. E.P. Ippen and C.V. Shank, Appl. Phys. Lett. $\underline{27}$, 488 (1975).
7. D.B. McDonald, D. Waldeck and G.R. Fleming, Opt. Commun. in press.
8. C.V. Shank and E.P. Ippen, *Ultrashort Light Pulses*, ed. by S.L. Shapiro, Topics in Applied Physics, Vol. 18 (Springer, Berlin, Heidelberg, New York 1977)
9. N.J. Frigo, T. Daly and H. Mahr, IEEE J. Quant. Elect. *QE13*, 101 (1977)
10. D.M.Kim, J.Kuhl, R. Lambrich and D. von der Linde, Opt. Comm. $\underline{27}$, 123 (1978)
11. R.J. Robbins, G.R. Fleming, G.S. Beddard, G.W. Robinson, P.J. Thistlewaite and G.J. Woolfe, J. Amer. Chem. Soc. in press
12. G.S. Beddard, G.R. Fleming, G. Porter, J. Rowe and M. Westby, in preparation
13. A. von Jena and H.E. Lessing, Appl. Phys. $\underline{19}$, 131 (1979)
14. H.J.Eichler, U. Klein and D. Langhans, Chem. Phys. Lett. $\underline{67}$, 21 (1979)
15. F.Franks and D.J. Ives, Quant. Rev. Chem. Soc. $\underline{20}$, 1 (1966)
16. V.A. Mikhailov, J. Struct. Chem. $\underline{9}$, 397 (1968)

Molecular Motions and Solvent Stabilization of Charges

K.G. Spears, K.M. Steinmetz-Bauer, and T.H. Gray
Department of Chemistry, Northwestern University, Evanston, IL 60201, USA

1. Introduction

The solution phase properties of ionic molecules are an important area of chemistry because of the common occurrence of charged species in reaction pathways and the large interactions of solvent with ionic molecules. In this work we report studies of rotational relaxation of a single ionic molecule in a variety of solvent environments.

Rotational diffusion times of large molecules have been measured by a number of workers with picosecond techniques [1-7]. The bulk solvent viscosity is generally proportional to the observed rotational correlation times for a given class of solvent molecules. Earlier work from this group [4] established the importance of special solvent interactions such as hydrogen bonding for ionic molecules having low steric hindrance to hydrogen bonding. The earlier work compared dipolar aprotic solvents and protic solvents for the dianion dye molecule, rose bengal. The application of Stokes-Einstein rotational diffusion calculations suggested that only for dipolar aprotics was this simple model a reasonable approximation. For rose bengal the correlation of viscosity, n, and orientation relaxation τ_{or}, led to τ_{or}/n of 200 ps/cp and 400 ps/cp for aprotics and alcohols, respectively.

The goal of this work was to use a simpler molecular structure that still maintained an exposed oxygen anion similar to the case of rose bengal. In Fig.1 we show rose bengal and the molecule selected for this work, resorufin. The rose bengal molecule is a dianion and its oblate spheroidal shape is quite different from the prolate spheroid shape of resorufin. Because the dipole axis of resorufin is along the molecular axis, one expects a single orientational decay characteristic of tumbling of the long axis. One might expect a clearer demonstration of hydrogen bonding effects in resorufin than in rose bengal. The positive counter ion in the salt of resorufin was a tertiary butyl ammonium (TBA) cation $N^+(C_4H_9)_4$. This was selected to ensure that the cation would not interfere with orientational decay because it is well solvated in dipolar aprotic solvents. The TBA cation allowed the addition of small cations capable of ion-pairing with resorufin in aprotic solvents.

2. Measurement Technique

The picosecond measurements were done by the method of time correlated photon counting. An argon ion laser (Control Laser) was mode-locked with a Harris Corp. mode-locker at 50 MHz. A synchronously pumped Rhodamine 6G dye laser was tuned with a 2 plate birefringent filter to 575 nm to provide excitation

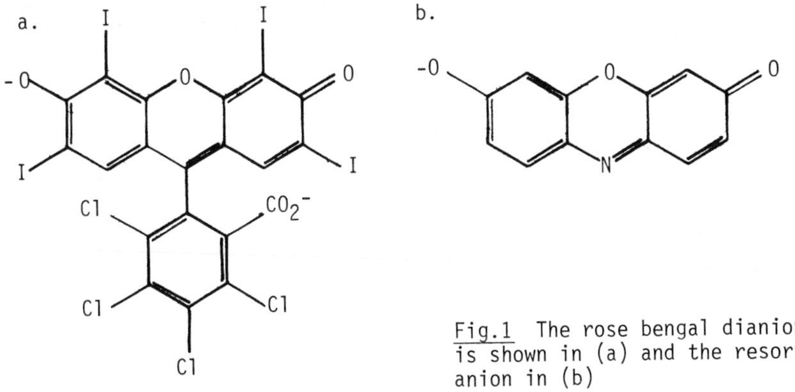

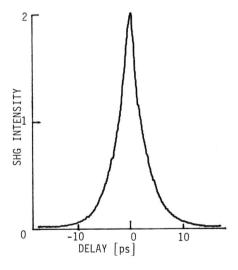

a.

b.

Fig.1 The rose bengal dianion is shown in (a) and the resorufin anion in (b)

pulses at 100 MHz. The dye laser cavity gain was adjusted by a wedged plate and the length was adjusted by a differential screw having micron resolution. Non-collinear SHG auto-correlation was done at 5 Hz by an oscillating prism to allow convenient adjustments, while a computer recording was done by translating a precision stage in steps as fine as 0.01 mm. An auto-correlation trace is shown in Fig.2 and it gives a FWHM pulse width of 1.9 ps assuming a single sided exponential pulse shape.

The time correlated photon counting methods in use by this group have been described previously [8,9]. An ortec 583 constant fraction discriminator and an Amperex XP2020 photomultiplier have improved the time response so that the observed FWHM is ~290 ps. In most of these experiments we used ~8 ps/channel resolution and no discernible change was evident over a sample taking interval (< 2 ps shifts in peak location).

The procedure of measurement was to record an instrument response signal by scattering from a water solution of centrifuged coffee creamer between

SHG INTENSITY vs **DELAY [ps]**

Fig.2 An auto-correlation trace that yields a pulse FWHM = 1.9 ps assuming a single sided exponential.

all data scans. A polarization vertical to the excitation and perpendicular to the excitation polarization was collected and a calculation of $I_{\parallel}$-$I_{\perp}$ gave the exponential decay function defined by τ_{or} [1-7]. The 3-5 ns lifetime (slightly solvent dependent) made it convenient to match the intensities of $I_{\parallel}$ and $I_{\perp}$ at 10 ns after excitation. The resultant curve was deconvoluted to yield an initial τ_{or} and a nonlinear least squares program optimized the subtraction and computed the final τ_{or}.

3. Results

The tentative results for pure solvents are contained in Table 1. These results do not yet represent extensive averages over multiple runs and are probably good to ± 10% for numbers ⩾ 75 ps. Our accuracy for correlation times < 50 ps is not yet established and we estimate ± 25%. Typical data for resorufin in DMSO are shown in Fig.3 along with an excitation pulse shape for comparison. This lifetime was 76 ps and the subtraction procedure clearly gives increased noise at longer times.

Table 1 Solvent dependent orientation decay times

Solvent	τ_{or} [ps]	τ_{or}/η [ps/cp]
MeOH	110	200
EtOH	235	220
2-PrOH	445	210
DMF	30	41
DMSO	77	41
DMSO + $Mg(ClO_4)_2$	110	60
DMSO + $Mg(C_2H_3O_2)_2$	120	64
DMSO + $Na(C_2H_3O_2)$	115	61

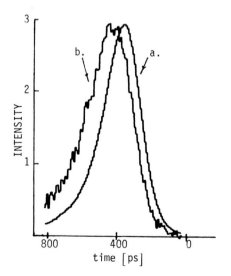

Fig.3 Curve (a) is the instrument response to a 1.9 ps pulse and curve (b) is $I_{\parallel}$-$I_{\perp}$ from polarized resorufin fluorescence with a τ_{or} of 76 ps

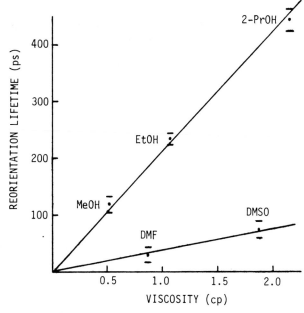

Fig.4 A plot of orientation decay times as a function of solvent viscosity

The data in Fig.4 show the correlation of viscosity and orientation relaxation. Clearly the dipolar aprotic solvents show a very different behavior than the alcohol solvents. The expected τ_{or} by a Stokes-Einstein calculation is about 60 ps/cp and Table 1 shows that resorufin in DMF and DMSO give τ_{or} similar to this value. One does not expect the Stokes-Einstein model to work very well, but the one surprising feature is that the observed slopes in ps/cp show a ratio of $\sim 5/1$ when comparing alcohols to aprotics. This is a very large effect when contrasted to the comparable 2/1 ratio for rose bengal. Hydrogen bonding appears to be more effective for the prolate shape of resorufin.

Additional preliminary experiments were done to establish the effects of counter ions in DMSO. Concentrations of salts were added at $\sim 10^{-4}$ M to $\sim 10^{-6}$ M solutions of resorufin. The salts were selected to provide large anions and small cations at a concentration too low to affect viscosity. The $Mg^{++}(ClO_4^-)_2$ and $Mg^{++}(C_2H_3O_2^-)_2$ gave values of ~ 60 ps/cp. These values represent an increase in orientation decay time of $\sim 50\%$ and strongly suggest the importance of the counter ion in a solvent separated ion pair. An effective increase in the "length of resorufin" by one DMSO molecule will increase τ_{or} by 20-25%. A completely solvated cation could easily add the equivalent of two DMSO molecules to the length and increase the τ_{or} by 40-50% if the contact time is long enough. The cations have a strongly bound first solvation shell and one does not expect significant second shell solvation in DMSO.

More data are being accumulated for these systems. Other techniques such as fluorescence analysis using SHG upconversion and transient absorption dichroism will be applied to the shorter time scales.

4. Conclusions

The orientation decay times of a prolate anionic molecule, resorufin, show a linear dependence on viscosity for a particular type of solvent. Resorufin in hydrogen bonding alcohols has $\sim$5 times longer orientation relaxation than in the dipolar aprotic solvents. The counter ions Mg^{++} and Na^{++} increase the orientation decay by $\sim$50% in DMSO.

5. Acknowledgments

The authors thank the National Science Foundation for their financial support of this work.

6. References

1. T.J.Chuang and K.B. Eisenthal, Chem. Phys. Letters 11(1971)368.
2. G.R. Fleming, J.M. Morris, and G.W. Robinson, Chem. Phys. 17(1976)91.
3. G.R. Fleming, A.E.W. Knight, J.M. Morris, R.J. Robbins and G.W. Robinson, Chem. Phys. Letters 51(1977)399.
4. K.G. Spears and L.E. Cramer, Chem. Phys. 30(1978)1.
5. G.B. Porter, P.J. Sadkowski and C.J. Tredwell, Chem. Phys. Letters 49(1977)416.
6. A. VonJena and H.E. Lessing, Chem. Phys. 40(1979)245.
7. S.A. Rice and G.A. Kenney-Wallace, Chem. Phys. 47(1980)161.
8. K.G. Spears, L.E. Cramer and L.D. Hoffland, Rev. Sci. Instr. 49(1978)255.
9. L.E. Cramer and K.G. Spears, J. Amer. Chem. Soc. 100(1978)221.

Picosecond Polarization Kinetics of Photoluminiscence of Semiconductors and Dyes

R.J. Seymour, P.Y. Lu, R.R. Alfano

Picosecond Laser and Spectroscopy Laboratory,
Physics Department, City College of New York,
New York, NY 10031, USA

We will describe a technique for simultaneously measuring photoluminescence from two orthogonal polarizations with picosecond resolution. We will show that this will enable one to determine both the total decay kinetics and the polarization exchange kinetics. Examples are given for both spin relaxation dynamics in solids and depolarization in dyes.

Referring to the case of analysis of circular polarized luminescence we can see the advantages of this technique. Here the two orthogonal circular polarizations are produced from spin up and spin down electrons in the conduction band. The rate equations governing the population change of spin up electrons (N_+) and spin down electrons (N_-), that is spin aligned parallel and antiparallel to the light excitation direction, after a delta excitation pulse $(t=o)$ are given by:

$$\frac{dN_+}{dt} = \frac{-N_+}{\tau} - \frac{N_+}{T_s} + \frac{N_-}{T_s} \tag{1}$$

$$\frac{dN_-}{dt} = \frac{-N_-}{\tau} - \frac{N_-}{T_s} + \frac{N_+}{T_s} \tag{2}$$

Where τ is the recombination time of the electrons in the conduction band and T_s is the spin alignment relaxation time of an electron in the conduction band. The sum of the equations (1) and (2) is the decay of the total electron population in the conduction band:

$$\frac{d(N_+ + N_-)}{dt} = \frac{-(N_+ + N_-)}{\tau} \tag{3}$$

which has a single exponentially decaying solution.

$$N_T(t) = N_+(t) + N_-(t) = N_T^0 e^{-t/\tau}, \tag{4}$$

where N_T is the total conduction band population and N_T^0 its initial value. The rate of change of the difference of the spin up and spin down populations is:

$$\frac{d(N_+ - N_-)}{dt} = \frac{-(N_+ - N_-)}{\tau} - \frac{2(N_+ - N_-)}{T_s} . \tag{5}$$

The solution is:

$$N_+(t) - N_-(t) = \left(N_+^0 - N_-^0\right) e^{-t/\tau} e^{-t/(T_s/2)}, \tag{6}$$

where N_+^0 and N_-^0 are the initial spin up and spin down populations, respectively. The degree of spin polarization as a function of time, $P_s(t)$, is obtained by dividing (6) by (4).

$$P_s(t) = \frac{N_+(t) - N_-(t)}{N_+(t) + N_-(t)} = \frac{N_+^0 - N_-^0}{N_+^0 + N_-^0} \; e^{-t/(T_s/2)}. \tag{7}$$

This percentage of the spin alignment decays with a single exponential time $T_s/2$.

The ability to separately and simultaneously determine the initial spin alignment, spin relaxation time, and total carrier decay time is proving important in the analysis of spin kinetics of GaAs. It has recently been shown that the transition probabilities vary as function of k much greater than had been previously assumed[1]. Furthermore, for excitation energies much larger than the bandgap there may be considerable loss of spin alignment during the initial momentum relaxation[2].

The method of transient photoselection uses a polarized light pulse to create an oriented population of excited state molecules. The oriented population can be studied by measuring the polarization kinetics of the fluorescence by time-resolved picosecond techniques. In particular, the dynamics of non-radiative energy transfer can be studied by photoselection. In the experiment, one measures the intensity I (t,ll) and I (t,⊥) at a given wavelength. Although the actual phenomenon may be more complicated we may use a simple theoretical model to describe the polarized fluorescence. For instance for spherically symmetrical molecules:

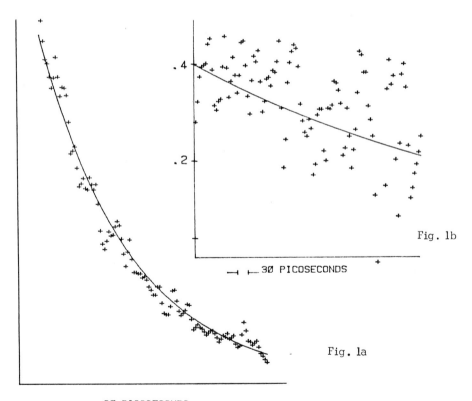

Fig. 1b

⊢⊣ ⊢— 3Ø PICOSECONDS

Fig. 1a

⊢⊣ ⊢— 3Ø PICOSECONDS

$$I(t,ll) = \left[1/3 + (4/15) \ P_2 \ (\cos \varphi) \ \exp \ (-6Dt) \right] R(t) \qquad (8)$$

and

$$I(t,\perp) = \left[1/3 - (2/15) \ P_2 \ (\cos \varphi) \ \exp \ (-6Dt) \right] R(t) \qquad (9)$$

where $R(t)$ is the probability that the fluorescent state remains excited at time t. $R(t) = e^{-K_F t}$ where $K_F = (K_R + K_{NR} + K_{ST})$, φ is the angle between the absorbing dipole and emission dipole; P_2 is the second degree Legendre polynomial; and D is diffusion rate constant due to random energy transfer walk, and possible Browian rotational motion; i.e. $D = D_{RW} + D_{ROT}$. For a totally asymmetrical molecule, extra exponential decay terms will appear in $I(t,ll)$ and $I(t,\perp)$. The degree of polarization is defined by[3]

$$p(t) = I(t,ll) - I(t,\perp) \ / \ I(t,ll) + 2I(t,\perp) \qquad (10)$$

and

$$p(t) = (2/5) \ P_2 \ (\cos \varphi) \ \exp \ (-6Dt). \qquad (11)$$

yields a single exponential decay in time with constant $\tau_D = 1/6D$. τ_D is determined by $p(t)$ which is a measure of the actual energy transfer time among the molecules. Fig. 1 shows the experimental results and fitted single exponential decays for DDI in glycerin. Figure 1a is the total luminescence fitted to a total decay time of 260 ps. The rotational lifetime is shown in fig. 1b with a fitted single exponential decay of 1000 ps.

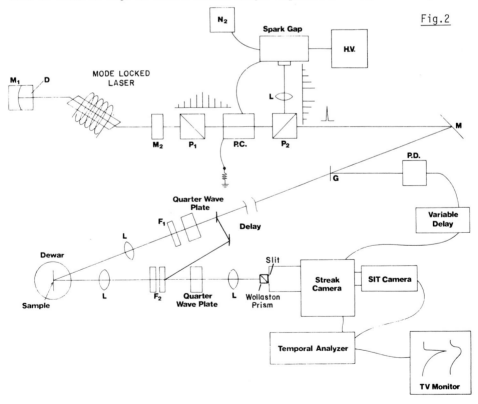

Fig.2

The experimental set-up for this technique is shown in fig. 2. It is for the simultaneous analysis of circularly polarized luminescence used in determining spin relaxation kinetics[4]. A ruby laser mode-locked by DDI in methanol provides the excitation. A spark gap and pockell cell is used to select a single pulse from the mode-locked train. The linearly polarized output of the laser single pulse selector is passed through a quarter-wave plate whose optic axis orientation is rotatable to produce either left or right circularly polarized light or linear polarized light. The laser pulse is then focussed onto the sample within a glass optical dewar.

The luminescence from the sample is collected by a lens, passed through a 1/4 wave plate, and filtered to remove the scattered excitation light. The luminescence is then refocussed onto the slit of a Hamamatsu C979 Temporal Disperser streak camera. Just before the slit of the streak camera is a Wollaston prism which separates the two polarizations with a 6° deviation and results in two focussed spots with orthogonal polarizations. There is some time spreading due to the prism but this is less than the streak camera resolution. This configuration provides for the simultaneous detection of both right and left circularly polarized luminescence kinetics from each excitation pulse.

The output of the streak camera is focussed onto a silicon intensified target (SIT) camera, Hamamatsu model C-1000. The camera output is digitized and stored in a two channel Hamamatsu C1098 Temporal Analyzer. This unit provides analog and digital outputs corresponding to the regions of the two orthogonal polarizations. This provides an independent time kinetic intensity profile for both the right and left circularly polarized luminescence on each laser shot.

References

1. E. J. Johnson, R. W. Davies, and A. Lempicki, Annual Report Contract No. F49620-78-C-0082 Air Force Office of Scientific Research July 31, 1979.
2. A. I. Ekimov and V. I. Safarov, JETP Lett. 13, 495 (1971).
3. R. D. Spencer and G. Weber, J. Chem. Phys. 52, 1654 (1970).
4. R. J. Seymour and R. R. Alfano, APL 37, 231 (1980).

This research is supported by NSF grant DMR-781-7805 and ENG 7920192 and AFOSR grant 80-0079.

Molecular Orientation Dynamics in Gels and Critical Mixtures

D.E. Cooper[1] and J.D. Litster

Department of Physics and Center for Materials Science and Engineering,
Massachusetts Institute of Technology, Cambridge, MA 02139, USA

1. Introduction

We have used pulses from a passively mode-locked dye laser (~1 psec, 5 kW)
to study the reorientation of dye molecules dissolved in gelatine gels of
concentrations from 2% to 10% gelatine. We have also used the Kerr effect
to measure orientational dynamics of nitrobenzene ($C_6H_5NO_2$) near the conso-
lute critical point with n-hexane (C_6H_{17}). These measurements determine the
local viscosity and interstitial pore size in the gel and the effect of critical fluctuations on molecular reorientation in binary critical mixtures.

2. Experiments in Gelatine

A gel consists of a crosslinked network of long polymer molecules with small
molecules of a fluid medium (usually water) trapped between the polymers.
Gels containing as much as 99.5% water can exhibit macroscopic shear rigidity,
which has led some biologists to propose that the water must be structured.

   Gelatine gels are weakly crosslinked by hydrogen bonds which are easily
broken thermally to cause a reversible gel-sol transition familiar to anyone
who has prepared Jello. In order to study the structure of the interstitial
water in both the gel and sol phases, we dissolved the laser dye oxazine 4
perchlorate (OX4) (concentration 10^{-4} to 10^{-3} molar) and measured the decay
of dichroism induced by a laser pulse. With a large molecule like OX4 in
water one would expect a hydrodynamic description to be valid; thus the dichroism should decay exponentially with a decay time given by the Debye-
Stokes-Einstein relation $\tau = \eta\nu/6kT$. Experiments in pure water between 20°C
and 95°C showed an exponential relaxation. After correction for the OX4
fluorescence lifetime (2.8 ns) we found the viscosity to vary as
$\eta = \eta_0 \exp(W/kT)$ with an activation energy W = 3.8 kcal/mole as expected
for water. The hydrodynamic volume of OX4 was $\nu = 3.8\times10^{-21}$ cm³. Having
established that τ scales with the viscosity, we carried out measurements
in gels from 20°C to 50°C (the gel-sol transition occurred at 28°C). We
found the reorientation times to be longer than in pure water and to increase
with gelatine concentrations. In any given gel, we found τ to scale with T
the same as in pure water (i.e. W = 3.8 kcal/mole within experimental error);
since W is a sensitive measure of local structure we conclude the water in
both gel and sol phases has identical local structure and viscosity to pure
water. The larger values of τ are caused by polymers disturbing the Stokes
velocity field around the OX4 molecule and can be used to determine the interstitial pore size. For a spherical molecule of radius a   reorienting

---

[1] Present address: Los Alamos Scientific Lab., Los Alamos NM 87545, USA

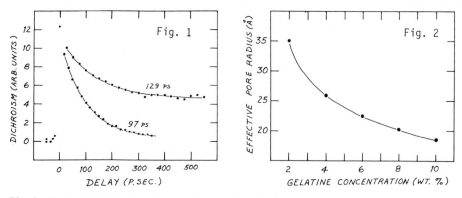

Fig.1 Induced dichroism decay of oxazine 4 dissolved in methanol in bulk (lower curve) and in Vycor glass (upper curve). The upper curve does not decay to zero as some dye is bound to the glass.

Fig.2 Effective pore sizes in various gelatine gels deduced according to discussion given in the text. The solid curve is drawn as a guide to the eye.

inside a spherical cavity of radius R, $\tau(\infty)/\tau(R) = 1 - (a/R)^3$; thus we expect $\tau(\text{water})/\tau(\text{gel}) \simeq 1 - (C/R_e)^3$, where R_e is an effective pore radius in the gel. This effect is illustrated in Fig.1 for Vycor glass of 40 Å pore size and was used to calibrate (giving C = 13.6 Å) our measurements in gels. The results are nearly independent of T and are shown in Fig.2 as a function of gelatine concentration.

3. Experiments in a Binary Critical Mixture

Nitrobenzene and n-hexane are soluble in any concentration above T_c = 20.3°C; below T_c there is a phase separation with one phase rich in $C_6H_5NO_2$, the other rich in C_6H_{17}. At T_c the osmotic compressibility, and consequently fluctuations in relative $C_6H_5NO_2$ concentration diverge. The static properties of the mixture have been extensively measured and can be quantitatively understood by renormalization group calculations in statistical mechanics. Our aim was to study the effect of critical fluctuations on a local scale as they influenced the orientation motion of the $C_6H_5NO_2$ molecules; we used a laser pulse to orient some of the molecules and measured the relaxation of the resulting birefringence. The reorientation time is shown in Fig.3 as a function of reduced temperature $\varepsilon = T/T_c - 1$. The orientation order induced by the laser pulse may be written as $Q_{\alpha\beta} = (1/2)\langle 3\zeta_\alpha\zeta_\beta - \delta_{\alpha\beta}\rangle$ where ζ_α, ζ_β are the Cartesian components of the polarization symmetry axis $\vec{\zeta}$ of the $C_6H_5NO_2$ molecule and the average is over a small but macroscopic volume. The induced birefringence is proportional to $Q_{\alpha\beta}$, and will relax as $Q_{\alpha\beta}(t) = Q_{\alpha\beta}(0) \exp(-t/\tau)$. From our experience with liquid crystal ordering [1], we expect the relaxation time to be $\tau = \nu\chi$ where χ is a weakly temperature dependent susceptibility ($\chi \sim 1/kT$) and ν is a transport coefficient proportional to the shear viscosity.

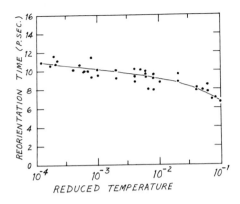

Fig. 3 Decay of induced birefringence in a nitrobenzene n-hexane critical mixture. The solid curve is a fit to (1)

Previous studies of molecular reorientation in critical mixtures [2] have been carried out by spectral analysis of depolarized light scattering in which the spectrum of interest must be deconvoluted from the instrumental resolution and a narrow polarized spectrum (from concentration fluctuations) whose intensity diverges at T_c. Our direct measurements in the time domain eliminate these difficulties. Our results were fit to the following theoretical [3] expression

$$\tau = \frac{\tau_0}{1 + \epsilon} \, e^{\frac{W}{1+\epsilon}} \, \epsilon^{-Y} + \tau^* \tag{1}$$

in which W (3.58 kcal/mole) is obtained from the shear viscosity far from T_c, and τ_0, τ^*, and Y were adjustable parameters. We found τ_0 = 0.15 ps, τ^* = 2.6 ps, and Y = 0.043. The meaning of the zero viscosity intercept τ^* is not clear (our data are fit only slightly less well by τ^*=0), but it is of the order one might expect for $C_6H_5NO_2$ as a free rotor; it may also account for nonsingular terms in ν.

The form of (1) comes from renormalization group (RG) calculations [3] for the shear viscosity η and the weak divergence comes from nonlinear coupling of local molecular motion to the critical concentration fluctuations. The value of Y is in excellent agreement with the RG predictions; in fact the accuracy of our data permit O(logarithmic divergence) < Y < 0.06. Although our value of τ is consistently 50% smaller than the light scattering results of BEYSENS and ZALCZER [2], we find an identical temperature dependence.

4. Conclusions and Acknowledgements

Our measurements provide a direct measurement of the interstitial fluid viscosity in gels and confirm that (inferred from other methods) there is nothing special about the interstitial water. Time resolved spectroscopy is a more direct way to measure molecular reorientation in critical fluids and could be used under circumstances of restricted geometry which are currently of interest.

This work was supported in part by the Joint Services Electronics Program (DAAG-29-78-C-0020) and the National Science Foundation (grants DMR-78-23555 and DMR-76-80895).

References

1. T.W. Stinson, J.D. Litster, and N.A. Clark, J. de Physique 33, C1-69 (1972)
2. D. Beysens and G. Zalczer, Phys. Rev. A18, 2280 (1978); I.L. Fabelinskii, V.S. Starunov, A.K. Atakhodzhaev, L.M. Sabiror, and J.M. Utarova, Opt. Comm. 20, 135 (1977); J.R. Petrula, H.L. Strauss, K. Q-H. Lao, and R. Pecora, J. Chem. Phys. 68, 623 (1978)
3. P.C. Hohenberg and B.I. Halperin, Rev. Mod. Phys. 49, 435 (1977)

Measurement of Excited State Absorption Cross Sections in Dye Solutions with Picosecond Light Pulses

A. Penzkofer and J. Wiedmann

Fakultät für Physik, Universität Regensburg,
D-8400 Regensburg, Federal Republic of Germany

1. Introduction

In excited state absorption measurements intense pump pulses promote molecules to excited states and weak probe beams induce transitions to higher lying states. The measurement of absolute excited state absorption cross-sections is complicated by the fact that the transmission of probe light depends on the number of excited molecules, on their orientational distribution and on the angle between the involved transition moments.

In the past several techniques have been developed for the measurement of absolute excited singlet state absorption cross-sections: i) orientation free absorption spectra were measured with pump and probe beam polarizations oriented under an angle of 54.7° [1]; ii) the excited state absorption was investigated after complete bleaching of the ground state [2]; and iii) the absorption measurements were compared with numerical simulations [3].

Here a further technique is described which avoids the problems of orientational anisotropy: An intense picosecond pump pulse (frequency ν_p) populates the S_1-state. After molecular reorientation (delay time $t_D \simeq 2\tau_{or}$) two probe beams are applied. One probe pulse at frequency ν_p measures the excited population and the other probe beam at frequency ν_L monitors the excited state absorption.

The method is applied to determine the absolute excited state absorption cross-sections $\sigma_e(\nu_L)$ of rhodamine 6G and rhodamine B in ethanol for a transition between the singlet states S_1 and S_4.

2. Description

In Fig.1 a simplified level scheme of the rhodamine dyes is presented. A pump pulse of frequency $\nu_p = 18\,960$ cm^{-1} (second harmonic of a modelocked Nd-glass laser; duration $\Delta t_p = 5$ ps) promotes the molecules to the S_1-band. Excited state absorption

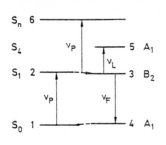

S$_n$ 6 ──────────

S$_4$ v$_P$ ── 5 A$_1$

          v$_L$

S$_1$ 2 ────── ─── 3 B$_2$

     v$_P$ v$_F$

S$_0$ 1 ────── ── 4 A$_1$

Fig.1 Energy
level diagram
for rhodamine
dyes.

at frequency v_P leads to transitions to level 6. A probe beam
of frequency v_L = 9 480 cm^{-1} excites molecules from the S$_1$ to
the S$_4$ level. This state is frequently termed as S$_2$ when two
weakly absorbing lower lying singlet states [4] are neglected.

After reorientation of the excited molecules in the S$_1$-state
the transmission of the probe beam at frequency v_L is given by

$$T_e = \exp \left[-\sigma_e(v_L) \int_0^\ell N_3(z)dz \right] \qquad (1)$$

$\sigma_e(v_L)$ is the isotropic S$_1$-S$_4$ excited state absorption cross-
section at frequency v_L.

The total population $\int_0^\ell N_3(z)dz$ of level 3 is monitored with
the probe beam of frequency v_P. Its transmission through the
sample is

$$T = \exp \left\{ -\sigma_{12} N\ell + [\sigma_{12} - \sigma_e(v_P)] \int_0^\ell N_3(z)dz \right\} \qquad (2)$$

$N = N_1(z) + N_3(z)$ is the total number density of dye molecules
in the solution. σ_{12} and $\sigma_e(v_P)$ are the isotropic ground state
and excited state absorption cross-sections at frequency v_P.
ℓ is the sample length.

Solving Eq. 2 for $\int_0^\ell N_3(z)dz$ and inserting the result into
Eq. 1 leads to

$$\sigma_e(v_L) = \frac{\ln(T_e)[\sigma_e(v_P) - \sigma_{12}]}{\ln(T) + \sigma_{12} N\ell} \qquad (3)$$

Eq. 3 is valid when: i) the excited state population $N_3(r)$ is
constant over the cross-sections of the probe beams, ii) the
stimulated emission cross-section $\sigma_{em}(v_L)$ at frequency v_L is
negligibly small. iii) the transfer to triplet states may be
neglected, iv) the probe beams do not affect the level populations (weak probe beam powers), and v) an isotropic orientational distribution in the S$_1$ state is established (delay of probe
beams $t_D \simeq 2\tau_{or}$).

120

3. Experiment

The experimental setup is depicted in Fig.2. A modelocked Nd-glass laser is used. A single pulse is selected with an electro-optical shutter and increased in energy with a Nd-glass amplifier. The second harmonic pump pulse is generated in a KDP crystal. A BK7 glass block is used for temporal separation of the fundamental and second harmonic pulses. The transmitted laser pulses at ν_p and ν_L are reduced in intensity and reflected back to the sample. They act as time delayed probe pulses and their transmissions are measured with detectors PD1-4.

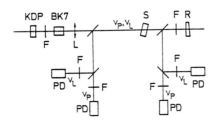

Fig. 2 Experimental arrangement. KDP, crystal for second harmonic generation; F, filters; BK7, glass block (length 10 cm) for temporal separations of probe beams ν_p and ν_L; S, dye sample; R, reflecting glass wedge; PD, photodetectors.

4. Results

The parameters of the analysed dyes [5] together with the determined excited state absorption cross-sections $\sigma_e(\nu_L)$ are listed in Table 1. The $\sigma_e(\nu_L)$-values are rather small. In case of rhodamine 6G, $\sigma_e(\nu_L)$ should be approximately equal to the peak S_1-S_4 absorption cross-section (transition to a position slightly above the S_4 potential energy curve [6]). The peak S_0-S_4 absorption in rhodamine 6G is a factor of two larger than the S_1-S_4 absorption ($\sigma(S_0$-$S_4) = 4.5\times10^{-17}$ cm^2 at $\nu=28600$ cm^{-1}). Our results agree with Dolan and Goldschmidt [2] who found $\sigma(S_1$-$S_4) \approx 1.5\times10^{-17}$ cm^2 for rhodamine 6G.

The measurement of absolute excited state absorption cross-sections at fixed frequencies allows the calibration of qualitative excited state absorption spectra which may be obtained with picosecond light continua [7].

Table 1

	Rhodamine 6G	Rhodamine B
concentration	1.65×10^{-5} M	2.9×10^{-5} M
solvent	ethanol	ethanol
σ_{12}	4.17×10^{-16} cm^2	2.1×10^{-16} cm^2
$\sigma_e(\nu_p)$	5×10^{-17} cm^2	5×10^{-17} cm^2
$\sigma_e(\nu_L)$	$(2 \pm 0.2)\times10^{-17}$ cm^2	$(4 \pm 1)\times10^{-18}$ cm^2

References

1. H.E. Lessing and A. von Jena, Chem. Phys. Lett. $\underline{59}$, 249 (1978).
2. G. Dolan and C.R. Goldschmidt, Chem. Phys. Lett. $\underline{39}$, 320 (1976).
3. A. Müller, J. Schulz-Hennig and H. Tashiro, Appl. Phys. $\underline{12}$, 333 (1977).
4. H. Jacobi and H. Kuhn, Z. Elektrochem. $\underline{66}$, 46 (1962).
5. W. Falkenstein, A. Penzkofer and W. Kaiser, Opt. Comm. $\underline{27}$, 151 (1978).
6. C. Orner and M.R. Topp, Chem. Phys. Lett. $\underline{36}$, 295 (1975).
7. A. Penzkofer and W. Kaiser, Opt. Quant. Electr. $\underline{9}$, 315 (1977).

122

Slow and Fast Response of the Optical Kerr Effect

J. Etchepare, G. Grillon, A. Antonetti, and A. Orszag

Laboratoire d'Optique Appliquée, Ecole Polytechnique - ENSTA
F-91120 Palaiseau, France

1. Introduction

The optical Kerr effect (O K E) results in a transient anisotropy induced in a material in which travels an intense polarized beam. The kinetics of the different mechanisms involved in this effect may be discriminated by the use of laser pulses of adequate duration. We present new results on, at least, two of the usually proposed mechanisms [1,2]. In a first part, we describe a double beam apparatus, working in the picosecond range, which demonstrates that many non saturated organic compounds exhibit a fast non resolved response superimposed upon the slower decay of molecular origin. In a second part, we present preliminary results obtained with a subpicosecond laser setup.

2. Theoretical background

The time dependent anisotropy $\Delta n\ (t)$ of the O K E can be expressed [3] by

$$\Delta n\ (t) = n_{2B}^i\ I_1\ (t) + \frac{n_{2B}^j}{\tau_j} \int_{-\infty}^{t} I_2\ (t')\ \exp - \frac{(t-t')}{\tau_j}\ dt' \qquad (1)$$

where two terms appear, regardless or not the associated relaxation time τ is shorter than the pump pulse width ; $n_{2B}^{i,j}$ represent the non linear indexes of refraction and I_1 the pulse intensity envelope. The phase lag between two-perpendicular polarizations of a beam at frequency λ_2, is

$$\Delta\phi\ (t) = \frac{2\pi L}{\lambda_2}\ \Delta n\ (t) \qquad (2)$$

where L is the length of the active medium. If the probe pulse I_2 is polarized at 45° with respect to the two above perpendicular polarizations, we can write the transmission TT (t) as :

$$TT\ (t) = N \int_{-\infty}^{+\infty} I_2\ (t-t')\ \sin^2 \frac{\Delta\phi(t')}{2}\ dt' \qquad (3)$$

where N is a normalization factor which takes into account the transmission of the probe pulses via the crossed polarizers.

If the relaxation times are short compared to the pulse width, and when the pulse energy is low enough to experience a phase lag $\Delta\phi < 1$, the transmission TT (t) becomes proportional to the third-order cross correlation of the pulses.

As the two pulses are generated from the same laser pulse, the above signal gives direct information on their width and shape.

When the active medium is characterized by, at least, a non linear index of refraction associated with a relaxation time τ^J longer than the pulse width Δt, a linear regression method yields the relaxation time.

3. Measurement of the OKE in the ten-picosecond regime

3.1. Description of the experiment

Our setup (fig.1) allows us to simultaneously record the transmissions of two non linear medium cells, and to measure

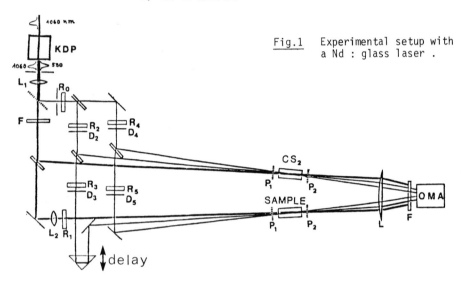

Fig.1   Experimental setup with a Nd : glass laser .

their ratio. The reference cell is filled with carbon disulfide : its molecular relaxation time is fast enough not to limit the response of the shutter and we have directly a precise knowledge of the shape and width of the actual pulses ; its Kerr constant is well established [4,5], so we can obtain accurate relative values of induced anisotropies. The principle of the experiment is to measure for each cell the ratio of two transmitted probe beams, one coming with variable time delay, the other being fixed in time, compared to the pump pulse. Therefore the relative transmission can be considered as independent of the energy fluctuations of the pulses.

The specific configuration used here employed a Nd^{3+} glass laser which delivers a train of pulses, at 1.06 μm, with a power density close to 200 MW/cm^2 ; the probe pulse, at 0.53 μm, obtained by SHG, have an energy kept low enough in order not to perturb the active medium. The four probe beams transmitted through the cells are frequency filtered before being imaged on an ISIT optical multichannel analyser target. The dynamic range of this apparatus is extended to about 10^5 by using appropriate optical densities.

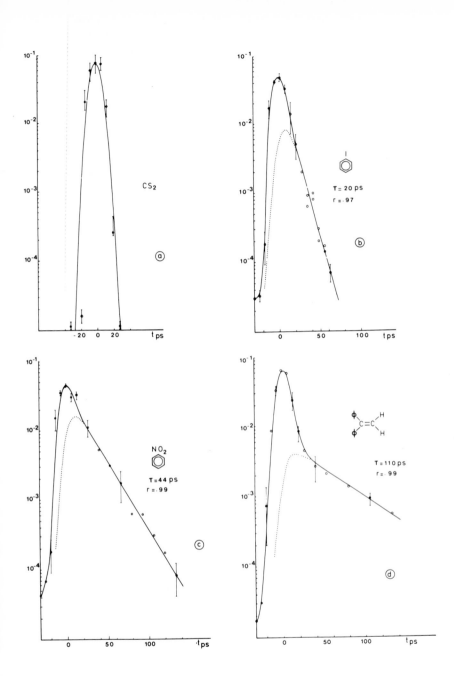

<u>Fig.2</u> Decay profiles of the transmitted Kerr signals.

When the molecular relaxation time is larger than the pulse width, the temporal profile of the transmitted light clearly reveals two components in all tested liquids. The table 1 gathers the essential of the results.

3.2. Discussion of the results

In a first step, the apparatus function is determined by filling the two cells with CS_2. The observed profile (fig.2a) fits a gaussian curve of 15 ps-width. The mean features of our results may be accounted for by the different figures (2b, 2c and 2d).

Table 1. Transmissions (in arbitrary units), molecular relaxation times and non linear indexes of refraction of some liquids.

compound	TT_{max} (relative)	τ_j [ps]	n_{2B}^j [10^{-13}esu]	n_{2B}^i [10^{-13}esu]
carbone disulfide	10 000	2.1 ± 0.3 [4]	200	/
iodobenzene	1300	20	52	67
nitrobenzene	1500	44	140	73
diphenylethylene	500	110	24	43

4. OKE in the subpicosecond regime

These experiments use subpicosecond (0.5 ps) amplified pulses [6]. Only two curves are presented here. Fig.3a represents the kinetics of 1-bromonaphtalene ;

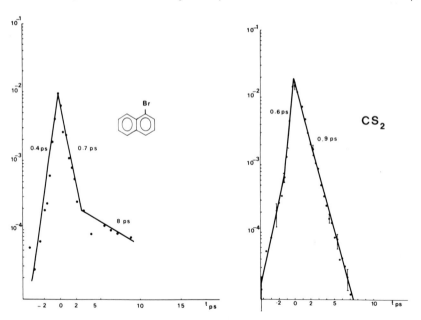

Fig.3.Decay profiles of OKE in the subpicosecond regime.

we observe still a fast decay response limited by the laser pulse width which demonstrates the existence of a very fast (of electronic origin) process. In fig.3b we have reproduced the kinetics of CS_2 ; its molecular relaxation time seems shorter than usually admitted and complementary experiments are in progress to measure more precisely this decay time.

5. Conclusion

We have clearly demonstrated the existence of at least two mechanisms with different kinetics in the OKE of several organic liquids. Moreover, the new results on CS_2 point out that its molecular relaxation time is probably shorter than 1.8 ps.

This work has been supported by a grant of the Direction des Recherches, Etudes et Techniques.

References

1. K. Sala and M.C. Richardson, Phys. Rev. A 12, 1036 (1975).

2. R.W. Hellwarth, Prog. Quant. Electr. 5, 1, (1977).

3. P.P. Ho and R.R. Alfano, Phys. Rev. A 20, 2170, (1979).

4. M.A. Duguay, E. Wolf, Progress in Optics XIV. North Holland (1976).

5. E.P. Ippen and C.V. Shank, Appl. Phys. Lett. 26, 92 (1975).

6. A. Migus, J.L. Martin, R. Astier and A. Orszag, This issue.

Part IV

Picosecond Relaxation Phenomena

Picosecond Measurements of Radiative and Radiationless Relaxation Processes in Aromatic Molecules

K.J. Choi, L.A. Hallidy, and M.R. Topp

Department of Chemistry, University of Pennsylvania,
Philadelphia, PA 19104, USA

1. Experimental Technology

Recently, measurements of subnanosecond transients in a number of carbonyl and heteroaromatic compounds, such as benzophenone, fluorenone
and acridine have begun to yield useful information about the effects
of solvents on radiationless relaxation processes. Picosecond techniques
based on solid-state lasers have been used extensively for these types
of measurement in a number of different laboratories. In our laboratory,
several experimental approaches not in wide use elsewhere have permitted
us a greater degree of experimental freedom.

Two principal techniques are available for time-resolved fluorescence
measurements in the picosecond regime. Although the streak camera is,
in principle, a superior device, one major drawback is the apparent
limitation to broad detection spectral bandwidths, which arises because
a standard monochromator is highly time-dispersive. Although it is
possible to circumvent this problem, we prefer for most purposes the
simplicity and high sensitivity of laser gating, via frequency conversion
(TRF) [1]. Spectral resolution is determined by a monochromator, effectively independently of the experimental time-resolution.

As a complement to this approach, we employ a method developed in our
laboratory in which two time-separated pulses are used to excite the
sample molecule to a higher singlet state [2]. Low-quantum yield, mid-
ultraviolet fluorescence is used to quantify the absorption of the
second pulse by the molecular S_1 state. The method provides a high
signal-to-noise probe of transient absorption events, but via fluorescence excitation (CTP approach).

Presently both TRF and CTP experiments are routinely carried out in
our laboratory with a high-repetition-rate Nd^{3+}-YAG laser. As we have
shown elsewhere [3,4], an injection-mode-locked ring-laser configuration
incorporating an optical Kerr-cell [5] delivers stable pulse-trains at
5 Hz, having average pulse-durations <8 ps.

2. Fluorescence Radiative Rate Constants

We are particularly interested in making measurements of fluorescence
radiative rate-constants as a means of state identification and for the
study of the effects of solvent environment [6]. So, in addition to
the usual study of spectrum and time, we require to calibrate the
absolute magnitude of our time-resolved fluorescence signals. This
is easier to interpret than for transient absorption measurements,
since one of the states involved, the ground state, is known precisely.

Without a knowledge of the magnitude of k_f, time-resolved fluorescence measurements can be very misleading.

The importance of the <u>direct</u> measurements can be seen if we consider three separate methods for measuring k_f. The most basic and, in some cases the most straightforward is summarised in the standard equation (1):

$$k_f = 2.88 \times 10^{-9} \frac{n_f^3}{n_a} \frac{\int F(\nu)d\nu}{\int \frac{F(\nu)}{\nu^3} d\nu} \cdot \int \frac{\varepsilon d\nu}{\nu} \qquad (1)$$

This requires three spectrum integrations, two of which, the fluorescence spectra, usually pose few problems. However, in several molecules of interest to us, absorption spectrum integration leads to the wrong answer!

In a molecule such as benzophenone in a non-polar, non-hydrogen-bonding solvent such as hexane, the weak n-π* state ($\varepsilon \sim 10^2$ M^{-1} cm^{-1}) is clearly separable from the stronger absorptions to the blue ($\varepsilon > 10^4$ M^{-1} cm^{-1}). In alcoholic and mildly acidic solvents, the n-π* absorption band is rapidly reduced to a weak shoulder barely resolvable from the tail of the intense and inhomogeneously broadened π-π* absorption bands.

In the case of acridine, an almost <u>total</u> overlap is observed. The absorption spectrum is nearly independent of solvent and, although the fluorescence <u>quantum-yield</u> varies more than three orders of magnitude, the fluorescence <u>spectrum</u> is only slightly different between polar and non-polar media. Although some guidance can be obtained from photo-chemical measurements, the assignment of the lowest excited singlet state is far from clear in solvents of intermediate character. In any case, the simple assignment of n-π* or π-π* character is too much of a generalisation and a much more quantitative description is required. It is therefore of paramount importance to obtain values of k_f for acridine, simply in order to know which state we are looking at. However, equation (1) is no longer useful since the last integral cannot be separately evaluated for the ^{1}n-π* state.

A second means of obtaining k_f values is to use equation (2):

$$k_f = k_d \, \Phi_F(\text{ref}) \frac{\int F(\nu)d\nu}{\int F_{\text{ref}}(\nu)d\nu} \qquad (2)$$

which incorporates more of a dynamic approach than equation (1). This can be readily calibrated by some reference compound, but it does require that k_d is a simple first-order rate constant and that the fluorescence spectrum is not a function of time. We have determined that this approach is permissible for acridine in simple solvents but not, for example, for fluorenone in alcohols.

The most direct approach requires the <u>absolute calibration</u> of the gated fluorescence intensity, and the value k_f is obtained from equation (3):

$$k_f = k_f(\text{ref}) \frac{f(\text{ref})}{f} \frac{F(\nu,\tau)}{F(\nu,\tau) \, (\text{ref})} \qquad (3)$$

Here we again use a calibration molecule. The f are correction factors for the fractions of the two spectra actually sampled while τ is some time after irradiation at which the fluorescence signal $(F(\nu))$ is assumed not to have decayed significantly. If sufficient time resolution is available, pulse deconvolution is unnecessary. The conditions are usually standardised so that the excited state populations are the same in both samples.

3. Acridine

Our data for acridine are presented in Table 1.

Table 1   Acridine singlet lifetime parameters

Solvent	Decay time (ps)		τ_{rad} (ns)	$\phi_F (\times 10^3)$
		lit		
Hexane	45	(33)	470	0.096
		(38)		
Benzene	54	-	163	0.33
DMF	59	-	215	0.28
MeCN	61	(75)	175	0.35
2-propanol	346	(120)	46	7.5
Ethanol	350	(250)	45	7.9
Methanol	377	-	37	10.2
Formamide	438	-	45	9.7
H_2O	-	(10^4)	(~ 30)	(320)

The important results were:

1. The decay time of singlet acridine in alcohols is nearly independent of the solvent at room temperature. This implies that, once the ^{1}n-π* and $^1\pi$-π* states are inverted by hydrogen bonding, the decay time is not sensitive to small changes in hydrogen-bonding strength. Thus the singlet decay time is not sensitive to the relative energies of $^1\pi$-π* and ^{3}n-π* states.

2. The longer decay time in aqueous solution may be understood in terms of a further inversion now of $^1\pi$-π* and ^{3}n-π* levels (analogously to the example of 9-fluorenone on going from hexane to benzene). The emitting acridine state can be unambiguously assigned as $^1\pi$-π* by its radiative lifetime of ~40 ns, although the fluorescence quantum yield is greatly different between aqueous and alcoholic solutions.

3. In non-hydrogen-bonding solvents, the decay times become much shorter and the radiative lifetimes much longer. Also, unlike the hydrogen-bonding situation both of these quantities are solvent sensitive. Thus, although in a general sense the ^{1}n-π* is now lowest in energy, we see strong evidence of

coupling with the $^1\pi$-π^* state as the relative energies are sensitive to solvent. The most obvious effect is the difference in the radiative life-time between hexane and polarisable solvents. Previously, this effect had only been studied indirectly from triplet quantum-yield measurements [7]. Again, simple decay time measurements would have been insufficient here.

4. Benzophenones

In the recent literature, we have reported the successful time resolution of the "non-fluorescent" first excited singlet state of biphenylene [8] using the CTP method. We have used the same approach to attack the perennial problem of benzophenone intersystem crossing. We have deter-mined that benzophenone, along with several of its derivatives, fluoresces at an acceptable level in the region 280-330 nm, when excited by wave-lengths longer than 350 nm, derived from a neodymium laser.

In benzophenone, as Table 2 shows, the singlet relaxation time is solvent dependent and, by comparison with data obtained in our laboratory and elsewhere, on a similar time scale to the thermalisation of excess vibrational energy. It is highly significant that we observe the singlet signal to persist for the same period as a transient feature in the time-resolved absorption spectrum obtained by GREENE et al. [9]. Further, our data indicate a monotonic decay of the singlet signal, showing that the fraction of singlet in the excited state population evolves constantly in time. Thus, a rapid singlet-triplet conversion is effectively ruled out

Table 2   The singlet decay time of benzophenone and related compounds

	Solvent	λ_p(nm)	τ_{SD}(ps)
Benzophenone (B)	hexane	530	16
	ethanol	530	12
	HFIP	1060	10
	H_3PO_4(85%)	TRF	950
2-hydroxy B	hexane	354 AC	6
	ethanol	354 AC	30
4-hydroxy B	hexane	1060	18
	ethanol	1060	27
	water	1060	<5
Anthrone	hexane	354 AC	9
Fluorenone	hexane	530 or TRF	120
	ethanol	530 or TRF	complex
Duroquinone	hexane	530	6

and we are left to consider the vibrational states of the singlet and triplet states when crossing does occur, i.e. when the excited molecule evolves towards predominantly triplet behaviour. Even in cases where time evolution of the radiative rate constant in hydrocarbons has provided evidence for the interception of incompletely vibrationally relaxed molecules, no specific hot-band features have been observed. This indicates that at best only low-frequency modes are populated after $\sim 10^{-12}$ s, so that it is questionable whether C=O stretch excitation would persist sufficiently long in benzophenone to sustain triplet "hot" band absorption. The conclusion appears to be therefore that, while inter-system crossing almost certainly involves non-thermalized singlet states, under our experimental conditions the process is occurring in the large-molecule limit.

Observations in different solvents, therefore, while initially exciting the molecule with different amounts of vibrational energy, probably represents a real difference in intersystem crossing, perturbed by strong hydrogen bonding to the carbonyl group.

Also presented in Table 2 are data for benzophenone derivatives. 2-hydroxybenzophenone has been the object of study in a number of groups, since it is a highly stable ultraviolet absorber [10,11]. The major reason for photostability is the rapid excited state quenching assisted by intramolecular hydrogen-atom (or proton) transfer. Time-resolved absorption measurements have been used to determine relaxation times for the triplet state in alcoholic and non-polar solvents to be 1.5 ns and 30 ps respectively [10]. However, there was some considerable doubt from these measurements as to the singlet relaxation time. Also, the possibility of the presence in comparable proportions of two distinct types of 2-hydroxybenzophenone in ethanol has been postulated on the basis of an observed two-stage decay profile in the transient absorption at 354 nm. (The validity of the two-state model of [10] must be questioned since excited singlet and triplet states of benzophenones should also absorb strongly near 350 nm). The two types of species would be either intra- or intermolecularly hydrogen bonded, as was proposed a decade ago by LAMOLA & SHARP [12]. Our results show in non-polar solvents a very rapid singlet decay while, in ethanol, the singlet relaxation time is both monotonic, and of approximately 30 ps duration. Thus, in the ethanol solution, we find strong evidence for a predominance of a single species which is different from that in a non-polar solvent, as well as a relaxation process which matches fairly well the primary relaxation in the transient absorption work. The picture that emerges, then is of a predominantly externally hydrogen-bonded 2-hydroxybenzophenone, which behaves similarly to 4-hydroxy-benzophenone in yielding a triplet state after 30 ps. The exact yield of triplet would depend on the rate at which the external hydrogen-bonds could be broken in S_1. Subsequently, the triplet relaxes again by hydrogen-bond breaking in 1.5 ns to the intramolecularly bonded form which rapidly yields ground state. The difference in rates of hydrogen-bond breaking can probably be explained on the basis of the known differerces in pKa of the carbonyl group.

The behaviour of 4-hydroxybenzophenone is even more solvent sensitive. In hexane, the situation closely resembles benzophenone itself, as our data confirm since the n-π^* and π-π^* absorption bands are well separated. In both ethanolic and aqueous solutions, the deprotonated form of 4-hydroxybenzophenone has been detected at long times by flash photolysis [13]. Our data have shown that the anion generation mechanism is significantly different between the two solvents. Thus, while in ethanol the singlet lifetime is ~30 ps, in aqueous solution the singlet is rapidly quenched (presumably by proton ejection) in <5 ps.

5. Vibrational Relaxation

Another important problem which is currently receiving much attention is the relaxation of excess vibrational energy following pulsed excitation. There are several stages in the relaxation process, which we may group as follows:

(a) coherence loss of the original set of levels
(b) intramolecular energy exchange
(c) intermolecular energy exchange, induced by collisions.

These processes may, or may not be separable in time.

One significant feature of equation (3) is that the product fk_f is shown to be a possible function of time. That is, the absolute intensity and spectrum of fluorescence depend on the states involved, including differences in the degree of vibrational relaxation. On a picosecond time scale, the non-uniqueness of a molecular fluorescence spectrum can readily be demonstrated. Thus, the time-evolution of the fk_f term becomes evident in a spectrally and temporally-resolved experiment on the early fluorescence of an aromatic hydrocarbon such as 3,4,9,10-dibenzpyrene [14]. This particular experiment used 8 ps pulses from a conventionally mode-locked Nd^{3+}-glass laser. The results are shown in Fig. 1. The fluorescence time profile

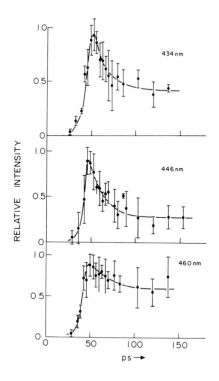

Fig. 1

was measured over ~150 ps at three different wavelengths: (434 nm) close to the electronic origin of the S_1-S_0 transition; (446 nm) at a minor peak in the time-integrated spectrum and (460 nm) at the position of the first principal vibrational band.

As far as could be determined, it was possible to extract the same initial relaxation time (~15 ps) from all three curves. However, from the ratios of the various maxima to asymptotes, a pronounced spectrum shape change is evident over the first few tens of picoseconds. The absence of any significant intensity to higher energies (>> kT) than the electronic origin ruled out the possibility of significant residual population of progression-forming vibrational modes. Since the molecule had been excited initially into the second excited singlet state, this was not unexpected. More recent experiments, designed to study the process of intramolecular energy migration have utilised excitation and probing of different vibrational levels of the same electronic state. The same conclusion was reached - that the originally populated modes relaxed into the bath faster than our time-resolution.

This work was supported in part by the Laboratory for Research on the Structure of Matter (DMR-7923647), the Regional Laser Laboratory and NSF grant CHE-78-35212.

References

1. L. A. Hallidy and M. R. Topp, Chem. Phys. Lett. <u>46</u> 8 (1977).

2. L.A. Hallidy, H.B. Lin and M.R. Topp, *Picosecond Phenomena*, ed. by C.V. Shank, E.P. Ippen and S.L. Shapiro, Springer Series in Chemical Physics, Vol. 4 (Springer, Berlin, Heidelberg, New York,1978) p. 314

3. K. J. Choi, L. A. Hallidy and M. R. Topp, J. Opt. Soc. Am. <u>70</u> 607 (1980).

4. K. J. Choi, M. R. Topp, this book, p.   .

5. A. J. Duerinckx, H. A. Vanherzeele, J. L. VanEck and A. E. Siegman, IEEE. J. Quant. Electr. QE-14 p. 983 (1978).

6. L. A. Hallidy, Ph.D. Thesis, University of Pennsylvania, 1980.

7. L. J. Noe, E. O. Degenkolb and P. M. Rentzepis, J. Chem. Phys. <u>68</u> 4435 (1978)

8. H. B. Lin and M. R. Topp, Chem. Phys. Lett. <u>64</u> 452 (1979).

9. B. I. Greene, R. M. Hochstrasser and R. B. Weisman, J. Chem. Phys. <u>70</u> 1247 (1979).

10. S. Y. Hou, W. M. Hetherington, G. M. Korenowski, and K. B. Gisenthal, Chem. Phys. Lett. <u>68</u> 282 (1979).

11. C. Merritt, G. W. Scott, A. Gupta and A. Yavrouian, Chem. Phys. Lett. <u>69</u> 169 (1980).

12. A. A. Lamola and L. J. Sharp, J. Phys. Chem. <u>70</u> 2634 (1966).

13. G. Porter and P. Suppan, Trans. Farad. Soc. <u>61</u> 1664 (1965).

14. K. J. Choi and M. R. Topp. Chem. Phys. Lett. <u>69</u> 441 (1980).

Vibrational Relaxation in Doped and Pure Molecular Crystals A Picosecond Photon Echo and CARS Study

W.H. Hesselink, B.H. Hesp, and D.A. Wiersma

Picosecond Laser and Spectroscopy Laboratory,
Department of Physical Chemistry, University of Groningen,
Nijenborgh 16, 9747 AG Groningen, The Netherlands

ABSTRACT

Using picosecond photon echoes a detailed study has been made of vibrational relaxation in the system pentacene in naphthalene at 1.5 K. The results indicate that low-frequency modes relax directly through phonon emission, while higher frequency modes decay through vibrational cascading. The role of the medium on the anharmonic intra-molecular vibration-vibration coupling is emphasized.

Delayed picosecond CARS experiments have been used to study relaxation of Frenkel-type vibrational excitons in pure naphthalene. The Raman lineshapes of these excitons are found to be Lorentzian and evidence is presented that this is caused by vibrational exchange narrowing.

The process of cooling of vibrationally hot large molecules in the condensed phase is presently not very well understood. Substantial progress however was recently made in the study of vibrational relaxation of small molecules in rare gas matrices [1]. The techniques used here are studies of hot band emission spectra and lifetime measurements using nsec pulsed dye lasers.

Vibrational relaxation in large molecules occurs on a picosecond time scale as evidenced by the weakness of the hot emission spectra. Relaxation times in these systems therefore can only be obtained by using psec light sources. Detailed understanding of vibrational relaxation requires not only determination of the relaxation time of the vibration excited, but also knowledge on the decay route. Information on the deactivation pathway can be obtained from hot emission spectra, as first shown by Rebane and co-workers [2], whereby we notice that the analysis of these spectra is far from trivial.

In this presentation we focus on the first question, namely the cooling rate of the vibration excited. We present detailed results here of psec photon echoes on mixed crystals of pentacene in naphthalene and p-terphenyl. Preliminary results of these experiments were reported in the first meeting on psec phenomena [3]. In addition we present results on the relaxation of vibrational excitons in pure crystals of naphthalene, obtained by psec delayed CARS experiments. It turns out that the vibrational transitions in these crystals are extremely narrow and evidence is presented that this occurs through vibrational-exchange narrowing. The laser set-up used to generate and detect picosecond photon echoes and delayed CARS has recently been described [4] in great detail. All crystals were grown in a Bridgman furnace whereby we notice that the naphthalene used for the CARS study was extremely purified by crystallization from ethanol, then treated with potassium metal and subsequently zone refined.

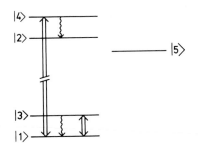

Fig. 1 Level scheme of relevant levels in pentacene and naphthalene

The level structure which is basic to understanding of the experiments is shown in Fig. 1. Level $|1\rangle$ is the groundstate, $|2\rangle$ the electronically excited state, $|3\rangle$ a vibrational state, $|4\rangle$ a vibronic state and $|5\rangle$ the lowest triplet state.

The psec photon echo is excited in the transition $\langle 4|\leftarrow|1\rangle$. Actually we have used the *accumulated stimulated* photon echo [5] to measure optical T_2

Table 1 Phase relaxation times at 1.5 K of the vibronic transitions of pentacene in naphthalene

PTC-h$_{14}$/NT-h$_8$		PTC-d$_{14}$/NT-h$_8$		PTC-h$_{14}$/NT-d$_8$	
ν^a(cm^{-1})	$\frac{1}{2}T_2$(ps)	ν^a(cm^{-1})	$\frac{1}{2}T_1$(ps)	ν^a(cm^{-1})	$\frac{1}{2}T_2$(ps)
136.7	2.3±0.4[b]	131.7	1.5±0.5[c]	135.9	7.5±2.5[c]
260.4	2.4±0.4[b]	252.6	3.7±0.7[b]	259.7	3.5±1[c]
307.6	19±2	288.7	14±2	307.4	20±2
346.4	8.5±1	314.7	13±1	344.3	14±4
448.5	15±1.5	415.4	30±4	447.4	19±2
521.6	1.5±0.2[c]	506.9	1.4±0.2[c]	520.3	1.7±0.2[c]
596.7	19.5±1.3	574.7	13±1.5	596.4	25±2
609.1	59±3	590.2	15±1.5	608.0	47±4
747.2	33±1.5	715.1	26±2	747.5	27±2
		834	11±2[d]		
		960	10±1.5		
		1197			
		1342			
		1367			
		1409	< 2		
		1418			
		1430			

[a] relative to the 0-0

[b] from linewidth and photon echo experiments

[c] from linewidth measurements

[d] decay showing beat pattern

Table II Phase relaxation times at 1.5 K of the vibronic transitions of pentacene in p-terphenyl at the 0_3 and 0_4 sites

0_3		0_4	
$\nu^a(cm^{-1})$	$\frac{1}{2}T_2$ (ps)	$\nu^a(cm^{-1})$	$\frac{1}{2}T_2$ (ps)
267.1	2.1 ± 0.5^b	267.8	2 ± 0.5^b
599.4	18 ± 3	599.4	10 ± 1
608.3	19 ± 3	606 9	32 ± 5
746.9	18 ± 3	744.9	31 ± 3

[a] relative to the 0-0

[b] from photon echo and linewidth measurements

on this transition. It can be shown that the intensity of this echo $I_e(t_{12})$ decreases with the first-second pulse separation (t_{12}) according to $I_e(t_{12})=I_e(o)e^{-4t_{12}/T_2}$. In a separate stimulated photon echo experiment on the vibronic transition involving the 747 cm^{-1} vibrational mode it was ascertained [4] that at low temperature (1.5 K) the pure dephasing contribution to optical T_2 was negligible or $T_2 = 2T_1$.

The results of experiments on three isotopic combinations of pentacene (PTC) in naphthalene (NT) are given in Table I. Note that $\frac{1}{2}T_2$ is the vibrational relaxation time at low temperature. In a separate similar experiment the relaxation times of several modes in different sites of pentacene in p-terphenyl were measured, the results of which are given in Table II.

The first thing we note is that there is an apparent erratic behaviour of relaxation times versus excess vibrational energy. This is most clear from a pictorial representation of the results, of which an example is given in Fig. 2. We note that a similar observation was made in the system porphin in n-octane by Voelker and MacFarlane [6] using the technique of high-resolution photochemical hole-burning [7].

We propose the following interpretation of the results. For vibrations of frequency less than twice the Debeye frequency (region A) *direct* relaxation through phonon emission occurs and the relaxation is fast. Note that

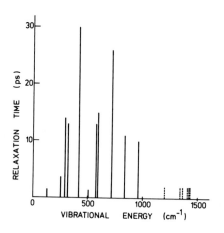

Fig. 2 Plot of the vibrational relaxation times versus excess vibrational energy of PTC-d_{14} in NT-h_8 at 1.5 K.

this implies that overtones of low-frequency modes also relax fast by phonon emission down the vibrational ladder. Assuming linear electron-phonon
coupling to be responsible for this direct phonon process, the vibrational
relaxation rate is predicted to increase linearly with the vibration quantum number [8]. From Table I we see that for the overtone of the 260 cm^{-1}
vibration this prediction is substantiated, which lends support to the suggested analysis. For fundamental vibrational frequencies in the range
$2\omega_D \lesssim \nu \lesssim 1200$ cm^{-1} (region B) vibrational relaxation is determined by *anharmonic* vibration-vibration coupling, whereby branching may occur. The re-
laxation is determined by both the "background" level structure and by the
effect of the medium on the vibrational anharmonicity. Note that this model
for relaxation in this sparse region implies that hot emission from vibrational levels other than the pumped one should be observed. For the vibrational levels we have studied this has not been verified sofar, however,
Van der Ende and Rettschnick [9] showed that by pumping in the 1426 cm^{-1}
level of pentacene in p-terphenyl, vibrational cascading does occur. The
observed guest and host isotope effect supports this model. While the guest
isotope effect is large there seems to be a distinct relation   between the
observed change in vibrational lifetime and frequency as shown in Fig. 3.
For levels that exhibit a large shift in vibrational frequency the relaxation time becomes longer and vice versa. This may be understood by realizing
that the number of states to which branching may occur decreases in the
first case while increasing in the second case. Although in region B, intramolecular anharmonic coupling determines the relaxation rate, the host (isotope) effect, and site effect witness the crucial importance of the medium
on the anharmonic intramolecular vibration-vibration coupling.

Finally in region C ($\nu \gtrsim 1200$ cm^{-1}) the vibrational level structure is
congested and strong vibrational mixing occurs, which leads to extremely
fast relaxation again. This idea is consistent with recent spectroscopic

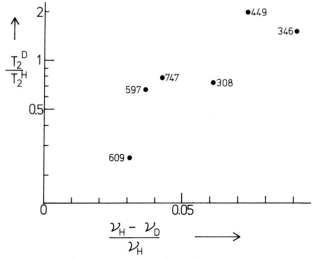

Fig. 3   Ratio of the dephasing times at 1.5 K of several vibronic bands in
$\overline{PTC-h}_{14}/NT-h_8(T_2^H)$ and PTC-$d_{14}/NT-h_8(T_2^D)$ plotted as a function of the
relative frequency shift upon deuteration. Indicated at each point are the
vibrational frequencies (in cm^{-1}) of PTC-h_{14}/NT-h_8.

observations on tetracene in a jet where strong anharmonic vibrational mixing occurs at 1000 cm^{-1} excess energy [10].

In conclusion of this part we note that the above description is also applicable to the understanding of the observed relaxation in porphin.

Vibrational relaxation (or dephasing) in the groundstate on the transition $<3|\leftrightarrow|1>$ of pure naphthalene crystals was studied by *delayed* psec CARS experiments. Very recently a similar type four-wave mixing experiment on a naphthalene vibration was performed by Decola, Hochstrasser and Trommsdorff [11]. The advantage of delayed psec CARS over the CW four-wave mixing, as employed by Decola et al, is that in the delayed CARS the non-resonant contribution to the signals is absent.

We have performed experiments on several vibrations in pure perproto- and perdeutero naphthalene. A typical delayed CARS-decay signal of the Ag-exciton level of the 766 cm^{-1} totally symmetric (ν_{8-}) mode of naphthalene-h$_8$ is shown in Fig. 4. The insert shows that the observed decay is purely exponential which indicates that the vibrational Frenkel exciton lineshape is purely Lorentzian. We note here that our experiments do not confirm the claim of Hess and Prasad [12] that this vibrational transition exposes a Gaussian line shape at 2 K.

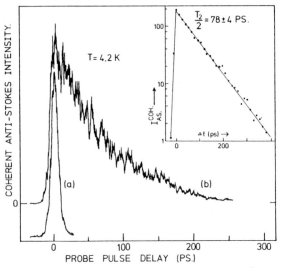

Fig. 4 Delayed CARS signal of the 766 cm^{-1} (ν_8) vibration of pure naphthalene.

The results obtained sofar are summarized in Table III. We first note that the present experiments do not reveal whether the observed relaxation times are due to vibrational dephasing or relaxation. However the drastic effect of perdeuteration on the relaxation times strongly suggests vibrational relaxation to be the dominant linewidth determining mechanism. The Lorentzian lineshape of the vibrational Frenkel type exciton suggests that this line is vibrationally exchange narrowed, as was also suggested by Decola et al [11]. Support for this idea is obtained by measuring the CARS relaxation at higher temperature where, as Fig. 5 shows, the lineshape be-

Table III CARS free induction decay times of some naphthalene vibrations

$\nu_{vib}(cm^{-1})$	polarization	temperature (K)	relaxation time (ps)
493(d_8)	aa	1.5	74±9
511(h_8)	aa	1.5	140±10
511(h_8)	bb	1.5	136±13
701(d_8)	bb	4.2	< 5
766(h_8)	bb	1.5	78±4
863(d_8)	aa	4.2	< 5
1021(h_8)	bb	1.5	22±2
1385(h_8)	aa	1.5	92±9
1385(h_8)	bb	1.5	98±14

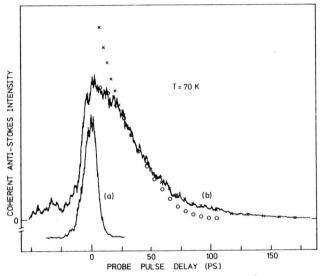

<u>Fig. 5</u> Delayed CARS signal of the 766 cm^{-1} (ν_8) vibration of pure naphthalene at 70 K. The open circles denote a fit assuming a Gaussian decay while the crosses attempt a fit to an exponential decay.

comes more Gaussian-like in the wings but still Lorentzian in the center. This indicates that at higher temperature the exciton becomes more locali- zed due to coupling to the lattice phonons. A similar observation on the triplet exciton lineshape in tetrachlorobenzene was made by Burland et al. [13]. For an exchange-narrowed line the Kubo theory [14] predicts the Lo- renztian homogeneous width (FWHM in Hz), to be $(\Delta^2\tau_c)/\pi$ where Δ (in rad. s^{-1}) is the amplitude and τ_c the correlation time of the stochastically fluctuating force. The driving force for exchange narrowing in this case is the nearest neighbour exchange which is at most a quarter of the factor group splitting. The factor group splitting is certainly less than 0.3 cm^{-1}

[12] which sets the lower limit for τ_c at 70 ps. We are presently pursuing CARS experiments on dilute isotopically mixed crystals of naphthalene in an attempt to directly determine the inhomogeneous Raman lineshape of an isolated naphthalene impurity (Δ). Note, however, that in this experiment the homogenous linewidth due to vibrational relaxation still may exceed the inhomogeneous linewidth. At any rate, the condition for exchange narrowing: $\Delta\tau_c \ll 1$ demands Δ to be less than 0.07 cm^{-1}, which seems quite reasonable.

Acknowledgement

We are grateful to Marianne A. Morsink for her courtesy to type this paper.

REFERENCES

1. F. Legay in *Chemical and Biochemical Applications of Lasers* (Academic Press, New York 1977) edited by C.B. Moore, Vol. II.
2. K.K. Rebane and P. Saari, J. Luminescence 12 (1976) 23.
3. W.H. Hesselink and D.A. Wiersma in *Picosecond Phenomena* edited by C.V. Shank, E.P. Ippen and S.L. Shapiro, pg. 192 (Springer-Verlag, Berlin, 1978).
4. W.H. Hesselink and D.A. Wiersma, J. Chem. Phys., July 15th issue, 1980.
5. W.H. Hesselink and D.A. Wiersma, Phys. Rev. Letters 43 (1979) 1991.
6. S. Voelker and R.M. MacFarlane, Chem. Phys. Letters 61 (1979) 421.
7. H. de Vries and D.A. Wiersma, Phys. Rev. Letters 36 (1976) 91.
8. A. Nitzan, S. Mukamel and J. Jortner, J. Chem. Phys. 63 (1975) 200.
9. A. van der Ende and R.P.H. Rettschnick, Proc. "Ultrafast Phenomena in Spectroscopy", Tallinn 1978.
10. A. Amirav, U. Even and J. Jortner, J. Chem. Phys. 71 (1979) 2319; Chem. Phys. Letters 71 (1980) 12; ibid. 72 (1980) 21.
11. P.L. Decola, R.M. Hochstrasser and H.P. Trommsdorff, Chem. Phys. Letters 72 (1980) 1.
12. L.A. Hess and P.N. Prasad, J. Chem. Phys. 72 (1980) 573.
13. D.M. Burland, D.E. Cooper, M.D. Fayer and C.R. Gochanour, Chem. Phys. Letters, 52 (1977) 279.
14. R. Kubo, Adv. Chem. Phys. 15 (1969) 101.

New Results on the Rapid Dynamics of Vibrational Modes of Organic Molecules in Liquid and Solid Solutions

A. Fendt, J.P. Maier, A. Seilmeier, and W. Kaiser

Physik Department der Technischen Universität München,
D-8000 München, Federal Republic of Germany

The population lifetimes of vibrational modes in the condensed phases are exceedingly short. With picosecond light pulses we are now in the position to measure directly these short time constants and obtain new information on the vibrational dynamics of polyatomic molecules. In our experimental investigations an excess population in certain vibrational modes is generated by an infrared pulse of 5 ps duration and a bandwidth of approximately 10 cm^{-1}. Tunable infrared pulses are produced in a single-path parametric system consisting of two or three $LiNbO_3$ crystals.

During and after the excitation process at the infrared frequency ν_1 the instantaneous degree of excess population is interrogated with delayed ultrashort probe pulses. Two probing techniques have been developed and are applied in the following investigations: (i) Spontaneous anti-Stokes scattering gives direct information which vibrational mode is excited and how fast the population of this mode relaxes to other vibrational states. Careful experimentation is required to measure the small anti-Stokes signals and to avoid disturbing nonlinear processes. The anti-Stokes signals are small on account of the small Raman scattering cross-section and the limited excess population generated with ultrashort infrared pulses. Investigations are restricted to neat liquids or highly concentrated systems. (ii) With probe pulses of appropriate frequency ν_2 transitions are made from the excited vibrational states to the first excited electronic state S_1. The fluorescence from S_1 is measured. Short time delays between the exciting IR-pulse and the probe pulse allow the observation of the time behavior of the vibrationally excited molecules. This probing technique has the advantage of high sensitivity making the investigation of highly diluted systems possible. On the other hand, the measured fluorescence signal does not tell us accurately the probed vibrational energy state. It will be shown that the combination of frequency dependent studies and time resolved measurements gives valuable new information on the vibrational excitation of the molecule and on the involved vibrational lifetimes.

In the following discussion we are mainly concerned with CH-stretching modes. These modes of high vibrational energy ($\sim 3000 cm^{-1}$) are known to be of major importance for the relaxation kinetics of electronic energy in polyatomic molecules. In recent years we have studied the population lifetimes of the v=1 state of CH, CH_2, and CH_3-stretching modes of numerous organic molecules. The

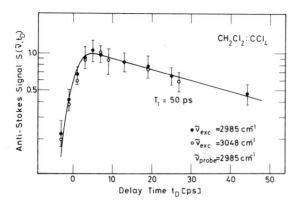

Fig. 1 Anti-Stokes Raman signal of the symmetric CH_2-mode versus delay time between excitation and probe pulse

lifetimes - varying between one and one hundred picoseconds - were found to depend critically on the individual vibrational states of the specific molecules. Two examples, CH_2Cl_2 and CH_3OH, should be discussed first.

In Fig.1 the anti-Stokes Raman signal of the symmetric CH_2-stretching mode of CH_2Cl_2 is presented versus time. Pumping directly the symmetric CH_2-stretching mode (ν_1) at 2985 cm^{-1} or the neighboring asymmetric CH_2-stretching mode (ν_6) at 3048 cm^{-1} gives the same results within the accuracy of our experiments. Obviously, there exists rapid energy exchange between these two modes. These findings are in agreement with numerical estimates of intra-molecular collision-induced vibrational transitions which suggest rate constants of $\sim 10^{12}$ s^{-1}. The longer time constant of 50 ps depicted by the curve of Fig.1 is connected with the transfer of vibrational energy to (Raman inactive) overtones and combination modes involving CH-bending components. The interaction with these latter modes is strongly affected by Fermi resonance. In Fig. 2 we show the infrared absorption spectrum and the Raman spectrum of CH_2Cl_2 depicting the frequency range of the ν_1 and ν_6 CH_2-stretching modes. The small peaks in the IR and Raman spectra at 2830 cm^{-1} are due to an overtone showing small Fermi resonance in this molecule. The relatively long time constant of 50 ps is due to this small Fermi resonance.

Drastically different is the situation in the CH_3OH where the asymmetric CH_3-stretching ν_2 = 2935 cm^{-1} is found to decay rapidly with a time constant of 1.6 ps (see Fig.3). This mode is Raman and IR-active and was excited by a resonant IR-pulse of ν = 2935 cm^{-1} The infrared absorption spectrum of Fig. 4 shows a complex structure around 2950 cm^{-1} suggesting strong Fermi resonance with overtones of half the energy value. Symmetry arguments support the notion of strong Fermi resonance. It appears from these data that the short population lifetime of the ν_2-stretching mode is caused by effective Fermi resonance to overtones.

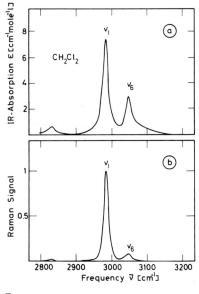

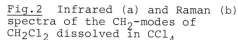

Fig.2 Infrared (a) and Raman (b) spectra of the CH$_2$-modes of CH$_2$Cl$_2$ dissolved in CCl$_4$

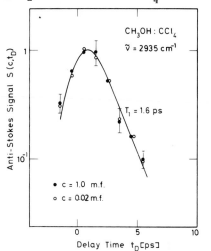

Fig.3 Anti-Stokes Raman signal of the symmetric CH$_3$-mode of CH$_3$OH versus time. Two mole fractions, 1.0 and 0.02 were measured.

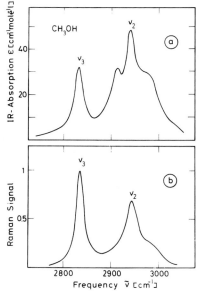

Fig.4 Infrared (a) and Raman (b) spectra of the CH$_3$ modes of CH$_3$OH dissolved in CCl$_4$

Now we turn to the two-pulse excitation of medium size molecules with large electronic dipole moment. The transition rate from the initial to the final state is a sum consisting of several terms. Each one is proportional to the product of the infrared dipole moment $\langle \chi_i^0 Q \chi_m^0 \rangle$ which connects the initial with the intermediate level (pulse of frequency ν_1) and the Franck-Condon factor $\langle \chi_m^0 \chi_f^1 \rangle$ which determines the transition from the intermediate to

147

the final state in the electronic state S_1 (pulse of frequency ν_2).
These two factors vary widely for different transitions:

$$M_{if} = \sum_m C_m <\chi_i^o \mid Q\chi_m^o><\chi_m^o \mid \chi_f^1>$$

The following points are of interest here. In the infrared range
around 3000 cm^{-1} the C-H stretching modes have prominent dipole
moments. The corresponding Franck-Condon (F-C) factors are known
to be considerably smaller than the F-C factors of the C-C ring
modes which couple quite well to the π-electrons of the chromophore. Frequently, the v=2 overtones of the C-C modes have the
largest F-C factors. These modes have their frequencies around
3000 cm^{-1} but with smaller infrared dipole moments. The total
transition via excited modes around 3000 cm^{-1} is a sum over
different components. The many modes with small infrared dipole
moments and small F-C factors give a broad frequency independent
contribution. The experimental results presented below bear out
these considerations.

We have investigated several molecules of the Coumarin family
which have diethylamino or dibutylamino groups attached to the
chromophore. The following data were taken with Coumarin 6 in
CCl_4. The infrared frequency of the pumping pulse was tuned between 2850 cm^{-1} and 3250 cm^{-1} while the probing pulse was held
constant at 18,940 cm^{-1}. First, the fluorescence was measured
when both pulses passed the sample simultaneously (time delay
$t_D = 0$). In Fig.5a, a distinct spectrum is depicted which has no
similarity with the infrared absorption spectrum taken of the
same solution in the same wavelength range (Fig.5c). IR data at
longer wavelengths suggest that the strong spectrum of Fig.5a

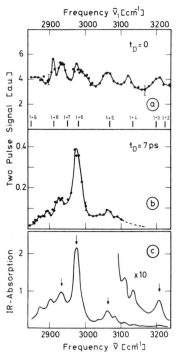

Fig.5 (a) Two-pulse spectrum of
Coumarin 6 in CCl_4 versus infrared
frequencies. No time delay between
the two pulses. (b) Two-pulse
spectrum measured with time delay
$t_D = 7$ ps. (c) Conventional IR
spectrum of the same solution. The
strongest bands belong to the CH_2
and CH_3 modes of the diethylamino
group

148

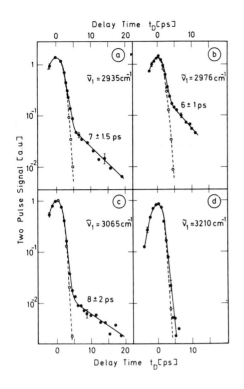

Fig.6 Two-pulse fluorescence versus time at four infrared frequencies. Longer time constants are observed when CH-stretching modes are excited. The broken curves belong to the simultaneously measured cross-correlation curves of the two pulses

results from infrared combination modes which comprise C=O vibrations. These modes are known to have favorable F-C factors. In Fig.6 we present time resolved measurements at various distinct infrared frequencies marked by arrows in Fig.5c. At the four frequencies ν_1 = 2935 cm^{-1}, 2976 cm^{-1}, 3065 cm^{-1}, and 3210 cm^{-1} we see a substantially different time behavior. At three frequency positions we find two time constants, a fast decay is followed by a slower exponential one with time values of approximately 7 ps. At the frequency ν_1 = 3210 cm^{-1}, on the other hand, only one fast decay is observed which is very close to the time resolution of the measuring system (the broken curves of Fig.6 represent simultaneously measured cross-correlation data of the two pulses). We note that CH, CH$_2$, and CH$_3$ stretching-modes are located at those frequencies where longer time constants are observed. At an infrared frequency of 3210 cm^{-1} we excite only combination modes which appear to decay very rapidly ($\leq$1 ps). To see more clearly the relation of the longer lifetimes with the CH-stretching modes we remeasured the infrared two-pulse spectrum with a fixed time delay of t_D = 7 ps. After this time, the population of the combination modes has disappeared and we look at the remaining vibrational excitation. The delayed infrared spectrum is depicted in Fig.5b. Now, we find a strong similarity with the one-photon infrared spectrum of Fig.5c. Obviously, the states which are interrogated after 7 ps are initially populated in relation to the infrared absorption.

149

The resemblance of the two spectra of Fig.5b and 5c is quite surprising. In particular, we find the same CH_3 to CH_2 ratio in the absorption and the two-pulse spectrum. One would have expected a smaller F-C factor of the CH_3 group at the end of the aliphatic chains (especially in butyl chains). In a tentative picture we suggest rapid intramolecular energy transfer within the chains attached to the chromophores. The excited system decays with one time constant of 7 ± 1 ps.

Similar experimental observations were made with a number of different molecules in various surroundings. Of special interest are results where the Coumarin molecules were incorporated into solid plastics. It is believed that the long-lived CH-modes decay via intramolecular processes.

References

A. Seilmeier, K. Spanner, A. Laubereau and W. Kaiser, Opt. Commun. 24, 237 (1978)
A. Fendt, W. Kranitzky, A. Laubereau and W. Kaiser, Opt. Commun. 28, 142 (1979)
A. Laubereau, S.F. Fischer, K. Spanner and W. Kaiser, Chem.Phys. 31,335 (1978)
J.P. Maier, A. Seilmeier and W. Kaiser, Chem.Phys.Lett.70,591 (1980

The Dynamics and Structure of Liquids Revealed by the Homogeneous and Inhomogeneous Broadening of Liquid Vibrational Transitions

C.B. Harris , H. Auweter[1], and S.M. George

Department of Chemistry, University of California and
Materials and Molecular Research Division, Lawrence Berkeley Laboratory,
Berkeley, CA 94720, USA

Vibrational lineshapes observed by isotropic spontaneous Raman scattering are broadened by homogeneous and inhomogeneous processes. Consequently, the vibrational correlation function must be expressed as a product of two correlation functions:

$$\phi(t) = \phi_H(t) \; \phi_{INH}(t) \tag{1}$$

The homogeneous vibrational correlation function, $\phi_H(t)$, yields information about the rapidly varying dynamical modulation processes which cause vibrational dephasing. The inhomogeneous vibrational correlation, $\phi_{INH}(t)$, yields information about the slowly varying modulation processes which create a distribution of environmental sites which establishes a distribution of vibrational frequencies. This distribution of vibrational frequencies causes effective vibrational dephasing because of destructive inferference among the individual frequencies in the distribution.

Since the mechanisms for homogeneous and inhomogeneous broadening processes are not well established, a series of experiments was performed to elucidate the physical basis of each process. In order to determine the mechanism of homogeneous broadening, the effect of vibrational mass and frequency on the homogeneous dephasing time was studied by investigating the consequence of isotopic substitution on the dephasing time of the symmetric CH_3-stretching vibration in methanol and acetone [1]. In order to explore the mechanism of inhomogeneous broadening, homogeneous linewidths of the symmetric CH_3-stretching vibration in a series of simple organic liquids were determined and then compared with their isotropic spontaneous Raman linewidths [2]. The experiments were based on a picosecond probing technique which is able to extract the homogeneous linewidth (i.e. the homogeneous dephasing time) from an inhomogeneously broadened vibrational band [3,4]. This technique overcomes the problem of inhomogeneous broadening and allows homogeneous linewidths to be determined in liquids with unknown inhomogeneity.

The effect of isotopic substitution on the homogeneous dephasing time of the symmetric CH_3-stretching vibration in methanol and acetone was negligible [1]. Both the binary collision model [5], and the hydrodynamic model [6] for vibrational dephasing predict substantial differences for the CH_3 and CD_3-stretching vibrations because of mass and frequency effects.

---

[1] Present address: Physikalisches Institut, Universität Stuttgart, Pfaffenwaldring 57, D-7000 Stuttgart 80, Federal Republic of Germany

The absence of an isotope effect is inconsistent with the predictions of the binary collision and hydrodynamic models, but is in approximate agreement with predictions given by the exchange theory [7], which proposes that dephasing of a high frequency vibration occurs as a result of low-frequency vibrations which are anharmonically coupled to the high frequency vibration. Although the results are qualitatively consistent with the exchange theory, temperature-dependent measurements are necessary to verify this vibrational dephasing mechanism.

Since the isotropic spontaneous Raman lineshape arises from the convolution of both the homogeneous and inhomogeneous broadening lineshape functions, the inhomogeneous broadening lineshape function can be obtained by deconvoluting the homogeneous Lorentzian lineshape function determined by the picosecond experiment from the isotropic spontaneous Raman lineshape function. The average experimental dephasing times, calculated homogeneous linewidths, isotropic Raman linewidths and deconvoluted inhomogeneous linewidths for all the liquids studied are compiled in Table 1. Some typical picosecond dephasing curves and isotropic spontaneous Raman lineshapes superimposed with their corresponding homogeneous Lorentzian lineshapes calculated from the measured dephasing times are shown in Figure 1. We found that symmetric CH_3-stretching vibrational transitions are inhomogeneously broadened to various degrees [2]. These results seriously jeopardize some of the conclusions of isotropic spontaneous Raman lineshape investigations which have assumed that vibrational linewidths were homogeneous in order to extract dynamical information.

The inhomogeneous broadening linewidths do not correlate with any liquid attraction parameter, with the possible exception of the ultrasonic absorption constant [8], which is the ratio of the observed ultrasonic absorption coefficient and the predicted classical ultrasonic absorption coefficient.

Table 1   Experimental dephasing times, calculated homogeneous linewidths, isotropic Raman linewidths and inhomogeneous broadening linewidths for the symmetric CH_3-stretching vibrations in the various liquids studied.

Molecule	ω (cm^{-1})	average experimental T_2 (psec)	Calculated homogeneous $\Delta\omega$ (cm^{-1})	Isotropic Spontaneous Raman $\Delta\omega$ (cm^{-1})	Gaussian Inhomogeneous $\Delta\omega$ (cm^{-1})
(1) 1,1,1-Trichloroethane	2938.5	2.6	4.1	4.3	0.9
(2) Methyl Iodide	2948	2.4	4.4	5.0	1.7
(3) Acetonitrile	2945	5.4	2.0	6.6	5.5
(4) Methyl Sulfide	2913.5	3.8	2.8	8.4	6.8
(5) Dimethyl Sulfoxide	2914.5	1.4	7.6	11.8	6.9
(6) Acetic Anhydride	2942.5	2.2	4.8	16.8	14.1
(7) Acetone	2925	3.0	3.5	16.5	14.6
(8) Methyl Formate	2961	2.2	4.8	17.5	14.8
(9) Pentane	2877	2.7	3.9	≈17	≈14.8
(10) Ethanol	2929	2.5	4.2	18.0	15.6
(11) Methanol	2836	2.4	4.4	18.8	16.3

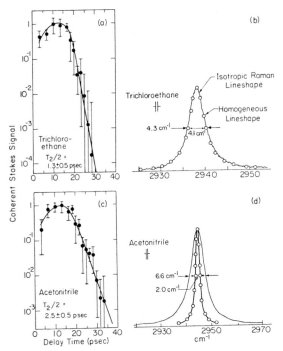

Figure 1   Experimental data for the symmetric CH_3-stretching vibration in trichloroethane and acetonitrile.   Coherently scattered Stokes signal as a function of probe pulse delay in (a) trichloroethane and (c) acetonitrile. Isotropic Raman lineshapes and homogeneous Lorentzian lineshapes calculated from the measured dephasing times in (b) trichloroethane and (d) acetonitrile.

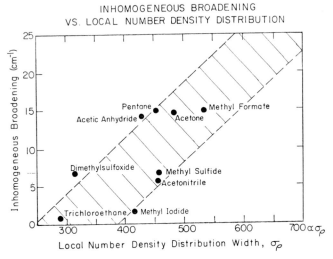

Figure 2   Graph of the inhomogeneous broadening linewidth versus the liquid's local number density distribution width.

Trichloroethane and methyl iodide, two linewidths which are broadened the least, have the largest ultrasonic absorption constants. This behavior is qualitatively consistent with Pinkerton's liquid classification scheme [8,9].

A correlation was established between the inhomogeneous broadening linewidths and the liquid's free volume [10], or σ_ρ, the distribution width of the liquid's local number density [11], which is related to the liquid's free volume through the isothermal compressibility. Given this correlation, which is shown in Figure 2, we have constructed a theory for inhomogeneous broadening. First, we propose that intermolecular forces shift vibrational frequencies. Second, we contend that the local number density establishes the intermolecular force that the local environment imposes on a molecular vibration. Since a local number density distribution with a width σ_ρ exists in the liquid, a distribution of intermolecular forces also exists. This distribution of intermolecular forces establishes a distribution of vibrational frequencies. Therefore, we suggest that the width of the distribution of intermolecular forces is measured by the inhomogeneous broadening linewidth and is proportional to the liquid's local number density distribution width. Thus the inhomogeneous broadening linewidth is dependent on the liquid's local number density distribution width.

The Kubo-Anderson stochastic theory of lineshape [12,13] provides a simple theoretical framework from which homogeneous and inhomogeneous broadening can be easily formalized and visualized. This approach treats the vibration as an oscillator with a frequency which is randomly modulated on fast (homogeneous) and slow (inhomogeneous) timescales. The random modulation, $\Delta\omega(t)$, for each timescale can be described in terms of two characteristic parameters: Δ, the amplitude of the frequency change during the modulation; and τ, the correlation time of the modulation. Using these parameters, the vibrational correlation function can be expressed as [14]:

$$\phi(t) = \phi_H(t)\ \phi_{INH}(t) \tag{2}$$

where:

$$\phi_H(t) = \exp\ [\ -\Delta_H^2\ (\tau_H^2\ [\exp\ (-t/\tau_H)\ -1])\ +\ t\tau_H] \tag{3}$$

and

$$\phi_{INH}(t) = \exp\ [\ -\Delta_{INH}^2\ (\tau_{INH}^2\ [\exp\ (-t/\tau_{INH})\ -1])\ +\ t\tau_{INH}] \tag{4}$$

Homogeneous broadening is caused by $\Delta\omega_H(t)$, the homogeneous stochastic modulation process, characterized by Δ_H and τ_H. Inhomogeneous broadening is caused by $\Delta\omega_{INH}(t)$, the inhomogeneous stochastic modulation process, characterized by Δ_{INH} and τ_{INH}. $\Delta\omega_H(t)$ and $\Delta\omega_{INH}(t)$ can be separated if they occur on different timescales. Particular forms for $\Delta\omega_H(t)$ are given by the binary collision model [5], the hydrodynamic model [6], the exchange theory model [7], or other models. In the limit that $\tau_{INH} > 1/\Delta_{INH}$, the inhomogeneous broadening linewidths which results from $\Delta\omega_{INH}(t)$ is proportional to Δ_{INH}. Since we have demonstrated that the inhomogeneous broadening linewidths correlate with σ_ρ, the distribution width of the liquid's local number density, we propose that Δ_{INH} is proportional to σ_ρ, for the inhomogeneous broadening of symmetric CH_3-stretching vibrations in liquids.

This treatment unifies the fast and slow modulation approaches to vibrational dephasing and demonstrates how isotropic spontaneous Raman scattering studies and picosecond coherent probing experiments can be used in conjunction to determine the linewidths for both the homogeneous and inhomogeneous broadening processes. Thus the homogeneous and inhomogeneous broadening of vibrational transitions can be studied and related to the dynamics and structure of the liquid.

References

1. C. B. Harris, H. Auweter and S. M. George, Phys. Rev. Lett. $\underline{44}$, 737 (1980).

2. S. M. George, H. Auweter and C. B. Harris, submitted to the Journal of Chemical Physics.

3. A. Laubereau, G. Wochner and W. Kaiser, Chem. Phys. $\underline{28}$, 363 (1978).

4. A. Laubereau and W. Kaiser, Rev. Mod. Phys. $\underline{50}$, 607 (1978).

5. S. F. Fischer and A. Laubereau, Chem. Phys. Lett. $\underline{35}$, 6 (1975).

6. D. W. Oxtoby, J. Chem. Phys. $\underline{70}$, 2605(1979).

7. C. B. Harris, R. M. Shelby and P. A. Cornelius, Phys. Rev. Lett. $\underline{38}$, 1415 (1977).

8. K. F. Herzfeld and T. A. Litovitz, Absorption and Dispersion of Ultrasonic Waves (Academic Press, New York, 1959).

9. J. M. M. Pinkerton, Proc. Phys. Soc. (London) $\underline{662}$, 129(1949).

10. H. Eyring and R. P. Marchi, J. Chem. Educ. $\underline{40}$, 562(1963).

11. D. A. McQuarrie, Statistical Mechanics (Harper and Row, New York, 1976).

12. P. W. Anderson, J. Phys. Soc. Japan $\underline{9}$, 316(1954).

13. R. Kubo in Fluctuations, Relaxation, and Resonance in Magnetic Systems, ed. by D. ter Haar (Plenum, New York, 1962).

14. W. G. Rothschild, J. Chem. Phys. $\underline{65}$, 455(1976); 65, 2958(1976).

This work was supported in part by the National Science Foundation and by the Division of Chemical Sciences, Office of Basic Energy Sciences, U.S. Department of Energy, under Contract No. W-7405-Eng-48. H. Auweter gratefully acknowledges support from the Deutsche Forschungsgemeinschaft for a postdoctoral fellowship.

Emission Risetimes in Picosecond Spectroscopy

G.W. Robinson and R.A. Auerbach

Department of Chemistry, Texas Tech University,
Lubbock, TX 79409, USA

1. Introduction

It is a common belief in picosecond spectroscopy that the measurement of a
risetime in an emission experiment provides information about a rate of transfer of energy or excitation. That this is not always true is easily seen.
Suppose, for example, that the rate of irreversible transfer from $D^* \to A$ is
reasonably fast for near neighbors, but is very slow for non-near neighbors.
If D^* species are prepared from D by a picosecond light pulse, those close to
A will transfer, leaving only those D^* that are far from A. If D^* diffusion
is very slow compared with the lifetime of D^*, the experiment is essentially
complete after this first fluorish of transfer. The emission risetime, but
low yield, will indeed provide a measure, averaged over near-neighbor distances and orientations, of the transfer. On the other hand, if the diffusion
of D^* is fast compared with the decay of D^*, then the system conditions are
continually "repaired" after each transfer, and the rise of A^* reflects only
the decay of A^* or of D^*, whichever is faster. If D^* decays faster than A^*
and its lifetime is dependent totally on the transfer dynamics, then again
information about the transfer can be obtained from measurements on the risetime of A^*. However, simply because the risetime of A^* matches the falltime
of D^* does not mean that one is measuring a transfer rate. If diffusion is
fast enough and the lifetime of D^* is less than that of A^*, then the risetime of A^* *must* equal the decay time of D^*, whatever the cause of the D^* decay.

In order to measure the temporal characteristics of transfer, either diffusion must be slowed down in some manner, or one has to be sufficiently
lucky and sufficiently knowledgeable to know that the dominant decay channel
for D^* is the transfer to A. Similar problems arise in the interpretation
of the decay of D^*. Intermediate cases lead to acceptor rise or donor decay
that are mathematical entanglements of the diffusion, the transfer, and the
decay of D^* and A^*. In addition to these complications, another one sets in
at early times because of the transient characteristics of diffusion at such
times [1,2]. Thus, monotonic emission measurements of the rise of A^* or the
decay of D^* can provide quantitative information about the elemental processes
of diffusion and transfer, but one must be alert to the fact that this is not
always the case.

These points are germane to picosecond experiments not only for energy/
excitation transfer [3], but also for inter- and intramolecular chemical
reactions [4] and charge-exchange processes [5,6] where bulk or local, translational or rotational, diffusion participates. The reactants (initial states)
mathematically behave as "donors", while the products (final states) play the
role of "acceptors". Even in crystals at low temperatures [7], the diffusion-
like motion of excitation can lead to the same precepts.

2. Rate Constants?

Another belief is that these transfer processes are susceptible to treatment by a set of kinetic equations. In their most simple form, these equations would be,

$$D^* + A \rightarrow A^* + D \qquad k_T \qquad\qquad\qquad (1a)$$

$$D^* \rightarrow D \qquad\qquad k_D \qquad\qquad\qquad (1b)$$

$$A^* \rightarrow A \qquad\qquad k_A \quad , \qquad\qquad\qquad (1c)$$

whose solutions, when fit to emission rise and fall data, provide the rate constants k_T, k_D, k_A. However, the kinetic equations contain no information whatsoever about the diffusion process, and in fact a rate constant k_T, independent of time, does not exist. This is because the temporal characteristics of energy transfer, a chemical reaction, or a charge exchange process depend on instantaneous spatial and orientational relationships between D^* and A, as well as on the instantaneous solvent structure. *Only for sets of subsystems where these factors are the same over a period of time Δt can one specify a k_T.* One must therefore treat each subsystem separately, calculate the rise and fall dynamics, and average over all subsystems to obtain the temporal properties that are being measured. This is equivalent to a master-equation approach[1] in chemical kinetics, which becomes more and more relevant as time-scales become shorter. Only in special cases, such as one where the diffusion of D^* is sufficiently fast and enough time has elasped, does k_T tend to a constant value. Fast diffusion is the "equalizer" that causes all excited donors to behave equivalently. Fast diffusion of the acceptors also "equalizes" the acceptors. But, generally excited acceptors can do only one thing, spontaneously decay, so this equalization for the excited acceptors is immaterial.

3. Fast/Slow Diffusion Limits

Historically, photochemists and radiation physicists have adopted two limits [9] for the attack of these problems: fast diffusion (the Stern–Volmer limit) and slow diffusion (the static transfer limit). In steady state photochemis-

---

[1]The coupled "master equations" [8] are,

$$\frac{dD_j^*}{dt} = -k_D D_j^* + \sum_k W(r_{jk})(D_k^* - D_j^*) - \sum_i X(r_{ji})D_j^*$$

$$\frac{dA_i^*}{dt} = -k_A A_i^* + \sum_\ell Y(r_{i\ell})(A_\ell^* - A_i^*) + \sum_j X(r_{ji})D_j^* \quad ,$$

where the letters $D_j^* \ldots$, $A_i^* \ldots$ represent probabilities that the jth donor, or ith acceptor is excited, and W, X, Y are distance-dependent elementary transfer rates for donor-donor, donor-acceptor and acceptor-acceptor, respectively. It has been assumed that the transfer is symmetrical between two given donors (or acceptors) and that acceptor → donor transfer is forbidden. One first calculates $D_j^*(t)$ and then $A_i^*(t)$ for all j and i and ensemble averages in order to obtain experimentally observable quantities. It is incorrect to assume averaged rates and solve "averaged rate equations". One actually should deal with five coupled equations, the other three describing D(t), A(t) and ρ(t), the photon density, but in the limit of small ρ(t) absorption, these are ignored.

try these two limits have been applicable in low viscosity and high viscosity solvents, respectively. However, in transient spectroscopy either limit may be important in the low viscosity case. Questions about how one can adjust experimental conditions in order to range between these two limits for the same problem are difficult ones, since the adjustment of diffusion coefficients by additives or by temperature changes can drastically affect other system parameters and cause interpretation to be difficult. Nonetheless, it is instructive to outline these methods here and to compare them with a more exact formalism, where diffusion is explicitly considered.

The kinetic equations for simple $D^* \rightarrow A$ transfer following instantaneous production of D^* in the limit of fast diffusion are,

$$\frac{dD^*}{dt} = -(k_D + k_T A)D^* \qquad (2a)$$

$$\frac{dA^*}{dt} = -k_A A^* + k_T D^* A \quad , \qquad (2b)$$

where k_D is the spontaneous decay rate of D^* with no A or A^* present, k_T is the transfer rate, and k_A is the spontaneous decay rate of A^*, no D or D^* present. The letters represent concentrations of the donor and acceptor species. The spontaneous rates contain the radiative plus all the nonradiative components: $k = k_r + k_{nr}^{(1)} + k_{nr}^{(2)} + \ldots$

When excitation densities are sufficiently low, A can be considered a constant and the solution to (2a) is trivial,

$$D^*(t) = D^*(0)\exp[-(k_D + k_T A)t] \quad , \qquad (3)$$

showing that D^* decays exponentially, but faster than the spontaneous rate because of the transfer to A.

Eq. (2b) is also quite easy to solve by using the trick,

$$A^*(t) = A_0^*(t)\exp[-k_A t] \quad , \qquad (4)$$

and solving for $A_0^*(t)$. This finally yields,

$$A^*(t) = \frac{k_T A D^*(0)}{k_A - K_D}\left[\exp(-K_D t) - \exp(-k_A t)\right] \quad , \qquad (5)$$

where $K_D = k_D + k_T A$, the decay rate of D^* in the presence of A. A similar result is obtained using more general methods [10] where account is taken of diffusion and the specific transfer law (Förster, r^{-4}, etc.) in the limit where transfer is just balanced by the rapid diffusion of new D^* species towards the acceptor.

In this fast diffusion limit, the exponential-difference temporal form is preserved independent of the character of the trapping law or any other details of the model[2]. Note from the form of (5) that when experimental emission data can be fit to a difference of two exponentials, by virtue of the fact that diffusion is sufficiently fast, the larger of k_A or K_D *always* describes the rising exponential and the smaller *always* describes the falling exponential[3].

3.1 Donor Dynamics — Slow Diffusion Case

In the limit of extremely slow D^* diffusion the rate equations must specifically take into account the dependence of the trapping rate on the radial distance between D^* and A. In the isotropic case and *no diffusion*, the rate equations may be obtained from the master equations in footnote 1 by omitting the donor-donor and acceptor-acceptor transfer terms. The solution for any particular D_j^* can be written down immediately,

$$D_j^*(t) = D_j^*(0)\exp[-k_D-\sum_i X(r_{ji})]t \quad . \tag{6}$$

Two limits give tractable results: the first where there is a large preponderance of donors and very few acceptors, and the second the reverse case.

In the first case, each donor interacts with only a single acceptor, and the sum in the exponential of (6) becomes a single term, dependent on the distance r_j between the acceptor and the jth donor,

$$D_j^*(r_j,t) = D_j(r_j,0)\exp[-k_D t]\exp[-k_T(r_j)t] \quad . \tag{7}$$

It will be assumed throughout that the initiating excitation is uniform, i.e. that $D^*(r,0) = D^*(0)$, a constant proportional to the energy absorbed from the light pulse. The overall donor decay is obtained by summing over j in a volume element Ω that contains only a single A. Replacing this sum by an integral gives for the overall donor decay,

$$D^*(t) = \Omega^{-1}D^*(0)\exp[-k_D t]\int_\Omega d^3r \; g(r)\exp[-k_T(r)t] \quad , \tag{8}$$

---

[2]In the limit of a single acceptor, i.e. very low acceptor concentrations, the general result [10] is the same as (5) except that $K_D=k_D$. However, as the acceptor concentration rises, and in the case of extremely fast diffusion, "equalization of donors" demands that the donor concentration far from an acceptor decays as,

$$D^*(0)e^{-k_D t}e^{-k_T At} \quad ,$$

and not as,

$$D^*(0)e^{-k_D t} \quad .$$

[3]In an absorption monitoring experiment, the ground state concentration of A first falls and then rises back to its original concentration according to $A(t) = A(\infty) - A^*(t)$. However, equation (5) is not altogether correct unless $A(\infty)>>A^*(t)$, for all t, in which case detection of the absorption is difficult. Thus, in making these kinds of studies, it is better to use emission monitoring techniques.

where $g(r)$ is a radial distribution function for D^* species surrounding an A. Note that as Ω becomes larger, an ever smaller fraction of D^* species decays by transfer to A, compared with the spontaneous decay of D^*. In the limit where there are no acceptors, Ω is the entire volume of the solution and the decay is entirely spontaneous. See Section V.C of [10].

In the case where there is a great preponderance of acceptors compared with donors, replacing the sum in the exponential of (6) by an integral yields,

$$D^*(t) = D^*(0)\exp[-k_D t]\exp[-At \int d^3r \ k_T(r)h(r)] \quad , \tag{9}$$

where $h(r)$ is the radial distribution function for A species surrounding a D^*, and in this case the integration extends over the entire volume of the system. Since the integral in the exponential of (9) is nothing more than an ensemble average $\langle k_T \rangle$ of $k_T(r)$, the decay of D^* has a form identical to the result (3) obtained through the kinetics formalism. Note that the value of $\langle k_T \rangle$ would decrease in a gas phase experiment as the pressure of A is reduced, or in a solution as the concentration of an inert diluent is raised.[4] *For a preponderance of acceptors, therefore, donor decay has the same mathematical form irrespective of the rate of D^* diffusion.* Donor diffusion merely carries D^* to a new location, which is no different from the old one. This result therefore falls within the framework of the italicized precept in Section 2.

3.2 Acceptor Dynamics — Slow Diffusion Case

Using the second equation in footnote 1, and ignoring the transfer terms, acceptor dynamics in the case of slow diffusion may be found by solving,

$$\frac{dA^*}{dt} = -k_A A^* + AD^*(0)\exp(-k_D t)\int d^3r \ k_T(r)g(r)\exp[-k_T(r)t] \quad , \tag{10}$$

in the case where there is a great preponderance of donors; or by evaluating,

$$A^*(t) = \frac{AD^*(0)\langle k_T\rangle}{k_A - k_D - A\langle k_T\rangle}\left[\exp(-k_D t - A\langle k_T\rangle t) - \exp(-k_A t)\right] \quad , \tag{11}$$

when there is a great preponderance of acceptors.

In special cases, an analytical expression can be obtained for (10). An example is Förster trapping where,

$$k_T(r) = k_D (R_0/r)^6 \quad . \tag{12}$$

If $g(r)$ is taken to be a constant, then (10) gives rise to a Dawson's integral expression for $A^*(t)$, as previously shown [10]. The risetime of $A^*(t)$ in this case of slow diffusion *is much faster* than in the fast diffusion limit, where

---

[4]Naturally, as one continues to dilute the concentration of A, the limit where there exists a $\langle k_T \rangle$ is no longer a valid concept. The exponential form of $D^*(t)$ gives way to a nonexponential decay, eventually evolving into a $t^{3/n}$ dependence at low concentrations ($<1M/\ell$) of A (n = exponent in inverse power law for transfer) [3,10].

$A^*(t)$ is a difference of exponentials. If $g(r)$ is peaked within the critical trapping radius R_0, the risetime of $A^*(t)$ becomes still faster, as shown in [10].

When there is a preponderance of acceptors, (11) applies for $A^*(t)$ dynamics. Interestingly, this is always a difference of exponentials. Analogous to the case of donor dynamics, the result (11) in the limit of slow diffusion mathematically matches the result (5) obtained from the kinetics equations. Thus, if $\langle k_T \rangle \gg k_A$ and k_D, a good way of obtaining $\langle k_T \rangle$ experimentally is to go to the limit where there is a preponderance of A and measure the risetime of A^* following pulsed excitation of the donor. Care must be exercised here, however, in the interpretation of the results if the concentration of A is *too* high. The phenomenon of *concentration quenching,* caused possibly by (excitation) diffusion [11] of A^* to deeper acceptor traps, such as impurities or dimers, might alter the dynamics of $A^*(t)$.

4. Conclusions

In this paper we have investigated the temporal properties of the initial and final states of a fast reaction, such as an energy exchange process. The initial states (D^*) of the system, prepared by a short pulse of energy, may be considered to be "donors", while the final states (A^*) may be thought of as "acceptors". In the paper, it has been assumed that both D^* and A^* can spontaneously decay, but that only D^* species are able to transfer energy. In the kinetics sense, this means that the rate of the back reaction has been assumed negligible compared with that of the forward reaction.

Since A^* can carry out only one function, spontaneous decay, diffusion of A^* is immaterial to the dynamics of the problem. On the other hand, D^* can either spontaneously decay or transfer energy to ground state acceptors (A). Thus, D^* diffusion does matter, since it can exchange D^* between inequivalent regions of the system (near or far from an A)[5].

The results of this paper and of a previous publication [10] indicate that when D^* diffusion is fast compared with the lifetimes of D^* and A^*, the dynamics of A^* is a simple difference of exponentials, one referring to D^* decay, the other to A^* decay. Except in cases where the D^* decay is totally dominated by the transfer, therefore to measure a transfer rate, D^* diffusion has to be slowed down. This may be accomplished in three ways: 1) additives, 2) lowering the temperature, or 3) diluting the donors. First of all, slowing diffusion may strongly decrease the quantum yield of transfer, causing detection to be difficult. In addition, using additives is a bad idea since they introduce indeterminate inhomogeneities in the temporal properties of the system. Temperature variations may change the rates of the processes being studied as well as diffusion rates.

This paper suggests that perhaps the best way of measuring transfer rates is to slow diffusion by donor dilution and to use a preponderance of acceptors. At sufficiently low donor concentrations, even though D^* may in fact

---

[5]Since ground state acceptors (A) can also do two things, 1) nothing, or 2) accept energy from a D^*, the diffusion rate of A is also important. It is therefore the relative diffusion between D^* and A that becomes the parameter of interest. In contrast, ground state donors can do but one thing only, nothing.

undergo rapid diffusion, this diffusion process is immaterial to the dynamics when there is a preponderance of A. The A^* risetime then yields the ensemble averaged transfer rate, providing the concentration of A is high enough so that the transfer rate is at least comparable to the faster of the two decay rates. However, in this case, concentration quenching considerations come into play, which may distort the temporal properties of A^*. Thus, in picosecond spectroscopy, the interpretation of emission risetimes in terms of transfer dynamics is not as straightforward as it may at first appear.

5. Acknowledgements

Support of this work by the R. A. Welch Foundation [K-099(e)] and the National Science Foundation [CHE77-21913] is gratefully acknowledged. We also wish to thank R. W. Zwanzig for helping to clarify this problem through discussions and collaboration.

References

1. J. T. Hynes, Ann. Rev. Phys. Chem. __28__, 301 (1977); B. J. Alder and T. E. Wainwright, J. Phys. Soc. Japan Suppl. __26__, 267 (1969).
2. S. W. Haan and R. W. Zwanzig, J. Chem. Phys. __68__, 1879 (1978).
3. M. Inokuti and F. Hirayama, J. Chem. Phys. __43__, 1978 (1965).
4. K. B. Eisenthal, Acc. Chem. Res. __8__, 118 (1975).
5. G. A. Kenny-Wallace, Acc. Chem. Res. __11__, 433 (1978).
6. R. A. Auerbach, J. A. Synowiec and G. W. Robinson, paper FB6, this conference.
7. D. C. Ahlgren and R. Kopelman, J. Chem. Phys. __70__, 3133 (1979).
8. A. Isihara, Statistical Physics (Academic Press, New York, London, 1971), p. 42 and p. 376.
9. J. B. Birks, J. Phys. B (Proc. Phys. Soc.) Ser. 2, __1__, 946 (1968).
10. R. A. Auerbach, G. W. Robinson and R. W. Zwanzig, J. Chem. Phys. __72__, 3528 (1980).
11. G. R. Fleming, private discussion.

Vibrational Relaxation in the Lowest Excited Singlet State of Anthracene in the Condensed Phase

R.W. Anderson

Xerox Corporation, Webster Research Center,
Rochester, NY 14644, USA

1. Introduction

Determination of the pathways of vibrational relaxation in excited state processes remains both a fundamental and experimentally difficult area in condensed phase systems. Indirect techniques such as homogeneous linewidth [1] and "hot" luminescence [2,3] measurements have been applied to low-temperature systems which exhibit sharp vibronic lines. Direct observations of the dynamics of vibrational relaxation have been accomplished by a number of techniques which are not restricted to systems at low temperatures or to those with sharp spectra. In general, these direct methods probe vibronic populations with either excited state absorption spectroscopy [4-6] or time-dependent emission studies [7-9].

While the ultimate goal in the study of vibrational relaxation is the characterization of the individual channels of deactivation, useful information may be gained by the measurement of the total decay rate of an initial optical excitation. In particular, the question of whether vibrational relaxation in a given system is competitive with the electronic relaxation processes frequently occurs in the picosecond domain. The observations reported in this paper are intended to explore a related issue - the effect of vibrational relaxation on the excited state absorption spectra of a large aromatic molecule in fluid solution.

2. Experimental

A laser photolysis apparatus using a mode-locked Nd^{+3}/glass laser is utilized to measure excited state absorption spectra in the 0-700 psec time range. Details of this apparatus are given elsewhere [10]. Briefly, 7 psec single pulses of the laser third harmonic at 355 nm are used for excitation while a laser induced continuum pulse is used as a broad band spectral probe. Transient spectra are obtained at separate times by introducing pathlength differences between the excitation and probe pulses.

3. Results and Discussion

The main result of the transient absorption studies is shown in Fig. 1. Transient spectra shown here were recorded for anthracene in a room temperature solution of 3-methylpentane at a concentration of 10^{-3}M. The normalized spectra shown are representative of the excited state absorption at long (t=700 psec) and short (t=3 psec) times relative to excitation.

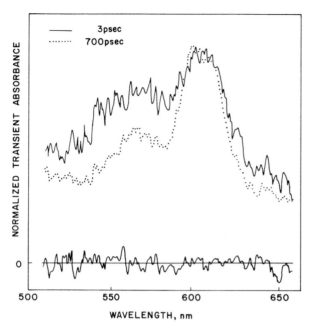

NORMALIZED TRANSIENT ABSORBANCE

— 3psec
······ 700psec

WAVELENGTH, nm

Fig. 1 Transient absorption spectrum of anthracene in 3-methyl-pentane
solution. Baseline spectrum taken with the u.v. excitation blocked.

Although these spectra are each averaged over several separate determina-
tions, the individual non-average results also clearly reveal the differences between the early and long time spectra. The early time spectrum is
measurably broader, with intensification evident in the vibronic regions
both to the blue and red of the 605nm maximum. Transient spectra taken at
intermediate times display a smooth evolution from the broadened form to
the long time shape over the time range 0-35 psec. A measure of the difference between the early and late time spectra is evident in the ratio of
the maxima at 605nm and 565nm. The ratio at long times (i.e. > 35 psec) is
constant, R_1, and the ratio at earlier times, R (t), is a function of the
level structure and dynamics involved. Analysis of the ratio R (t) according to a simple model yields the kinetic result shown in Fig. 2. The model
includes two assumptions: (i) only two states are being monitored (initial
and final) each with a characteristic spectrum and (ii) the population in
the initial state evolves to the final state via first order kinetics. A
straightforward consequence of this model is that a plot of $- \ln[R_1 - R(t)]$
as a function of time should yield a straight line whose slope is the rate
constant of the decay process. The data shown in Fig. 2 are fit to a line
with a slope of 12 psec^{-1}. The earliest data point in time is not included
in this fit since it falls in the time region where significant corrections
for pulse overlap would be required. Although the points beyond $\sim$ 10 psec
also represent an average over the interrogation pulse shape, this was not
included in the above analysis.

The transient spectra were recorded in the 510-650 nm wavelength region
which coincides with the well known $S_4 \leftarrow S_1$ ($^1A_{-1g} \leftarrow {}^1B_{+2u}$) transition in anthracene [11]. This region was emphasized for two reasons: i) the trans-
ient absorption in this region is characterized by possible vibronic struc-

164

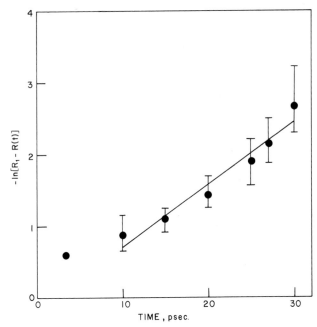

Fig. 2 Time dependence of the ratio of the maxima at 605nm and 565nm. (see text).

ture and ii) this region contains minimal excited triplet absorption even at long times [11]. Thus it is reasonable to assume that any dynamic spectral changes are due to the redistribution of the initial excess vibronic excitation. Effects due to rotational relaxation are ruled out by the insensitivity of the results to the relative polarization of the excitation and probe pulses.

Although the 355 nm excitation produces a modest (ca. 1500 cm^{-1}) excess vibronic excitation above the $S_1 \leftarrow S_0$ origin, the level structure of anthracene is dense enough [7] to allow this excitation to be partitioned among many vibrational modes. In solution the partitioning is facilitated by the broadening of the vibronic lines to ca.500 cm^{-1} (FWHM). Nevertheless in terms of the Franck-Condon levels the excitation may be accounted for in terms of three prominent modes ($\sim$1150cm^{-1}, $\sim$1430cm^{-1} and a combination at 1430cm^{-1} + 430cm^{-1}).

FLEMING et al. [12] have analyzed theoretically the effect of various vibronic excitations on excited state absorption spectra. While their results are of a general nature,they emphasize that the shape of vibronic bands in excited state absorption spectra are sensitive to several variables including the number of modes in the system, the number of modes excited and the temperature. In general, they conclude that excited state absorptions arising from unrelaxed vibronic states will be broader to both higher and lower energies relative to the spectrum from the relaxed state. Recently, SCHRÖDER and co-workers [5] have reported an excellent example of this phenomenon in the $T_n \leftarrow T_1$ spectra of naphthalene in the gas phase.

165

Since the pulse duration used in the anthracene transient absorption experiments is a significant fraction of the measured 12 psec decay time, it is not unexpected that the spectral broadening at early times is small. It is concluded, nevertheless, that this observation represents unrelaxed vibronic excitation in anthracene in fluid solution which decays with a time constant of approximately 12 psec. This time is consistent with the findings of HOCHSTRASSER and WESSEL [7] on anthracene doped naphthalene mixed crystals at 2K. In their experiments hot luminescence was observed for a period of ca. 70 psec with excitation into the phonon wing of the 400 cm^{-1} Franck-Condon mode, while no evidence was found for vibrationally unrelaxed states following excitation with 2950 cm^{-1} excess energy (i.e., rapid deactivation).

In conclusion, the observation of unrelaxed vibronic states in several condensed phase systems has recently been reported [6,7,9]. The results on anthracene reported here emphasize that these effects may extend well into the picosecond range, easily competitive with important electronic relaxation processes in polynuclear aromatic systems. Thus, excited state absorption spectroscopy serves as a useful tool to explore both the electronic as well as the vibrational composition and dynamics of excited systems.

References

1. S. Voelker and R. M. Macfarlane, Chem. Phys. Letters 61 (1979) 421
2. K. K. Rabane and P. Saari, J. Lumin. 12/13 (1976) 23
3. R. M. Hochstrasser and C. A. Nyi, J. Chem. Phys. 70 (1979) 1112
4. S. J. Formosinho, G. Porter and M. A. West, Proc. Roy. Soc. A 333 (1973) 289
5. H. Schröder, J. H. Neusser and E. W. Schlag, Chem. Phys. Letters 54 (1978) 4
6. B. I. Greene, R. M. Hochstrasser and R. B. Weisman, J. Chem. Phys. 70 (1979) 1247; B. I. Greene, R. M. Hochstrasser and R. B. Weisman, Chem. Phys. Letters 62 (1979) 427
7. R. M. Hochstrasser and J. E. Wessel, Chem. Phys. 6 (1974) 19
8. B. Kopainsky and W. Kaiser, Opt. Comm. 26 (1978) 219
9. K. Choi and M. R. Topp, Chem. Phys. Letters 69 (1980) 441
10. R. W. Anderson, D. E. Damschen, G. W. Scott and L. D. Talley, J. Chem. Phys. 71 (1979) 1134
11. D. Bebelaar, Chem. Phys. 3 (1974) 205
12. G. R. Fleming, O. L. J. Gijzeman and S. H. Lin, J. Chem. Soc. Faraday Transactions II 70 (1974) 1074

Picosecond Studies of Electronic Relaxation in Triphenylmethane Dyes by Fluorescence-Upconversion

G.S. Beddard, T. Doust, and M.W. Windsor[1]

Davy Faraday Research Laboratory, The Royal Institution,
21 Albemarle Street, London W1X 4BS, United Kingdom

INTRODUCTION

The electronic relaxation in solution of triphenylmethane dyes, TPM(Fig. 1), such as crystal violet (CV) and malachite green (MG), has been studied in several ways and is found to show a significant dependence on solvent viscosity. The fluorescence quantum yield, Q, is very small (<0.1) in fluid solvents, but increases to about 30 % in extremely viscous media [1-3]. Both fluorescence yield studies and picosecond spectroscopic kinetic measurements [4-8] of the decay of excited state absorption (ESA) and of the rates of ground state repopulation (GSR) show that increased

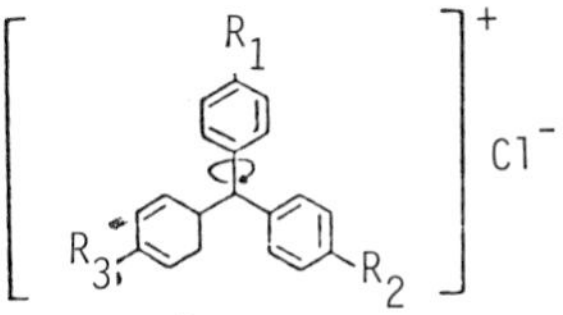

Fig. 1

Fig. 1 The triphenylmethane dyes crystal violet CV [$R_1 = R_2 = R_3 = N(CH_3)_2$] and malachite green MG [$R_1 = H$, $R_2 = R_3 = N(CH_3)_2$]

solvent viscosity leads to reduced rates of electronic relaxation. The cations of TPM dyes assume a 3-bladed propeller configuration in solution. The model proposed by Forster and Hoffmann (FH) [1] envisages that the excited molecule undergoes a conformational change involving synchronous rotation of the phenyl rings to new conformations that have enhanced rates of non-radiative decay. The ring rotation is driven by steric repulsion between ortho hydrogen atoms on adjacent phenyl rings and is hindered by solvent viscosity. Their model predicts a $Q \propto \eta^{2/3}$ dependence that is supported by some measurements [1,3,8], but the time dependence of excited state relaxation of $\exp(-at^3)$, also predicted by the model, has not been observed. Picosecond GSR studies of CV by Magde and Windsor[4] indicated that internal conversion is the dominant mechanism, but their data did not distinguish between several possible functional descriptions of the recovery. Ippen et al. [6] studied GSR for MG and found a single exponential recovery for $\eta < 1$ poise and a double exponential for $\eta > 1$ poise. A single exponential decay of fluorescence for MG was reported by Yu et al. [7], but Hirsch and Mahr[8] observed a double exponential decay. However, the lifetimes of the faster decays of Hirsch and Mahr agree both with the lifetimes of Yu et al. and the slower component of GSR reported by Ippen et al. Very recently Cremers and

---

[1] Permanent address: Department of Chemistry, Washington State University, Pullman, WA 99164, USA

Windsor[9], in a detailed study by picosecond flash photolysis, found that a sum of many decaying exponentials was required to fit their GSR and ESA kinetic data. In their model excitation produces in the excited state a replica of the ground-state conformational distribution. In high viscosity solvents the conformations are fixed and each conformation exhibits an independent and different rate of decay, thus leading to an overall multi-exponential decay. In solvents of lower viscosity, conformational changes can occur on the time scale of the relaxation and reduced viscosity leads to a transfer of excited molecules into the faster non-radiative decay channels. This model reconciles the kinetic data of Hirsch and Mahr and of Ippen et al., but the disparity between the kinetic and quantum yield measurements remains to be explained. Although a large amount of kinetic data has now been accumulated on the viscosity-dependent relaxation of CV via studies of ESA and GSR, to date no time-resolved studies of CV fluorescence have yet been made. We present here a picosecond kinetic study of the solvent-dependent relaxation of both CV and MG studied by laser upconversion of the fluorescence emission.

EXPERIMENTAL

A synchronously-pumped jet-stream dye laser producing pulses of 4 ps duration at 595 nm was used to excite the fluorescence of solutions of CV and MG in various viscous solvent mixtures. Heating of the sample volume was avoided by the use of a jet stream for the more fluid solutions and a sample cell rotated eccentrically about an axis normal to the plane of the cell for the more viscous solutions. The fluorescence was upconverted to about 310 nm by mixing in a $LiIO_3$ crystal with a portion of the excitation beam obtained by use of a suitable beam splitter. By varying the time delay of this latter beam with a delay line controlled by a stepping motor, the fluorescence decay profile can be mapped out. The fluorescence signal was observed over the range 650-800 nm by angle tuning the $LiIO_3$ crystal. CV samples were purified samples provided by Cremers and Windsor. Commercial samples of CV not specially purified, gave results that did not differ from those for the purified material. The MG used was a biological stain sample of 98 % purity. All experiments were performed at 21.5°C. Solution viscosities were also measured at this temperature with an Ostwald viscometer which was calibrated against oils of standard viscosity. Viscosity data were reproducible to ± 2 %.

RESULTS

Our results are summarized in Figures 2-4. Fig. 2 is a logarithmic plot of the decay of CV fluorescence in a glycol/water solution of viscosity 11P. The time axis corresponds to 3 ps per channel. Note that the decay is very close to being exponential except at the earliest times. A computer fit to a double exponential decay gives a major component with a lifetime of 150 ps and a minor (15 % amplitude) contribution from a faster decaying component with a lifetime of 41 ps.

We studied the fluorescence decay of CV in solutions ranging in viscosity from 0.8 P to 50 P. The results for both the long and short decay components are summarized in a log-log plot in Fig. 3. Within experimental error both fit quite well to a dependence of the lifetime on the 2/3 power of the viscosity in agreement with the FH model.

Similar data were obtained for MG, but the dual exponential character of the decay was more pronounced than for CV even in the more fluid solvents. Figure 4 shows the kinetics in a glycerol/water mixture of viscosity 11.5 P together with a dual exponential fit to the data. There is a long component of 153 ps and a shorter component (45 % amplitude) of 58 ps. The viscosity dependence of the MG fluorescence shows an

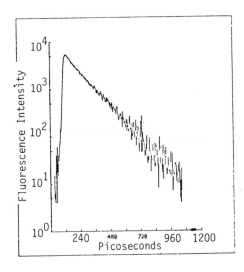

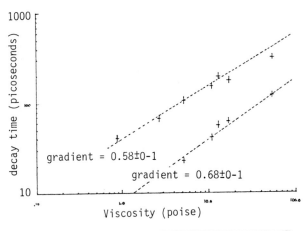

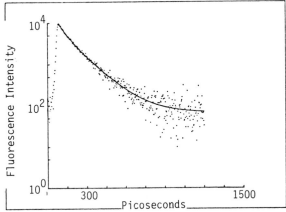

Fig. 2 Decay of crystal violet fluorescence in glycerol/water solution of viscosity 11 poise t 21.5°C

Fig. 3 Log-log plot of fluorescence decay times of crystal violet versus solvent viscosity

Fig. 4 Decay of malachite green fluorescence in glycerol/water solution of viscosity.11.5 poise at 21.5°C. The solid line is a dual exponential fit to the experimental data

approximately 1/3 power dependence of both lifetimes on viscosity in agreement with the data of Hirsch and Mahr.

We also studied the wavelength dependence of the fluorescence decay of CV over the range 650-800 nm. To within 1% we were unable to observe any variation in the decay times over this range.

DISCUSSION

Our results can be understood in terms of the model of Cremers and Windsor [9]. Excitation produces in the excited state, a replica of the ground-state population distributed over various angular conformations of the phenyl rings. In high viscosity solvents ring twisting is very slow and this distribution does not change on the time scale of fluorescence emission. The fluorescence decay is multi-exponential because, owing to the convergence of the upper and ground state potential energy surfaces with increasing ring angle, the rate of internal conversion also increases with increasing ring angle. Thus a different rate of fluorescence decay obtains for each angular conformation, determined by the internal conversion rate appropriate to each value of ring angle. In low viscosity solvents, the rate of twisting of the phenyl rings becomes rate controlling and the decay approximates to a single exponential. We do not at present understand the differences between the behavior of CV and that of MG. Probably these differences are connected with differences in the relative disposition of the upper and lower potential energy surfaces for the two molecules.

The weaker dependence of the decay time on viscosity for MG implies that ring rotation by the solvent is less hindered than in the case of CV. This is consistent with the presence in MG of one unsubstituted phenyl ring, i.e., one ring does not carry a $-N(CH_3)_2$ group in the para position, leading to a lower degree of π bonding to the central carbon atom and therefore less resistance to twisting.

ACKNOWLEDGEMENTS

This work was supported by the Science Research Council. GSB is grateful to the Royal Society for a Jaffe Research Fellowship and T. Doust for a SRC Studentship. M.W.W. acknowledges support from NSF grant PCM79-02911 and from the U.S. Army Research Office under grant DAA G29-76-9-0275.

References

1. Th. Forster and G. Hoffmann, Z. Physik. Chem. NF 75 (1971) 63.
2. C. J. Mastrangelo and H. W. Offen, Chem. Phys. Letters 46 (1977) 588.
3. L. A. Brey, G. B. Schuster and H. G. Drickamer, J. Chem. Phys. 67 (1977) 2648.
4. D. Magde and M. W. Windsor, Chem. Phys. Letters 24 (1974) 144.
5. B. A. Bushaw, Ph.D Thesis, Washington State University (1975).
6. E. P. Ippen, C. V. Shank and A. Bergman, Chem. Phys. Letters 38 (1976) 611.
7. W. Yu, F. Pellegrino, M. Grant and R. R. Alfano, J. Chem. Phys. 67 (1977) 588.
8. M. D. Hirsch and H. Mahr, Chem. Phys. Letters 60 (1979) 299.
9. D. A. Cremers and M. W. Windsor, Chem. Phys. Letters 71 (1980) 27.

Application of a Picosecond Dye Laser to the Study of Relaxation Processes in S-Tetrazine Vapour

J. Langelaar, D. Bebelaar, M.W. Leeuw, J.J.F. Ramaekers, and R.P.H. Rettschnick

Laboratory for Physical Chemistry, University of Amsterdam, Nieuwe Achtergracht 127, 1018 WS Amsterdam, The Netherlands

Picosecond pulses from a cw dye laser (sodium fluorescein) synchronously pumped by a mode-locked argon laser [1] have been used to pump several rovibrational bands in the $^1A_g \rightarrow {}^1B_{3u}$ electronic transition of s-tetrazine ($C_2N_4H_2$; fig. 1).

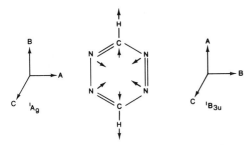

Fig. 1 s-tetrazine with inertial axes in ground and excited state.

The resonance fluorescence was observed via a high—resolution monochromator (0.12 nm/mm) with two exit slits to perform either spectrally or time—resolved measurements. The time resolution of the used single photon counting (SPC) equipment allows to measure decay times down to 100 ps without the need of deconvolution. The single photon counting technique is proved to be very useful for low—light—level fluorescence—lifetime studies in molecular vapours. The usefulness of the SPC is due to its linearity, high-dynamic range and good signal to noise ratio in combination with an extremely high sensitivity. Work is in progress to improve considerably the time resolution of the SPC system.

Under collision-free conditions the observed decaytimes vary from 400 ps for the 16 b^2 level up to 1470 ps for the 16 a^1 level; for the emission from the 0^0-level a decaytime of 820 ps was observed.

The results shown in fig. 2 are obtained for excitations at the maximum of the rotational contour of the absorption bands. It is worth noticing that no simple dependence of the non--radiative decay on the vibronic level energy is observed as in the case of benzene [2].

In fig. 3 the rotational contour of the 0_0^0 absorption is shown, while the rotational contour of the emission band is presented in fig. 4 for different excitation wavelengths; the different excitation lines are indicated by arrows in fig. 3. For the zero point level as well as for all vibronic levels studied we observed a decrease of the decay time with increasing

171

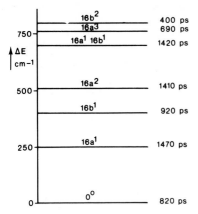

Fig. 2 Observed SVL lifetimes for several vibronic levels.

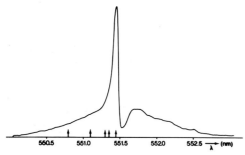

Fig. 3 Low resolution spectrum of the $\overline{0_0^0}$-absorption band of s-tetrazine at a pressure of 1 Torr.

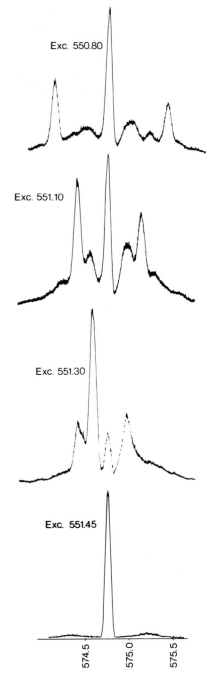

Fig. 4 Change of the rotational ▶ contour of the 0^0-emission at different excitation wavelengths indicated in fig. 3.

172

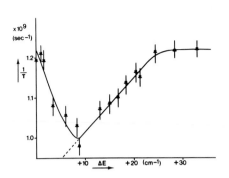

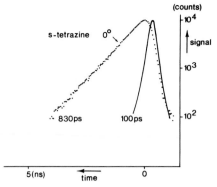

Fig. 5 Change of the decay rate of the 0⁰ level as a function of the energy difference with respect to the central Q-branch.

Fig. 6 Typical exponential decay curve of the 0⁰-emission detected in the Q-branch. The profile of the laserpulse(8ps) as seen by the detection system is shown as well

rotational energy. A typical result is given in fig. 5. The mode-locked laser linewidth of 60 pm (1 pm ≃ 1GHz ≃ 0.03 cm⁻¹) restricts each electronic excitation to a limited set of rotational levels. In fig. 6 a decay curve obtained after excitation in the Q-branch is shown; the multi-exponential decay is a result of the overlap of many rotational transitions and therefor shows a smooth exponential behaviour.

To study the influence of the rotational energy on the radiationless processes in s-tetrazine vapour, experiments resulting in a single exponential decaytime are desirable. The experimental conditions which have to be fulfilled are: - single rotational level excitation
                       - laserpulseduration << observed decaytime
                       - no collisional processes.
All experiments presented in this paper fulfill the last two requirements. The effect of the excitation bandwidth on the decay pattern is presently in study. Part of the high—resolution excitation P-branch spectrum obtained with a laser linewidth of 1 MHz is shown in fig. 7, together with the calculated spectrum.
From these results it is concluded that in some spectral regions rovibronic level excitation in s-tetrazine vapour at room temperature is possible. With a transform—limited laser puls of 720 ps FWHM (linewidth 0.02 cm⁻¹) a well—resolved single rotational level fluorescence spectrum is obtained. Preliminary single rovibronic level data show no significant difference with the decay times presented in fig. 5. Thus an increase of radiationless processes with rotational energy can be concluded. We believe that the radiationless process involved is most probably a result of rotational predissociation.

We also have studied intermolecular energy transfer for s-tetrazine with different collision partners in the one collision limit. The cross section for vibrational relaxation has been obtained for 8 vibronic levels with six different collision partners. It is concluded that V-T transfer plays an important role. Furthermore a high selectivity is observed for specific intra molecular relaxation channels. For these relaxation channels the observed branching ratio's are dependent on the collision partner.

The rotational relaxation rates are up to four times the gaskinetic value.

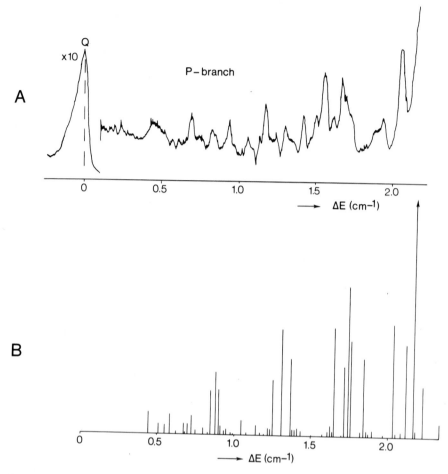

Fig. 7 A Part of the high-resolution excitation spectrum of the P-branch of the 6_0^1 transition in s-tetrazine.
Fig. 7 B Calculated P-branch spectrum for J < 20 for the same transition assuming an asymmetric rotator (for rotational constants see ref.3).

Acknowledgement

The investigations were supported in part by the Netherlands Foundation for Chemical Research (SON) with financial aid from the Netherlands Organization for the Advancement of Pure Research (ZWO).

References

1. J. de Vries, D. Bebelaar, J. Langelaar, Opt. Comm. *18*, 24 (1976)
2. K.G. Spears and S.A. Rice, J. Chem. Phys. *55*, 5561 (1971)
3. V.A. Job and K.K. Innes, J. Mol. Spectr. *71*, 299 (1978)

Dynamics of Reaction and Intramolecular Decay of Excited 3,5-Dinitroanisole: A Model for Ultrafast Radiationless Deactivation of Excited Nitroaromatics

F.L. Plantenga and C.A.G.O. Varma

Department of Chemistry, State University,
Gorlaeus Laboratories, P.O. Box 9502
2300 RA Leyden, The Netherlands

In a previous investigation we noticed that intersystem crossing from the excited singlet states of 3-nitroanisole and of 3,5-dinitroanisole (3,5-DINA) into their lowest triplet state T_o is completed on a subnanosecond timescale. The lifetime of the triplet state varies over more than three orders of magnitude as a function of the concentration H_2O, in aqueous solvent mixtures. An explanation of the variation in lifetime has been offered in terms of a model, in which the C-NO$_2$ torsional motions are inhibited, by incorporating the excited molecule in a cluster of H_2O molecules [1]. Further investigations have confirmed the importance of the structure of hydrogen bonding solvents in lengthening the triplet lifetime. It became evident that the structure in these solvents affects the rate of dissociation of a hydrogen bond between solvent and excited solute [2]. An e.s.r. study of the negative ion of 3,5-DINA in aqueous solution reveals that only a single NO$_2$ group in the anion forms a hydrogen bond [2]. As a consequence of the hydrogen bond, the extra charge in the anion gets practically localized on this particular NO$_2$ group. Since the first (π-π*) excited singlet state S_1 of 3,5-DINA is a charge transfer state with a dipole moment of 8.5 Debye, we assume that accumulation of negative charge on the NO$_2$ group in this state leads also to the formation of only one bond between one of these groups and hydrogen bonding solvents. A semiempirical quantum chemical calculation, which includes all valence electrons, shows that π_o, the highest occupied π orbital, is delocalized over the whole molecule, whereas π_1*, the first empty π orbital, is highly localized on the NO$_2$ groups [2].

Predissociation of the C-NO$_2$ bond has been held responsible for the absence of any emission from excited nitroaromatic compounds, which have their first excited singlet state more than 21000 cm^{-1} above the ground state S_0 [3]. In the case to be discussed here, predissociation can be ruled out, because triplet state reactions give high yields in substitution products with the C-NO$_2$ bonds preserved [4]. We will present evidence that the ultra fast radiationless decay of nitroaromatics arises, when the excitation is strongly localized on the NO$_2$ group. It will be shown that the efficiency of internal conversion from S_1 and of intersystem crossing from T_o to S_o in 3,5-DINA are reduced appreciably, when the excited singlet molecule reacts with H_2O.

An excite and probe technique shown in Fig. 1 has been, applied in our study of the kinetics of population and of decay of the lowest triplet state.

175

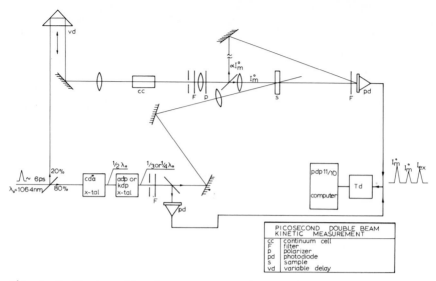

PICOSECOND   DOUBLE BEAM
KINETIC   MEASUREMENT
cc | continuum cell
F | filter
p | polarizer
pd | photodiode
s | sample
vd | variable delay

Fig. 1   Double beam kinetic spectrometer for the detection of
         photoinduced transient species

Part of a single, 6 psec, 1.06 μ fundamental laser pulse is either tripled
or quadrupled in frequency prior to excitation of the sample. The other
part is guided along an optical delay line into a liquid cell, where a
spectral continuum is generated for the measurement of transient absorptions
in the sample.   Interference filters are inserted in the continuum beam to
limit the bandwidth of the probing light. A transient digitizer is used to
transfer the signals from the photodiodes to a computer. The excitation
and probe pulses are practically colinear and are considered to have the
same Gaussian temporal profile. The excitation pulse arrives at the sample
at time t = 0 and its intensity $I_{ex}(t)$ is given by:

$$I_{ex}(t) = \frac{A\ ex}{\sigma\sqrt{2\pi}}\ \exp\{-t^2/\ \sigma^2\} \tag{1}$$

The intensity of the unattenuated probe pulse $I_p^{(o)}(t)$, delayed by an amount
Δt with respect to the excitation pulse, is given by:

$$I_p^{(o)}(t) = \frac{A_p}{\sigma\sqrt{2\pi}}\exp\{-(t-\Delta t)^2/\ \sigma^2\} \tag{2}$$

In (1) and (2) the width of the pulse is represented by 2σ and the area under the pulse envelope is denoted by A_{ex} and A_p respectively. A single
photodiode determines the time integrals $U_1(\Delta t)$ and U_2 of respectively the
transmitted and the unattenuated probe pulse.

$$U_1(\Delta t) = \frac{A_p}{\sigma\sqrt{2\pi}}\ \int_{-\infty}^{+\infty} 10^{-D(t)}\ \exp\{-(t-\Delta t)^2/\ \sigma^2\}dt \tag{3}$$

$$U_2 = \int_{-\infty}^{+\infty} I_p^o(t)dt \tag{4}$$

where in (3) the transient optical density is denoted by D(t). Another photodiode determines the time integral V of the excitation pulse.

$$V = \int_{-\infty}^{+\infty} I_{ex}(t)dt \qquad (5)$$

With the aid of (3)-(5), the following quantity is determined.

$$R(\Delta t) = \frac{1}{V} \, ^{10}\!\log \left\{ \frac{U_2}{U_1(\Delta t)} \right\} \qquad (6)$$

If $\Delta t \gg 2\sigma$, we have $D(\Delta t) = R(\Delta t)$ and $D(\Delta t) \neq R(\Delta t)$ otherwise.

In the manner just described we have found that the solution of 3,5-DINA in CH_3CN has a broad transient absorption band around 470 nm, which decays single exponentially with a rate constant of $1.28 \times 10^9 s^{-1}$. Because the solution has a fluorescence quantum yield of less than 10^{-4}, an identification of the transient species with the lowest excited singlet state would mean a radiative decay rate $k_r < 10^5 s^{-1}$. This is even too small for a transition from the first $^1(n-\pi^*)$ state, which is seen in absorption from S_o with a maximum extinction coefficient of 10^2. Therefore the transient absorption will be considered to arise from the triplet state T_o. A complete deuteration of 3,5-DINA does not change the lifetime of T_o. If the excitation did extend over the aromatic ring, a deuterium effect on the lifetime should have been found, just like in the case of aromatic hydrocarbons [5]. The absence of a deuterium effect indicates that the triplet state involves a $(n-\pi_1^*)$ local excitation of the NO_2 groups, in which an electron from lone pair orbital n on one of the two NO_2 groups is promoted to the π_1^* orbital. If we take into account, that the exchange integral, which determines the singlet-triplet energy separation for this electron configuration, is smaller than for the $(\pi_o-\pi_1^*)$ configuration, we come to the $^1(n-\pi_1^*)$ state as the lowest excited singlet state of 3,5-DINA in non hydrogen bonding solvents. The strong localization of the $(n-\pi_1^*)$ excitation must cause an appreciable deformation of the NO_2 group involved, relative to the geometries in S_o and in $^1(\pi_o-\pi_1^*)$. Despite a separation of more than 2.5 eV from the ground state, the $^3(n-\pi_1^*)$ state has a lifetime as short as ca. 800 psec. This is probably controlled by strong non adiabatic coupling to the ground state, arising from intersection of the two potential energy surfaces close to the bottom of the triplet state surface [6]. Such a crossing may occur when the molecular distortion is large.

Excitation of 3,5-DINA with the 353 nm laser pulse creates primarily S_1, i.e., the $^1(\pi_o-\pi_1^*)$ state, independent of the particular solvent choosen. Both the decay of this state and the population of the triplet state proceed on a picosecond time scale. In Fig. 2 we present the growth of the triplet absorption for solutions in H_2O and in two mixtures of CH_3CN and H_2O. For long delay times the quantity $R(\Delta t)$ is seen to reach an asymptotic value, which represents the maximum in optical density as a function of time and which will be denoted by $D(t_m)$. Fig. 3 shows $D(t_m)$ as a function of the molefraction H_2O in the solvent.
In order to understand the behaviour shown in Figs. 2 and 3, we propose a kinetic scheme in Fig. 4, which includes a reaction of H_2O with the primary excited state S_1 and with T_o.

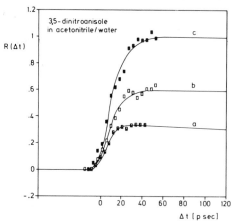

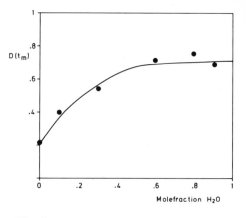

Fig.2 Growth of triplet absorption of 3,5-DINA in mixtures of
CH_3CN and H_2O
a) molefraction H_2O 0.00
b) molefraction H_2O 0.13
c) molefraction H_2O 0.56

Fig.3 Increase of maximum
transient optical density of
3,5-DINA in mixtures of CH_3CN
and H_2O with molefraction H_2O

For the triplet quantum yield $\emptyset_T$ of 3,5-DINA, we have found from energy
transfer experiments $\emptyset_T$ < 0.1 in apolar media, $\emptyset_T$= 0.45 in acetonitrile and
$\emptyset_T$= 0.9 in water. In the case of a solution in CH_3CN, the population of
the triplet state T_o involves only the rate constants k_1 and k_2. Using (6)
and $\emptyset_T = k_2/(k_1+k_2)$ as constraint, a numerical simulation of $R(\Delta t)$ yields
$k_1 = 10^{11}s^{-1}$ and $k_2 = 8 \times 10^{10}s^{-1}$. The large value of k_2 indicates, that
intersystem crossing proceeds directly from $^1(\pi_o-\pi_1{}^*)$ to $^3(n-\pi_1{}^*)$ due both
to strong spin orbit coupling and to strong non adiabatic coupling between
these states. The latter being the dominant factor, because of the distortion in the $^3(n-\pi_1{}^*)$ state. The relative weak spin orbit coupling of
the lower $^1(n-\pi_1{}^*)$ state with the $^3(n-\pi_1{}^*)$ state cannot account for the observed intersystem crossing rate. In order to get $\emptyset_T$= 0.45, the intersystem crossing rate of the primary excited state must be comparable to the
rate of relaxation to $^1(n-\pi_1{}^*)$. The fact that internal conversion to
$^1(n-\pi_1{}^*)$ does not dominate the decay of $^1(\pi_o-\pi_1{}^*)$ has to be associated
with the close spacing between these states. From gasphase and solution
spectra, we estimate the separation ΔE_s to be less than 2000 cm^{-1}. The
density of energy dissipating states will be low at such a small energy
gap and consequently the relaxation rate will be slow. This picture readily explains the low triplet quantum yield observed in apolar solvents.
Because the $^1(\pi_o-\pi_1{}^*)$ state is a charge transfer state, it will go up in
energy when the polarity of the solvent is reduced, thereby increasing
ΔE_s to almost 6000 cm^{-1} in cyclohexane. As a consequence of the enhanced
density of dissipating states, the relaxation to $^1(n-\pi_1{}^*)$ dominates in the
apolar solutions. The $^1(n-\pi_1{}^*)$, state will subsequently decay mainly
through internal conversion to S_o, because of the molecular distortion and
the weak spin orbit coupling to $^3(n-\pi_1{}^*)$.

Before we turn to the population of the triplet state of 3,5-DINA in
hydrogen bonding solvents, we have to mention that the ground state molecule does not form hydrogen bonds and that the hydrogen bonded triplet
state T_o, HB decays with rate constant K_8, which in water amounts only to
$0.72 \times 10^6 s^{-1}$ [1]. Obviously, the hydrogen bond brings the $^3(\pi_o-\pi_1{}^*)$ sta-

178

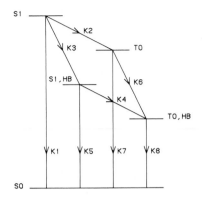

S1

K2

K3 ——— TO

S1,HB

K6

K4

——— TO,HB

K1 K5 K7 K8

S0 ———

Fig. 4 Scheme representing kinetics involved in triplet states of 3,5-DINA, where HB indicates hydrogen bonded states

Fig. 4 Scheme representing kinetics involved in triplet states of 3,5-DINA, where HB indicates hydrogen bonded states

te, involving a delocalized excitation, below the $^3(n-\pi_1^*)$. When this happens, the lowest excited hydrogen bonded singlet state S_1,HB should also be considered as a $(\pi_0-\pi_1^*)$ excitation. Just like in the case of the triplet states, the direct decay to the ground state will be slower from $^1(\pi_0-\pi_1^*)$ i.e. $k_5 << k_1$. Since T_0 involves excitation of an NO_2 lone pair electron to π_1^*, the affinity to form hydrogen bonds in this state will be less than in the ground state. We may therefore put $k_6= 0$.

Knowing $\emptyset_T$ for the solutions in CH_3CN and in H_2O, the ratio $\varepsilon_1/\varepsilon_2$ of the extinction coefficients of respectively T_0 and T_0, HB is readily determined from the corresponding values $D(t_m)$. The obtained value $\varepsilon_1/\varepsilon_2= 0.67$ is considered as constant in the series of solutions in mixtures of CH_3CN and H_2O. With the knowledge already at hand, we are able to simulate the experimental curves $R(\Delta t)$ in Fig. 2 numerically, by using (6) and the scheme in Fig. 4. In the simulation k_3 and k_4 are treated as variable parameters, once k_5 has been choosen. Acceptable simulations are obtained only, when k_5 is negligible. The values of the various rate constants are summarized in table 1.

Table 1 Rate constants for the decay channels of excited 3,5-DINA in solvent mixtures of CH_3CN and H_2O

Mole fractions CH_3CN/H_2O	$10^{-11}k_1$ sec^{-1}	$10^{-11}k_2$ sec^{-1}	$10^{-11}k_3$ sec^{-1}	$10^{-11}k_4$ sec^{-1}	$10^{-11}k_5$ sec^{-1}	$10^{-11}k_6$ sec^{-1}	$10^{-11}k_7$ sec^{-1}	$10^{-11}k_8$ sec^{-1}
1.00/0.00	1.0	0.8	–	–	–	–	0.01	–
0.87/0.13	1.0	0.8	1.0	0.8	0	0	0.01	0.02
0.44/0.56	1.0	0.8	4.0	0.8	0	0	0.01	<<0.02

It appears that in the region of molefractions H_2O between 0.00 and 0.4, the reaction of 3,5-DINA in state S_1, i.e. in $^1(\pi_0-\pi_1^*)$ with H_2O is one of the rate determining steps in the population of T_0, HB. In this region it is found that $k_3=k_d[H_2O]$ with $k_d= 3 \times 10^{10}$ l/mole s. The value of k_d is of the magnitude of a diffusion rate constant. When the molefraction H_2O is larger than 0.4, the relatively slow intersystem crossing from S_1, HB becomes the rate determining step in the growth of the triplet absorption.

179

By substituting the H atom on carbon atom number 4 of the ring in 3,5-DINA by a t.butyl group, we have been able to rotate both NO_2 groups about $60°$ around the $C-NO_2$ bond [7]. It should be realized that at a rotation of $90°$, one would encounter a perfectly localized $(n-\pi*)$ excitation on the NO_2 groups and that the first $(\pi-\pi*)$ transition from the ground state would go up in energy compared to the planar molecule. The rate constants for population and decay of the triplet state given in table 2 allow some conclusions about the energy level scheme of 4-t.butyl-3,5-DINA.

Table 2. Triplet quantum yield and rate constants for population and decay of the triplet state of 4-t.butyl-3,5-DINA

Solvent	$\emptyset_T$	$10^{-11}k_1$ sec^{-1}	$10^{-11}k_2$ sec^{-1}	$10^{-6}k_7$ sec^{-1}	$10^{-6}k_8$ sec^{-1}
cyclohexane	<0.2	1.0	<0.5	>100	
CH_3CN	0.43	0.5	0.5	12	
CF_3CH_2OH	0.45	0.5	0.5		1.0

It seems that the primary excited state is a $^1(n-\pi_1*)$ state localized on an NO_2 group, which crosses efficiently in polar solvents to the $^3(\pi_0-\pi_1*)$ state. Due to the charge transfer character of this state, a hydrogen bond will be formed subsequently and will enhance the triplet lifetime in hydrogen bonding solvents. The enhancement arises from an increase of the gap ΔE_T with $^3(n-\pi_1*)$ from which spin orbital coupling to ground state has to be borrowed vibronically. In apolar solvents the $^3(n-\pi_1*)$ state becomes the lowest triplet state. Since this has less spin orbit coupling with the primary (lowest) singlet state, the triplet quantum yield is less than in polar solvents.

Acknowledgement

The investigations were supported by the Netherlands Foundation for Chemical Research (S.O.N.) with financial aid from the Netherlands Organization for the Advancement of Pure Research (Z.W.O.).

References

1. C.A.G.O. Varma, F.L. Plantenga, C.A.M. van den Ende, P.H.M. van Zeyl, J.J. Tamminga and J. Cornelisse, Chem. Phys. 22, 475 (1977).
2. F.L. Plantenga and C.A.G.O. Varma, in preparation.
3. E. Lippert and J. Kelm, Helv. Chim. Acta 61, 279 (1978).
4. J.J. Tamminga, Thesis, Leyden, (1979).
5. W. Siebrand, in Organic Molecular Photophysics, ed. J.B. Birks (John Wiley, London, 1975), Vol. 1.
6. R. Englman, Non Radiative Decay of Ions and Molecules in Solids (North Holland, Amsterdam, 1979).
7. F.L. Plantenga, J.P.M. van der Ploeg, C.A.G.O. Varma and H.J.V.H. Geise, in preparation.

Study of the Intersystem Crossing in Ir and Rh Complexes with the Use of Subpicosecond Laser Pulse

T. Kobayashi and Y. Ohashi

The Institute of Physical and Chemical Research,
Hirosawa, Wako, Saitama 351, Japan

1. Introduction

There is currently great interest in the photochemical and luminescent properties of transition metal complexes. Because of the interest, $T_n \leftarrow T_1$ absorption spectra [1] and the formation time constant [2] of the lowest triplet
state (T_1) for dichlorobis (1,10-phenanthroline) iridium (III) chloride in 95%
v/v DMF-H_2O or in pure H_2O solution were measured with the use of a N_2 laser
and a mode-locked ruby laser [1,2]. The smaller intersystem crossing rate
constant of cis-Ir[Cl_2(phen)$_2$]Cl in 45% v/v DMF-H_2O mixed solvent (37 ps) than
in 95% v/v DMF-H_2O mixed solvent (26 ps) was interpreted in terms of the increase
crease in $^3(\pi\pi^*)$ character and the simultaneous decrease in $^3(d\pi^*)$ character
in the lowest triplet state of the model [2]. Presiously the rise times of
the luminescence of cis-[RhX$_2$(bpy)$_2$]X (X = Cl, Br) where bpy = 2,2'-bipyridine
were reported to be 350-630 ns in solution at room and liquid nitrogen temperatures [3]. The authors of the present paper and DEMAS[4] independently noticed
that the rise times reported are not the intrinsic ones but are the time constants of the detection photomultiplier system which are determined by the stray
capacitance of the detection system and output resistance.

In the present paper, the rise time constant of the triplet state of the Ir
complex in ethanol-methanol (v/v = 4:1) was measured with the use of subpicosecond spectroscopic technique.

2. Experimental

The time dependence of the absorbance change in picosecond time scale was
observed with the use of an Ar laser pumped dye laser. Subpicosecond pulses
(0.5 ps width) from a passively mode-locked rhodamine 6G dye laser operated at
612 nm were cavity dumped at a 10-Hz rate [5]. These pulses were amplified with
three-stage dye amplifiers pumped by the frequency-doubled output of a Nd/YAG
laser to powers of 1-4 GW[6]. The output of the amplifier was divided to form
two separate beams. The first beam was used to pump the sample after passing
through a variable optical delay line driven by a stepping motor and then
focused into a KDP crystal which generated a second harmonic at 306 nm. The
output second harmonic was focused on the sample cell. The second beam was
focused into a cell containing H_2O, CCl_4, or EtOH, generating a broad band subpicosecond continuum which was used to monitor the sample. The monitor pulses
were split into two, probe and reference pulses, and the former pulses were
imaged through the region of the sample excited by the pump pulses while the
latter pulses were imaged out of the excited region of the sample. Both pulses
were refocused on the entrance slit of monochromator to which an OMA was
attached.

3. Results and Discussion

3.1 Time-resolved absorption spectrum of the Ir complex.

Figures 1 and 2 show the picosecond time-resolved absorption spectra of cis-$[IrCl_2(phen)_2]$ Cl (complex A) at 2 ps, 7 ps, 54 ps, and 75 ps, after the 0.5 ps pulse excitation at 306 nm. The time-resolved spectrum of the Ir complex immediately after a 10 ns pulse excitation at 337.1 nm resembles that after 75 ps excitation [1]. The spectrum obtained by the nanosecond spectroscopy is attributed to $T_n \leftarrow T_1$ absorption since the lifetime of the transient spectrum of the Ir complex in deoxygenated solution at 19°C is 60 ns which is equal, within experimental error, to the phosphorescence lifetime of the same sample at the same temperature. The time-resolved absorption spectrum of the Ir complex at 2 ps after subpicosecond pulse excitation has its maximum around 600 nm while that at 75 ps after excitation has a maximum at 540-560 nm.

From the experimental results it is concluded that there is a precursor (P) of the lowest triplet state (T_1) and the isosbestic wavelength of P and T_1 is located at 590-570 nm. The absorption spectrum of $T_n \leftarrow T_1$ is very broad and its maximum is located at 540-560 nm. The band width of the $T_n \leftarrow T_1$ absorption spectrum is greater than 6,000 cm^{-1}. In order to assign P, we observed the rise curve of $T_n \leftarrow T_1$ absorption spectrum at 550 nm. There is only weak $S_n \leftarrow S_1$ absorption at 550 nm. From the rise curve, the time constant of $P \rightarrow T_1$ is obtained to be 26 ± 10 ps.

The excitation of the Ir complex in ethanol-methanol (v/v = 4:1) by 306 nm results in the population of higher excited singlet states of $^1(\pi\pi*)$ character. There are several singlet states of $^1(d\pi*)$, $^1(\pi\pi*)$, and $^1(dd)$ characters and several triplet states of $^3(d\pi*)$, $^3(\pi\pi*)$, and $^3(dd)$ characters between the initially populated singlet state (S_n) of $^1(\pi\pi*)$ character and the phosphorescent state. The overall time necessary for radiationless relaxation from the excited singlet state to T_1 is 26 ± 10 ps. The bottleneck of the relaxation process can be either the internal conversion within the singlet or triplet manifold, or the intersystem crossing from the singlet to the triplet manifold. The bottleneck is considered to be the intersystem crossing process because of the reason discussed below.

The time constant for intersystem crossing of the Ir complex is 21-37 ps in aqueous or mixed solvent of water and dimethylformamide (DMF). The rise time observed for $T_n \leftarrow T_1$ absorption in the present study for the Ir complex in ethanol-methanol solution is within the range of the observed time constant given above. Therefore it is reasonable to assign the spectrum to be a $T_n \leftarrow T_1$ transition. The dielectric constant of ethanol-methanol is 26 while those of water, 45% DMF-H_2O and 95% DMF-H_2O are 80, 61, and 39, respectively[1]. By changing solvent, a variation in the lowest triplet state formation rate is found. As was previously reported by the authors, T_1 of the Ir complex in 95% v/v DMF-H_2O is mainly of $^3(d\pi*)$ character. As the solvent polarity increases, the $^3(d\pi*)$ state shifts to higher energy and mixes with the $^3(\pi\pi*)$ state. The observed time constant of intersystem crossing in ethanol-methanol is closer to that for the sample in 95% DMF-H_2O than that for the sample in 45% DMF-H_2O. This is reasonable since the dielectric constant of ethanol-methanol is closer to that of 95% DMF-H_2O than 45% DMF-H_2O.

S_1 is of $^1(d\pi*)$ character and is located between the $^3(d\pi*)$ and $^3(\pi\pi*)$ states in 95% v/v DMF-H_2O and also shifts to higher energy with an increase in solvent polarity. The dielectric constants of ethanol-methanol (v/v = 4:1), and 95% DMF-H_2O are less than those of 45% DMF-H_2O and H_2O, therefore the character of the T_1 state in the Ir complex is more of $^3(d\pi*)$ character in the former solvents than in the latter solvents. Hence the rate of intersystem crossing is enhanced because of the increased contribution of $^1(d\pi*)$ - $^3(d\pi*)$ spin-orbit interaction which is greater than $^1(d\pi*)$ - $^3(\pi\pi*)$ interaction since the former contains one-center integrals while the latter does not.

From the above discussion the observed time constant of 26 ± 10 ps is reasonably assigned as that of the intersystem crossing from S_1 and T_1; therefore P can be attributed to S_1. This assignment can also be supported by the consideration of the spectrum of P. In this solvent, both S_1 and T_1 states are of $d\pi*$ character. The exchange interaction between S_n and T_n with $d\pi*$ character is much smaller than that between S_1 and T_1, because of the widespread higher $\pi*$ orbitals related to S_n and T_n. On the other hand, by the interaction between the lowest $^3(d\pi*)$ and a $^3(\pi\pi*)$ nearby lying, T_1 in ethanol-methanol (principally of $^3(d\pi*)$ character) may probably be stabilized by $(1-2) \times 10^3$ cm^{-1}, than S_1. Therefore the energy separation between $S_n \leftarrow S_1$ and $T_n \leftarrow T_1$ transition is $(1-2) \times 10^3$ cm^{-1}. The observed absorption maxima of $S_n \leftarrow S_1$ and $T_n \leftarrow T_1$ transitions are 16.7×10^3 cm^{-1} and 18.2×10^3 cm^{-1}, respectively. The difference between the two transitions is therefore 1.5×10^3 cm^{-1}. This is consistent with the estimated energy separation from $d\pi*-\pi\pi*$ interaction. The energy separation was estimated for $[IrCl_2(5,6-Mephen)_2]Cl$ in alcohols to be 800-1300 cm^{-1} [7].

The observed strong $S_n \leftarrow S_1$ and $T_n \leftarrow T_1$ transitions are primarily assigned to the transitions between electronic states of the same character ($d\pi* \leftarrow d\pi*$, or $\pi\pi* \leftarrow \pi\pi*$).

3.2 Time-resolved absorption spectra of the Rh complexes

The time-resolved spectra were observed for $[RhCl_2(bpy)_2]Cl$ (complex B) at the delay time of 15 ps excited by 306 nm subpicosecond pulse and $[RhBr_2(bpy)_2]Br$ (complex C) at the delay time of 1 ps and 17 ps. The spectra of the Br complex are shown in Fig. 3. The absorption spectra of the complexes resemble $T_n \leftarrow T_1$ absorption spectra observed with the use of nanosecond laser spectroscopy apparatus. This means that the transient spectra of B and C complexes at 15 ps or 17 ps after subpicosecond pulse excitation is due to $T_n \leftarrow T_1$ transitions. The rise time of the transient absorbance change for complex B was estimated to be 2.4 ± 0.6 ps from the rise curve. This is obtained without deconvolution of excitation and probing light pulses, so this time constant is longer than the real one for the growing-in process of the $T_n \leftarrow T_1$ absorption. The time-resolution limit of the system for the Cl complex sample is estimated to be about 1.1 ps. This is obtained by taking into account the decrease in the time resolution because of the difference in the group velocity between the excitation and interogation light pulses.

The apparent rise time obtained for complex C from the observed rise curve is 1.3 ± 0.4 ps. The estimated resolution time of complex C sample is 0.8 ps.

From the experimental results, the growing-in time of the lowest triplet state of complexes B and C is slightly shorter than 2.4 ps (Cl complex) or 1.3 ps (Br complex) in ethanol-methanol (4:1) mixed solvent at 19°C.

For the complexes B and C, the 306 nm, 337.1 nm, and 347.2 nm excitations are almost exclusively into ligand localized $^1(\pi-\pi*)$ states and the phosphorescent state is a localized $^3(d-d)$ ligand field state[8]. Thus the relaxation processes between excitation and emission involves both changes in multiplicity and in orbital parentage. As determined experimentally the effective rate for relaxation to the emitting level must be $>5 \times 10^{11}$ s^{-1} for the

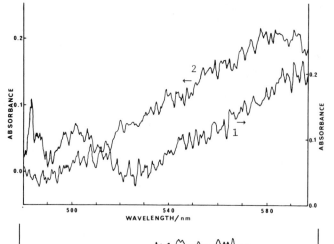

Fig. 1 The pico-second time-resolved absorption spectra of cis[IrCl$_2$(phen)$_2$]Cl in ethanol-methanol (v/v = 4:1) at 2 psec (curve 1) and 7 psec (curve 2) after 306 nm sub-picosecond pulse excitation at room temperature.

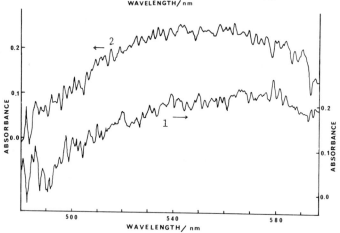

Fig. 2 The pico-second time-resolved absorption spectra of cis[IrCl$_2$(phen)$_2$]Cl in ethanol-methanol (v/v = 4:1) at 54 psec (curve 1) and 75 psec (curve 2) after 306 nm sub-picosecond pulse excitation.

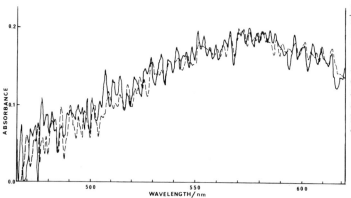

Fig. 3 The pico-second time-resolved absorption spectra of [RhBr$_2$(bpy)$_2$]Br at the delay time of 1 psec (———) and 17 psec (----) after subpicosecond pulse excitation at 306 nm.

Cl complex and $>1 \times 10^{12} s^{-1}$ for the Br complex in ethanol-methanol (4:1) mixed solvent.

There are two probable energy relaxation paths populating the emitting level in B and C complexes.

$$^1(\pi\pi^*) \to {}^3(\pi\pi^*) \to {}^3(dd) \quad (1) \qquad {}^1(\pi\pi^*) \to {}^1(dd) \to {}^3(dd) \quad (2)$$

In the previous paper, pathway (1) was considered to be dominant. However this was based on erroneous experimental results. If pathway (1) is the dominant one, the $^1(\pi\pi^*) \to {}^3(\pi\pi^*)$ intersystem crossing rate in the complexes would have to exceed 5×10^{11} s^{-1} to account for the present data.

While this might be possible because of the presence of the nearby heavy Rh and Cl or Br atoms it seems highly unlikely that the rate of $^1(\pi\pi^*) \to {}^3(\pi\pi^*)$ type intersystem crossing is enhanced from ordinary 10^{3-6} s^{-1} to 5×10^{11} s^{-1} by a factor of more than 10^5. In pathway (2) however, the intersystem crossing is completely localized in the dd manifold of the Rh heavy atom. In other words, the matrix element in the equation which represent the transition probability of $^1(dd)$ to $^3(dd)$ is expressed in terms of one-center integrals on the heavy atom.

An example of the above consideration is the $^4T \to {}^2E$ ($^4(dd) \to {}^2(dd)$) intersystem crossing rate in low atomic number Cr (III) complexes which is as high as 4×10^{10} s^{-1} [9,10]. Thus one would expect this rate to be extremely fast. If our assignment to pathway (2) is correct then the $^1(\pi\pi) \to {}^1(dd)$ internal conversion and the $^1(dd) \to {}^3(dd)$ intersystem crossing processes have rates of 5×10^{11} s^{-1} in complex B and of 1×10^{12} s^{-1} in complex C.

T.K. would like to express sincere gratitude to Dr. C.V. Shank for his hospitality during T.K.'s stay at Bell Labs., for his kindness to put the picosecond spectroscopic equipment at T.K.'s disposal and for his fruitful discussion.

References

1. Y. Ohashi and T. Kobayashi, Bull. Chem. Soc. Japan, 52, 2214 (1979).
2. Y. Ohashi and T. Kobayashi, J. Phys. Chem., 83,551 (1979).
3. Y. Ohashi, K. Yoshihara and S. Nagakura, J. Molec. Spect., 38, 43 (1971).
4. J.N. Demas, J. Phys. Chem., in press.
5. C.V. Shank, E.P. Ippen, A. Migus, and R.L. Fork, to be published.
6. E.P. Ippen and C.V. Shank, *Ultrashort Light Pulses*, ed. by S.L. Shapiro, Topics in Applied Physics, Vol. 18 (Springer, Berlin, Heidelberg, New York, 1977) p. 83
7. R.J. Watts, G.A. Crosby, and J.L. Sansregret, Inorg. Chem., 11, 1474 (1972)
8. J.N. Demas and G.A. Crosby, J. Amer. Chem. Soc., 92, 7262 (1970)
9. A.D. Kirs, P.E. Hoggard, G. Porter, M.G. Rockley, and M.W. Windsor, Chem. Phys. Letters, 37, 199 (1976)
10. S.C. Pyke and M.W. Windsor, J. Amer. Chem. Soc., 100, 6518 (1978)

Vibrational Energy Transfer in Pure Liquid Hydrogen Chloride

D. Ricard and J. Chesnoy

Laboratoire d'Optique Quantique du C.N.R.S., Ecole Polytechnique
F-91128 Palaiseau Cédex, France

Vibrational energy transfer has already been studied in a number of diatomic molecules in the liquid phase. In the case of H_2 and D_2 the results could be interpreted in terms of isolated binary collisions throughout the density range extending from gas to liquid [1].

Hydrogen chloride is an interesting material to study from several points of view. It forms a polar but non associated liquid. Vibrational relaxation in gas phase HCl has been extensively studied by C. B. MOORE and his collaborators. Among other things, they concluded that rotation plays a major role in the v = 1 level deexcitation [2] and they observed that the relaxation constant
$$k = 1/\rho T_1$$
where ρ is the number density and T_1 the v = 1 level lifetime passes through a minimum near room temperature and increases when the temperature is further reduced [3]. They attributed this effect to the formation of Van der Waals dimers.

In the work we report here, we have been mainly concerned with vibration to translation (V-T) energy transfer but we will also touch upon the topic of vibration to vibration (V-V) energy transfer.

Principle and experimental

Our measurements have been performed using an infrared double resonance technique whose principle rests on the fact that, due to anharmonicity, the transition frequency ω_2 between the v = 1 and v = 2 levels is smaller than the transition frequency ω_1 between the v = 0 and v_1 = 1 levels. For example, near the boiling point, at 188°K, $\omega_1 \simeq 2785$ cm^{-1} and, if we assume that the anharmonicity constant has the same value as in the gas phase, $\omega_2 \simeq 2680$cm^{-1}.

The measurement proceeds in two steps. A first picosecond pulse of frequency ω_1 or close to ω_1 excites some of the molecules to the v = 1 level. A second delayed picosecond pulse of frequency ω_2 probes the population of the v = 1 level. We measure the attenuation of the probe pulse as a function of the delay between the two pulses.

The absorption coefficient at frequency ω_2 is given by the formula

$$\alpha(\omega_2) = \sigma_{01}(\omega_2)(\rho_0-\rho_1) + \sigma_{12}(\omega_2)(\rho_1-\rho_2) + \ldots \tag{1}$$

where σ_{ij} is the absorption cross section for the v = i to v = j transition and ρ_i is the population of the v = i level.

When the excitation rate is small, the population of levels higher than $v = 1$ is negligible and (1) takes the simpler form

$$\alpha(\omega_2) \simeq \sigma_{01}(\omega_2)\rho + [\sigma_{12}(\omega_2) - 2\sigma_{01}(\omega_2)]\rho_1 \qquad (2)$$

The first term is the normal absorption coefficient at frequency ω_2. The second term is its change due to excitation of some of the molecules to the $v = 1$ level.

If τ_0 denotes the normal transmission of the sample cell at frequency ω_2 and τ its transmission after excitation, then

$$\ln(\tau_0/\tau) = \sigma_e N_1/S \qquad (3)$$

where N_1 is the total number of excited molecules, S is the beam cross section and τ_e an effective absorption cross section

$$\sigma_e = \sigma_{12}(\omega_2) - 2\sigma_{01}(\omega_2)$$

Thus, $\ln(\tau_0/\tau)$ plotted versus the delay decays exponentially as N_1.

The experiment starts with a single laser pulse at 1.06 μ delivered by a mode-locked neodymium glass source. The two infrared pulses are derived from this laser pulse by single pass parametric amplification in lithium niobate crystals as was first developed by LAUBEREAU and KAISER [4]. Liquid HCl is contained in an 80 μm thick cell placed inside a variable temperature cryostat. The beams are focused on the sample cell. The probe beam then passes through an aperture (to make sure that we probe a homogeneously excited zone) and a 5 cm^{-1} bandpass monochromator. The energy of the probe pulse is measured by InSb photovoltaic detectors.

When the excitation rate is small, the experimental data shows the expected exponential decay, with a lifetime T_1 of the order of 1 nanosecond. But when the excitation rate is too large, there appears a fast decay component at short delay times. This prompts us in examining the possible causes of error in the measurement of T_1. There are two of them.

The first one is heating. During the deexcitation of the $v = 1$ level, energy is transferred from vibration to translation, leading to sample heating. Since T_1 is temperature dependent, this could cause an error. For example, an excitation rate of 1 % leads, after complete deexcitation, to a 5.5°K heating.

The second one is vibration to vibration energy transfer. When the excitation rate is too large, the $v = 1$ level may also be deexcited by V-V processes such as

$$HCl(v=1) + HCl(v=1) \underset{k_2}{\overset{k_1}{\rightleftarrows}} HCl(v=2) + HCl(v=0) + \Delta E \qquad (4)$$

which are much faster than V-T transfer. These lead to population of higher vibrational levels at the expense of the $v = 1$ level. Although the expression of the transmission is now slightly more complicated than given by (3), it is easy to convince oneself that the fast decay component mentioned above is due to V-V transfer.

V-V transfer can also alter the measurement of T_1. When the excitation rate is small, ρ_1, the $v = 1$ level population, decays exponentially. But when the excitation rate is larger, the $v = 2$ level population becomes non negligible, more processes have to be taken into account, and the net result is a non exponential decay for ρ_1.

These two possible causes of error are favored by a large excitation rate. So, in order to get meaningful results in the measurement of T_1, one must use a low excitation rate.

Results and discussion

V-V transfer may be a cause of error in the measurement of T_1, but it is also an interesting physical phenomenon. V-V processes of the type

$$HCl(v=1) + HCl(v=1) \underset{k_2}{\overset{k_1'}{\rightleftarrows}} HCl(v=2) + HCl(v=0) + \Delta E \tag{5}$$

lead to an equilibration of the populations of the vibrational levels. Being much faster than V-T transfer, the excitation rate

$$x = (\rho_1 + 2\rho_2 + ...)/\rho \tag{6}$$

remains constant during the equilibration. If we can neglect levels higher than $v = 2$, ρ_1 and ρ_2 then obey simple equations

$$\frac{d\rho_2}{dt} = k_1\rho_1^2 - k_2\rho_0\rho_2 \tag{7}$$

$$\frac{d\rho_1}{dt} = -2\frac{d\rho_2}{dt} \tag{8}$$

which can be solved numerically. By observing the fast decay component, one can then get the value of the rate constants k_1 and k_2 (which are related by $k_1/k_2 = \exp(\Delta E/k_B T)$).

Working deliberately with a larger excitation rate than for T_1 measurements, we could get, at 173°K in the liquid phase, an estimate

$$k_2 \sim 2 \times 10^{-12} cm^3 s^{-1}$$

This corresponds, under low excitation conditions, to a $v = 2$ level lifetime of about 25 picoseconds. This liquid phase result is to be compared with the gas phase value obtained at room temperature by LEONE and MOORE [5] : $k_2 = 2.8 \times 10^{-12} cm^3 s^{-1}$.

Let us turn now to vibration to translation transfer. We have measured T_1 at four different temperatures in the liquid phase and at one temperature in the solid phase, always under vapor pressure. The results are given in Table 1. Notice that, in the liquid phase, T_1 decreases when the temperature increases.

Table 1

Temperature (°K)	148	173	198	223	248
Measured T_1 (10^{-9}s)	1.6	2.1	1.6	1.3	1.0
Calculated T_1 (10^{-9}s)		2.4	2.9	3.3	3.6

Let us compare our results to gas phase data. To do so, we will use a model which was recently developed by DELALANDE and GALE [1]. This model assumes that relaxation is due to isolated binary collisions, that the molecules interact through an isotropic model potential consisting of a hard core of diameters σ and an attractive part of depth ε and finally that deexcitation takes place on the infinitely repulsive part. Then, a simple calculation leads to the result

$$T_1^{-1} = \rho g(\sigma) k_0(T) \qquad (9)$$

where $g(r)$ is the radial distribution function and k_0 depends only on temperature. Therefore, at a given temperature, this implies

$$\frac{T_1^{liq}}{T_1^{gas}} = \frac{\rho^{gas}}{\rho^{liq}} \frac{g(\sigma)^{gas}}{g(\sigma)^{liq}} \qquad (10)$$

At low densities one has

$$g(\sigma)^{gas} = \exp(\varepsilon/k_B T) \qquad (11)$$

For densities corresponding to the liquid phase, one has, to a high degree of accuracy

$$g(\sigma)^{liq} = (1 - \eta/2)/(1 - \eta)^3 \qquad (12)$$

where η, the packing fraction, is $\eta = \pi \rho \sigma^3/6$. Then, with $\sigma = 3.3$ Å and $\varepsilon = 360°K$ (deduced from viscosity data), we calculate, from gas phase data, the values of T_1, shown in the last row of Table 1.

There is order of magnitude agreement with the measured T_1, but we do not reproduce the observed temperature dependence. The model we have used, which gives very good agreement in the case of H_2, seems to fail in the case of HCl. First, we may question the hypothesis of an isotropic intermolecular potential. In our case, the pair distribution function and the deexcitation probability per collision are certainly both orientation dependent. But we may also questionize the isolated binary collision hypothesis. We may very well have a contribution of the long-range attractive part of the potential.

Since this work seems to be very promising , we are presently pursuing it.

References

1. C. Delalande and G. M. Gale, J. Chem. Phys., 71, 4804 (1979)
2. H. L. Chen and C. B. Moore, J. Chem. Phys., 54, 4072 (1971)
3. P. F. Zittel and C. B. Moore, J. Chem. Phys., 59, 6636 (1973)
4. A. Laubereau, L. Greiter and W. Kaiser, Appl. Phys. Lett., 25,87 (1974)
5. S. R. Leone and C. B. Moore, Chem. Phys. Lett., 19, 340 (1973)

Investigation of Subpicosecond Dephasing Processes
by Transient Spatial Parametric Effect in Resonant Media

T. Yajima, Y. Ishida, and Y. Taira

Institute for Solid State Physics, The University of Tokyo.
Roppongi, Minato-ku, Tokyo 106, Japan

The determination of dephasing time T_2 (or transverse relaxation time) as-sociated with excited states of materials is of importance for its signifi-cant influences on light-matter interaction. The values of T_2 are, however, very short generally in condensed matter, being often in a range of subpico-second and femtosecond range. Although these extremely short values of T_2 associated, in particular, with electronic excited states have hitherto been measured mostly by nonlinear spectroscopic methods in the frequency domain [1-5], it is also desirable to determine these values by observing the phe-nomena directly in the time domain to complement and compare with the results by the frequency-domain method. Conventional coherent transient methods are, however, hard to be applied at present for the relaxation study in this ex-treme time region due to the limitation of present laser performance.

We therefore proposed a transient nonlinear optical means based on a de-generate third-order mixing in $\vec{k}$-space called spatial parametric effect which provides always some information, on T_2 under a variety of experimental con-ditions [6]. Our previous study has given a basic theory and an experimental demonstration using picosecond (10^{-11} s) pulses from the second harmonic of a mode-locked Nd:YAG laser. This paper presents a new experimental result with short subpicosecond (0.2 - 0.3 ps) dye laser pulses and its interpre-tation based on a more advanced theory.

The basic feature of the nonlinear optical process considered here is as follows. Two noncollinear light pulses with the wave vectors $\vec{k}_1$, $\vec{k}_2$ and the same frequency ω are incident on an one-photon resonant material to produce new light components with the wave vectors $\vec{k}_3 = 2\vec{k}_1 - \vec{k}_2$ and $\vec{k}_4 = 2\vec{k}_2 - \vec{k}_1$ at ω through the third-order nonlinear effect. The correlation traces of this process, i.e., the time-integrated output energy of one of the light beams with $\vec{k}_3$ and $\vec{k}_4$ as a function of the time separation between two input pulses, then reflect the relaxation effect for sufficiently short pulses. A density matrix analysis based on a two-level atomic model incorporating phenomeno-logical relaxation times, T_1 (longitudinal) and T_2, showed that when t_p (pulse width) $\ll T_2$, the trace shows an exponential decay determined only by T_2, the decay rates being $2T_2^{-1}$ and $4T_2^{-1}$ for homogeneously and inhomogeneously broadened transitions, respectively [6]. When $t_p \gg T_2$, the correlation trace shows dominantly the characteristics of the incident light pulse rather than relax-ation effect, but even in this case, we can estimate an upper limit of T_2.

In general, the correlation trace becomes a convolution of the pulse shape and the relaxation effect, whose relevant theoretical result should be compared with the measured one in order to get accurate value of T_2 or its upper limit.

For a broadly inhomogeneous transition satisfying the condition, $\Delta\omega_i$ (inhomogeneous width) $\gg T_2^{-1}$ and t_p^{-1}, the theoretical auto-correlation trace for the $\vec{k}_3$ beam is given by the formula

$$I_3(\tau_d) = \int_{-\infty}^{\infty} S^2(t, \tau_d)dt$$

$$S(t, \tau_d) = KE_1^2E_2 \int_{-\infty}^{t'}\int_{-\infty}^{t''} \{F(t'- \tau_d)F(t''- \tau_d)F(t'+ t''- t)$$

$$\times \exp[(t''- t')/T_1 + 2(t'- t)/T_2]\}dt''dt' \qquad , \qquad (1)$$

where $I_3(\tau_d)$ is the output light energy, $\tau_d = t_1 - t_2$ is the separation between the centers of two input light pulses with the electric fields of peak amplitudes E_1, E_2 and of the same waveform $F(t)$. Eq.(1) can be derived from the more general formula given by (9) of [6]. (Note that in [6] $\vec{k}_3$ is defined as $\vec{k}_3 = 2\vec{k}_2- \vec{k}_1$ instead of $\vec{k}_3 = 2\vec{k}_1- \vec{k}_2$.) A result of computer calculation of (1) for a Gaussian input pulse and for $T_1 = T_2$ is shown in Fig.1. It is seen that the distorsion of the correlation trace due to the relaxation effect can be detectable for the condition $T_2 \gtrsim 2t_p$. Then $2t_p$ would be adopted as an upper limit of T_2 in case of undetectable distorsion.

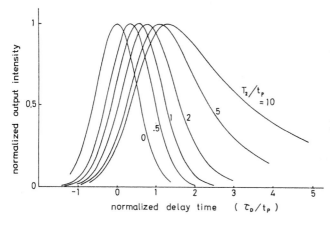

Fig.1
Calculated correlation traces for the spatial parametric effect showing the T_2 dependence. A Gaussian pulse of width t_p and a relation $T_1 = T_2$ have been assumed.

Figure 1 reveals also that there occurs a remarkable shift of the peak of the correlation trace, which is a more sensitive measure of the relaxation effect than the distorsion of the trace and therefore can effectively serve for the determination of T_2. This shift is generally dependent on both T_1 and T_2, and its calculated behavior is shown in Fig.2. If T_1 is known by other means, one can determine the value of T_2 from the shift even when $T_2 \ll t_p$. Even if T_1 is unknown we can know an upper limit of T_2 much shorter than t_p. T_1 is usually interpreted as the population relaxation time between the two levels constituting the optical transition. But in a material with very fast spectral cross relaxation, it would have to be replaced by an effective longitudinal relaxation time $T_1' = (T_1^{-1} + T_3^{-1})^{-1}$, where T_3 is the cross relaxation time, as infered from the theory of frequency-domain spectroscopy [1]. In this case, T_1' is much shorter than T_1 and becomes much close to T_2. The exact transient theory incorporating the cross relaxation

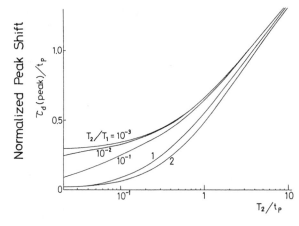

Fig.2
Fig.2
Calculated peak shift of
the correlation trace of
the spatial parametric
effect for a Gaussian
pulse showing the T_1 and
T_2 dependences.

effect is still left for further study. The shift behavior of the correlation trace is also slightly dependent on the pulse shape. However, this effect was found to provide no significant influence on the estimation of T_2 from the comparison between the cases of Gaussian and sech^2 pulses.

The transient spatial parametric effect considered here covers a wide range of phenomena continuously varying from usual nonlinear optical mixing to coherent transient process. The photon echo in the low field limit has also been included as a limiting case. For the purpose of relaxation study the present method serves as a flexible means in the sense that we can always observe an output light signal giving some information on T_2 for both homogeneous and inhomogeneous transitions, for both $t_p \ll T_2$ and $t_p \gg T_2$ and for both situations of input pulses separated and overlapped in time.

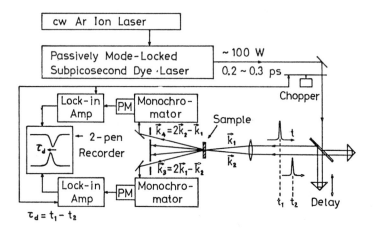

Fig.3 Experimental arrangement for the study of subpicosecond spatial parametric effect.

192

The whole experimental system for the present study is shown in Fig.3. The subpicosecond light source used is a passively mode-locked dye laser pumped by a cw argon ion laser. The laser consists of four cavity mirrors and a single dye jet stream of 0.2 mm thickness without any other dispersive elements. A mixed dye solution containing both the amplifying (rhodamine 6G, 2×10^{-3} M) and mode-locking (DODCI, 8×10^{-6} M) dyes in ethylene glycol was used. The laser system is similar to that used by Diels et al. [7], but laser parameters and performances are somewhat different. The laser produced a continuous train of nearly transform-limited pulses with a width of typically 0.2 - 0.3 ps and a peak power of about 100 W at around 6100 Å. With the careful adjustment of various laser parameters, we have obtained a shortest pulse of down to below 0.15 ps under the optimum condition.

The samples selected for relaxation study are several dyes in solution and in solid sheet with typical concentrations of $1 \sim 5 \times 10^{-4}$ M. The laser wavelength is within the width at nearly the half-maximum of the broad absorption band corresponding to the $S_0 \to S_1$ electronic transition of these dye molecules. Two input laser beams making an angle of $2°$ were focused by a lens of f=15 cm into a sample cell of 1 mm thickness. The detection system has been arranged to be capable of observing the two output beams with $\vec{k}_3$ and $\vec{k}_4$ at the same time for the reason described later. For the observation of output light signal, special care was taken for the spatial overlapping of the two beams at the sample and the elimination of stray scattering light component. The final correlation trace was obtained on a pen recorder with the use of a lock-in amplifier and a light chopper. In spite of the low power of the light source, fairly good signal to noise ratio was obtained due to the resonance enhancement effect, the efficient signal averaging and the high laser beam quality.

The measurements were first made for only the $\vec{k}_3$ beam. The observed correlation traces showed nearly symmetric character for all the samples examined, (i) DODCI in ethanol or ethylene glycol, (ii) brilliant green in ethanol, (iii) pinacyanol chloride in ethanol, and (iv) 1,3'-diethyl-2,2'-quinolythia carbocyanine iodide in solid vinyl-acetate. This implies that all the traces are pulse-width limited without detectable distorsion due to the relaxation effect. The pulse widths deduced from these third-order correlation traces (the correlation width divided by a factor 1.23 or 1.30 for Gaussian or $sech^2$ pulses, respectively) were consistent with those obtained by the conventional SHG correlation method. In the light of the above theory, the upper limit of T_2 estimated from only the shape of the correlation traces then becomes 0.5 ps for a typical pulse width of 0.25 ps.

In order to get the result of higher time resolution, we next tried to measure the peak shift of the correlation trace. This shift can conveniently be determined as the half value of the relative shift of the two peaks of the correlation traces associated with the $\vec{k}_3$ and $\vec{k}_4$ output beams measured simultaneously, because the two correlation traces produce the shifts of equal magnitude with opposite sign for the same direction of optical-delay scanning. Among the samples mentioned above, (i) and (iv) showed clear peak shift, but others did not show detectable shift. An example of the recorder traces exhibiting the shift is shown in Fig.4. In determining the shift, a residual small shift of instrumental origin has been eliminated carefully. From the measured shift and the theoretical result of Fig.2, the values of T_2 are estimated to be $\lesssim 0.05$ ps, $\lesssim 0.04$ ps and $\lesssim 0.01$ ps for the samples (i), (iv) and others, respectively. The individual comparison between these results and those by other means is not strictly capable at present because of the

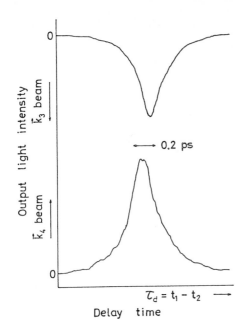

Fig.4
Measured correlation traces for the $\vec{k}_3$ and $\vec{k}_4$ beams produced by the spatial parametric effect exhibiting a relative delay of the two peaks. The sample is DODCI in ethanol.

unavoidable different conditions of samples and wavelengths. Nevertheless, it is notable that the present results indicate directly in the time domain the existence of very short T_2 far below 0.1 ps which has often been estimated from the frequency-domain measurements. In order to have full understanding of the observed results, it is required to proceed a more systematic experimental study and also a theoretical refinement, especially regarding the cross relaxation effect.

In summary, we have shown that the transient spatial parametric effect in resonant media is observable even with a low-power, short subpicosecond laser source. The shape and the shift of the correlation traces provide a new useful means of estimating the dephasing time T_2 of down to below 0.1 ps coupled with a theory developed here.

References

1. T. Yajima, H. Souma, Phys. Rev. *A17*, 309 (1978)
2. T. Yajima, H. Souma and Y. Ishida, Phys. Rev. *A17*, 1439 (1978)
3. J.J. Song, J.H. Lee and M.D. Levenson, Phys. Rev. *A17*, 1439 (1978)
4. H. Souma, Y. Taira and T. Yajima, *Picosecond Phenomena*, ed. by C.V. Shank, E.P. Ippen and S.L. Shapiro (Springer, Berlin, Heidelberg, New York, 1978) p. 224
5. H. Souma, T. Yajima and Y. Taira, J. Phys. Spc. Japan *48*, No. 6 (1980)
6. T. Yajima and Y. Taira, J. Phys. Soc. Japan *47*, 1620 (1979)
7. J.-C. Diels, E.W. Vyn Stryland and D. Gold, *Picosecond Phenomena*, ed. by C.V. Shank, E.P. Ippen and S.L. Shapiro (Springer, Berlin, Heidelberg, New York, 1978) p. 117

Optical Properties of Coherent Transients which are Stimulated by Picosecond Pulses at Three Frequencies

C.V. Heer

Department of Physics, Ohio State University,
Columbus, OH 43210, USA

1. Introduction

The coherent transient effects such as optical nutation, free induction decay, photon echoes, pulses formed by probe pulses, Carr-Purcell echoes,
phase conjugation, etc. can be expected to be generated by picosecond pulses.
Some of these effects have been observed already [1]. Many of these nonlinear effects are due to coherent space phase as well as time-phase information which is stored in the off diagonal density matrix elements and become observable via the electric polarization. In a somewhat more subtle
manner space phase information can be stored in the diagonal density matrix
elements and lead to phenomena such as forward or backward phase conjugation,
probe pulse effects, etc. Thus the decay of both the diagonal and off diagonal density matrix elements can be measured. This paper is primarily concerned with the optical properties of the waves which are generated by picosecond pulses at three frequencies ω_1, ω_2, ω_3 between energy states a, b, c,
d which are shown in Fig. 1. Three wave effects are considered since the sum
frequency is an allowed electric dipole transition for gases, isotropic liquids, and cubic solids.

2. Theory

Coherent transients with excitation by pulses at three frequencies can be easily studied by a method which was used in a recent paper [2]. A four-level
resonant interaction is considered and for simplicity the atomic or molecular

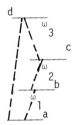

Fig.1 States a, b, c, d and the
near-resonant frequencies ω_1, ω_2,
ω_3 are defined in this figure.
ab, bc, cd, ad are electric dipole
transitions and ac, bd are electric
quadrupole transitions.

states are denoted by a, b, c, d. For an electric dipole interaction $V(t) =$
$- \vec{P} \cdot \vec{E}$ and the interaction in the rotating wave approximation can be written
as

$$\hbar^{-1}V(t) = |b)(a|v_1\, e^{i\Psi_1} + |c)(b|v_2\, e^{i\Psi_2} + |d)(c|v_3\, e^{i\Psi_3} + H.\ c. \tag{1}$$

195

where the eikonal phase $\Psi_n = \phi_n + (\dot{\phi}_n - \omega_n)t$ and $-\not{h}\, v_1 = (b|\vec{P}|a)\cdot\hat{u}_1\, E_1$, etc., relates the electric dipole matrix element and the electric field amplitude. The optical properties of the waves are included in $\phi_n(\vec{r}_\alpha)$ where $\vec{r}_\alpha$ is the molecular position. The gradient of the phase yields the wave vector at $\vec{r}$

$$\vec{k}_n = \text{grad } \phi_n \tag{2}$$

and for plane waves $\phi = \vec{k}\cdot\vec{r}$. If $E_n(t)$ describes the amplitude of each wave, the time evolution can be obtained readily for square pulses in the absence of damping. With these assumptions the time evolution of the density matrix is

$$\sigma(t) = U(t,t')\, \sigma(t')\, U^+(t,t') \tag{3}$$

Since the electric dipole moment $\vec{d}$ of the radiating molecule follows from $\vec{d} = \text{Tr } \vec{P}\, \sigma(t)$ only $\sigma(t)$ is needed in the discussion. Other pulse shapes and damping require a numerical solution. The assumption of square pulses and no damping permits a discussion of the essential features of the coherent effects.

The evolution is described by

$$U(t,t') = A(t)\, e^{-i\zeta(t-t')} A^+(t') \tag{4}$$

for a pulse which is described by the ζ matrix where a, b, c, d labels

$$\zeta = \begin{bmatrix} 0 & v_1^* & 0 & 0 \\ v_1 & \Delta_1 & v_2^* & 0 \\ 0 & v_2 & \Delta_2 & v_3^* \\ 0 & 0 & v_3 & \Delta_3 \end{bmatrix} \tag{5}$$

the rows and columns, and $\Delta_1 = \omega_{ba} - \omega_1 + \dot{\phi}_1$, etc. The operator $A(t) = \exp i\delta(t)$ where

$$\delta = \Psi_1\, |b)(b| + (\Psi_1 + \Psi_2)\, |c)(c| + (\Psi_1 + \Psi_2 + \Psi_3)\, |d)(d|$$

keeps track of the space and time phase in the matrix elements. The four roots λ_k of ζ in the following expansion completes the theory for a square pulse,

$$e^{i\zeta t} = \sum_k e^{i\lambda_k t} \prod_{k\neq\ell} (\zeta - \lambda_\ell)/(\lambda_k - \lambda_\ell) \tag{6}$$

and are discussed in detail in [2].

3. Optical Nutation

If the amplitudes of the three pulses are turned on simultaneously, then the density matrix element $\sigma_{da}(t)$ oscillates at the sum frequency $\omega_N = \omega_1 + \omega_2 + \omega_3$, σ_{db} at $\omega_2 + \omega_3$, σ_{ca} at $\omega_1 + \omega_2$, σ_{ba} at ω_1, etc. Although these may be equally large σ_{db} and σ_{ca} can only generate an electric dipole via $\vec{p}_e e^{i\vec{k}\cdot\vec{r}}$ and are characteristic of quadrupole transitions. The amplitude of σ_{da} etc. are determined by the roots λ_k and (4). The macroscopic electric polarization is determined by an average over the Δ_n and optical nutation can occur during the picosecond pulse. Since each molecule can have a different Δ_n, this is in a

196

sense an average over the λ_k in (6) and the exponential term is quite sensitive to this average. The wave generated at the sum frequency ω_N has a space phase which is given by $\phi_N = \Sigma \phi_n$. Phase matching is necessary in thick samples and for plane waves

$$\vec{k}_N = \vec{k}_1 + \vec{k}_2 + \vec{k}_3.$$

For physically thin samples the phases can be added and for spherical waves with radii of curvature z_n, the sum frequency curvature follows from

$$1/z_N = \Sigma \, (\omega_n/\omega_N)(1/z_n).$$

4. Free Induction Decay with Probe Pulse Waves

At the end of the picosecond pulse at time t' all matrix elements $\sigma(t')$ have non-zero values and information is stored until decay occurs. If the average over Δ_n of $\sigma_{da}(t')$ exp $-i\Delta_F(t-t')$ is non-zero, then free induction decay occurs. Even though this average can decay in a very short time interval the $\sigma_{da}(t')$ may decay much more slowly during the interval t"-t'. A probe pulse which is applied at time t" can be used to measure the decay during this interval. Thus a probe pulse ζ_S at time t" with frequencies ω_2 and ω_3 generates via $\sigma_{da}(t")$ a time dependent $\sigma_{ba}(t)$ which creates a new pulse at frequency ω_1 with a phase ϕ_1

$$\phi_1 = \phi_1^I + \phi_2^I + \phi_3^I - \phi_2^S - \phi_3^S$$

If the probe pulse is not collinear with the original pulse, this new wave occurs in a new direction. Even the decay of σ_{ca} becomes observable by the σ_{ca} (t") and oscillates at frequency ω_1 with a phase $\phi_1 = \phi_1^I + \phi_2^I - \phi_2^S$. Again if the pulse and probe at ω_2 are non-collinear plane waves, the new wave is in a different direction. If the original pulse is formed by plane waves and the probe pulse is a diverging wave, the radii of curvature are related by $1/z_1 = -(\omega_2/\omega_1)(1/z_2^S)$ when the indices of refraction are near unity. Thus the new wave comes to focus. The pulse duration is determined by an average over Δ_n and is similar to free induction decay.

5. Photon Echoes and Probe Pulse Waves

If two pulses are used which are described by matrices ζ^I and ζ^{II}, then at the end of the second pulse matrix elements like $\sigma_{da}(t')$ are non-zero. Radiation from the macroscopic system occurs for a time related to the average over the Δ_n. If the spread in Δ is sufficiently large, then at a time which corresponds to the interval between pulses, this average is no longer small and the macroscopic system radiates coherently and is the echo. The echo phase at the sum frequency follows from $\sigma_{da}(t')$ and is $\phi_E = \Sigma(2\phi_n^{II} - \phi_n^I)$. Echoes also occur at the single frequencies. Again focusing effects can occur. A probe pulse can be used to generate new waves and the new waves have there largest amplitude for a probe pulse at the time of the echo. Thus a probe pulse with frequencies ω_2 and ω_3 can generate from $\sigma_{da}(t')$ via $\sigma_{ba}(t)$ a new wave at frequency ω_1 with a phase which is determined by

$$\phi_1 = \Sigma(2\phi_n^{II} - \phi_n^I - \phi_n^S) + \phi_1^S.$$

$\sigma_{ca}(t')$ can be observed by a probe pulse at frequency ω_2. $\sigma_{ba}(t)$ then depends on $\sigma_{ca}(t')$ and a new wave is generated at frequency ω_1 with a phase which follows from the previous expression with n = 1,2. $\sigma_{db}(t')$ when stimulated by ω_3 generates a new wave at ω_2 via $\sigma_{cb}(t)$.

6. Phase Conjugate Waves

Information is stored in the off–diagonal density matrix elements and has been considered in the previous discussion. It is also possible to store information in the diagonal density matrix elements. This is apparent by returning to (1) and writing

$$v_1 = A + B\, e^{i\alpha}$$

Then the roots λ_k which occur in (6) depend on the product

$$v_1 v_1^* = A^2 + B^2 + 2AB\, \cos \alpha$$

and contain a term $\cos \alpha$ as a function of position $\vec{r}$. For plane waves $\alpha = (\vec{k}'' - \vec{k}') \cdot \vec{r}$. This spatial information is stored in the roots λ_k and occurs in the diagonal as well as the off-diagonal elements of the density matrices. During stimulation with a pulse or a pair of pulses the diagonal density matrix elements take on values $\sigma_{mm}(r,t')$ at the end of the pulse. A probe pulse can now be used to observe these diagonal matrix elements. The probe pulse generates a new wave off this phase grating and the optical features are similar to those observed in phase conjugation in either the forward or backward direction by four wave mixing.

6. Carr-Purcell Echoes

Since picosecond pulses can be produced as a sequence of pulses, Carr-Purcell photon echoes can be produced. If the first pulse has strength $\tfrac{1}{2}\pi$ and at time T later a sequence of pulses with period 2T are started, an echo will form at interval T after each pulse. If R_m is an operator which generates the m th echo, $R_{m-1} = U_0(t, t_{2m-1})\, U_m(t_{2m-1}, t_{2m-2})$, then the density matrix [3] for the single frequency aspect is given by

$$\sigma_{ab}(t) = (a|R_{m-1}|b)\; \sigma_{ba}(t_{2m-2})\; (a|R_{m-1}^+|b)$$

The anomalous optical and polarization effects become apparent in this expression. σ_{ab} evolves from $\sigma_{ba}(t_{2m-2})$ and the complex conjugate phase appears. The echo phase follows from $\phi_{E1} = 2\phi_2 - \phi_1$, $\phi_{E2} = 2\phi_3 - 2\phi_2 + \phi_1$, etc. If the sequence of pulses makes a small angle θ with the first pulse, the first echo occurs at $-\theta$, the second at $+\theta$, the third at $-\theta$, etc. If the first pulse is diverging and the sequence is a plane wave the first echo comes to focus, the second echo diverges, the third comes to focus, etc. Since σ_{ba} depends on the phase conjugate of the earlier pulse, random processes which introduce phase shifts in the time domain tend to cancel and the Carr-Purcell decay can be much slower than the single echo decay as the interval T is increased. These ideas can be extended to excitation at three frequencies.

7. References

1. W. H. Hesselink and D. A. Wiersma, Phys. Rev. Lett. 43 1991 (1979); S. C. Rand, A. Wokaun, R. G. DeVoe, and R. G. Brewer, Phys. Rev. Lett. 43 1868 (1979); D. E. Cooper, R. D. Wieting, and M. D. Fayer, (preprint); R. G. DeVoe and R. G. Brewer, Phys. Rev. Lett. 20 2449 (1979).

2. C. V. Heer and R. L. Sutherland, Phys. Rev. A19 2026 (1979). Some

essential features of this expansion (6) become apparent by considering resonance and making all $\Delta_n = 0$. The four roots follow from

$$\lambda_{\pm}^2 = \tfrac{1}{2} \, (v_1^* v_1 + v_2^* v_2 + v_3^* v_3) \pm [\tfrac{1}{4} \, (v_1 v_1^* + v_2 v_2^* + v_3 v_3^*)^2 - v_1^* v_1 v_3^* v_3]^{\tfrac{1}{2}}$$

Direct use of the expansion theorem (6) yields

$$e^{i\zeta t} = (\lambda_+^2 - \lambda_-^2)^{-1} \, [(\lambda_+^2 \cos\lambda_- t - \lambda_-^2 \cos\lambda_+ t) - i\zeta(\lambda_-^2 \lambda_+^{-1} \sin\lambda_+ t - \lambda_+^2 \lambda_-^{-1} \sin\lambda_- t)$$

$$+ \, \zeta^2 \, (\cos\lambda_+ t - \cos\lambda_- t) + i\zeta^3 \, (\lambda_+^{-1} \sin\lambda_+ t - \lambda_-^{-1} \sin\lambda_- t)],$$

where ζ^2 and ζ^3 follow from (5) by matrix multiplication. Thus the matrix
element from state d to a depends only on ζ^3,

$$(d| e^{i\zeta\tau}|a) = i \, (v_1 v_2 v_3) \, (\lambda_+^2 - \lambda_-^2)^{-1} \times [\lambda_+^{-1} \sin\lambda_+ \tau - \lambda_-^{-1} \sin\lambda_- \tau].$$

3. C. V. Heer, Phys. Rev. A13 1908 (1976).

Part V

Picosecond Chemical Processes

Picosecond Electron Relaxation and Photochemical Dynamics

G.A. Kenney-Wallace, L.A. Hunt, and K.L. Sala

Department of Chemistry, University of Toronto,
Toronto, M5S 1A1, Canada

Electron scattering, localization and subsequent optical excitation and electron transfer are processes that underlie a wide range of picosecond phenomena [1] which in
turn encompass the domains of quantum electronics, molecular physics, chemistry
and biology. In this paper we will focus on electronic transport in liquids and present
the first results pinpointing the identity of the sites for electron localization in polar
liquids, describe new results on isotope effects in a novel laser-induced electron
transfer mechanism involving strong electron-phonon coupling, and finally make some
brief remarks on short pulse measurements in tunable dye laser spectroscopy with
synchronously pumped laser systems.

First, consider some of the physical processes that accompany the injection of
a low energy electron, via laser-induced photoionization (Fig.1a), or photodetachment, or direct ionization with a ps pulsed e^--beam, into three different classes of
materials, namely crystals, amorphous semiconductors and liquids. The usual density of states diagram for intrinsic semiconductors can be translated into a simple
energy level diagram (Fig.1b), illustrating the conduction band, or continuum of extended states, and the valence band, or deeply localized states. At subpicosecond
times, frozen into the liquid is a significant degree of short-range order. Microscopic
regions of the liquid resemble a lattice (supporting only phonon propagation) superimposed on the longer range, random molecular network typical of an amorphous semiconductor. Carriers or hot quasifree electrons (e_{qf}^-) relaxing through the conduction

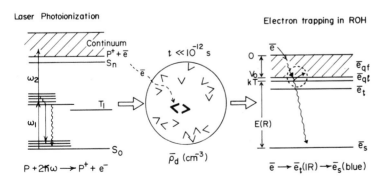

Fig.1   Photoionization and electron trapping sequence.

band by scattering through the liquid or amorphous solid, exhibit a high mobility ($\mu \geq 1$ cm^{2}V^{-1}s^{-1}) which drops by orders of magnitude upon localization. Bound by 1-2 eV, trapped electrons (e_t^-) exhibit intense optical absorption spectra [2]. Quasi-localized states (e_{ql}^-) close to the mobility edge (V_0) are within kT of the continuum states and provide temporary periods of localization for the fast carriers. Transitions between these extended and localized electron states are accompanied by radiative and nonradiative processes, and, along with electron localization mechanisms, have become the subject of intense current study for both fundamental and applied reasons. The emission from electrons in alkali halide crystals forms the basis of mode-locked, colour centre lasers [3], the high mobility of carriers in a-Si is the basis for ps electronics [4], while electrons in liquids display potential as both high mobility, short lifetime carriers and ultrafast saturable absorbers.

1. Electron Localization

The cluster model of electron solvation in polar liquids [2,5] has proposed a two-step process: (1) trapping of the excess electron by a preexisting site and (2) local medium reorganization and growth of the cluster of molecules to give a configuration-ally relaxed e_s^-. The latter step is characterized by prominent spectral shifts to the visible as the IR absorbing trapped electrons acquire a larger cluster of molecules and an increased binding energy. In n-ROH step 2 is correlated to molecular rotation.

But what of the former localization step? We had previously proposed [5] that these sites were configurational and density fluctuations frozen-in the molecular liquid on the subpicosecond time scale of multiple electron scattering events [6]. Furthermore, hydrogen-bonded dimers, trimers and tetramers could be the sites in alcohols (ROH).

Consider a unit volume element V in the liquid (centre, Fig.1) in which the average number density of OH dipoles is given by $\bar{\rho}_d$, cm^{-3}. In pure n-ROH, $\bar{\rho}_d$ ranges from 1.5×10^{22} to 3×10^{21} on going from methanol to decanol. The probability of a given alcohol multimer M_n being present in V is proportional to ρ_d, the intermolecular potentials $U_{ij}(r)$ and the equilibrium constant for nM $\rightleftharpoons M_n$. By diluting the system such that ρ_d is decreased by $\sim 10^3$, the probability of locating a species M_n in V becomes vanishingly small for $n > 1$. Since the transient negative ion ROH$^-$ is not observed, only the translational diffusion of ROH molecules to form M_n^- can provide a mechanism whereby e$^-$ stabilization in alcohol sites can occur. If this diffusion is slow compared to ion recombination, stable trapped electron species cannot be formed, hence no increase in the ps optical absorption of the system will be seen.

The single pulse, stroboscopic ps e-beam and sampling techniques were employed with a time resolution of 10 ps: full details of these and the ft NMR ^{13}C spin lattice (T_1) relaxation times have appeared elsewhere [5,7].

Figure 2a illustrates the amplitude of the e_t^- absorption signals for $\bar{\rho}_d = 5 \times 10^{20}$ cm^{-3} n-butanol (C_4) diluted in cyclohexane (top trace), n-hexane (centre) and iso-octane (lower trace). Note that the kinetic behaviour is quite different, manifesting growth and spectral shifts attributable to further solvation processes of e_t^- by ROH in cyclohexane and (to a lesser extent) in n-hexane, but only decay of e_t^- in isooctane, implying that ion recombination is predominant. These differences arise from

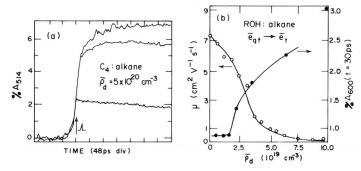

Fig.2   Electron absorptions in ROH:alkane.

variations in the local molecular structures of the fluid, and variations in V_0 and μ, or the electron-alkane scattering potential.

When the amplitude of the initial absorption for C_4:n-hexane against $\bar{\rho}_d$ in Fig. 2b we observe a critical value of $\bar{\rho}_d$ above which the absorption increases and the kinetics of the e_t^- signal move from decay to growth ($e_t^- \to e_s^-$). The sudden increase in absorption is in a mirror image relationship to the sudden loss of carrier mobility, shown here for propanol:isooctane [8].

Measurements of the T_1 (s) values of the carbon-atom mobility in the alcohol chains over a wide range of $\bar{\rho}_d$ identified the critical $\bar{\rho}_d$ as an alcohol "nucleation" threshold. In the extreme narrowing limit, the longer the T_1 time, the faster the rotational correlation time of the atom [7]. Hydrogen-bonding, via the OH, anchors the motion of the adjacent ^{13}C atom and thus a study of the T_1 times of isotopically* labelled $CH_3(CH_2)_n{}^*CH_2OH$ as a function of $\bar{\rho}_d$ can follow the onset of the formation of hydrogen bonded clusters ($nM \rightleftharpoons M_n$).

When the relaxation rates T_1^{-1} (s^{-1}) are plotted in Fig.3 against χ (mole fraction of ROH) and $\bar{\rho}_d$ for two alcohols in two alkanes, the T_1^{-1} values appear relatively constant until for C_4, as $\bar{\rho}_d$ approaches 3×10^{19} cm^{-3}, a clear change (I) in slope is seen. This occurs at $\sim 10^{20}$ for t-butanol (t-C_4), in n-hexane, and a further change in slope (II) is seen at $\bar{\rho}_d \sim 10^{21}$ cm^{-3}. From these nmr and other data [7], we can assign I to the onset of dimer formation in C_4 and cyclic trimer formation in t-C_4. By $\bar{\rho}_d \sim 10^{20}$ cm^{-3}, the equilibrium $2M_2 \rightleftharpoons M_4$ is established for C_4 and subsequent increases in $\bar{\rho}_d$ lead to larger clusters and a broad distribution of cluster sizes. In t-C_4, the $3M \rightleftharpoons M_3$ equilibrium dominates until at II, the population density of n-mers (where $n > 3$) becomes the dominant feature. Comparison of the ps optical data with the T_1 data is striking. The absorbance $A(\chi)$ normalized to the $A(\chi = 1)$ value reveals exactly the same profile (I,II) for each alcohol. The optical data for C_4 in cyclohexane are plotted in lieu of n-hexane, to illustrate how even the factor of ~ 3 observed spectroscopically in the threshold $\bar{\rho}_d$ for e^- trapping by ROH in the two alkanes appears in the intercepts of the NMR data too.

In summary, we have shown for the first time that (1) the preexisting M_n electron trap in alcohol:alkane systems is a dimer in n-butanol and a cyclic trimer in t-butanol, and (2) the critical $\bar{\rho}_d$ at which such sites become availabe depends on

205

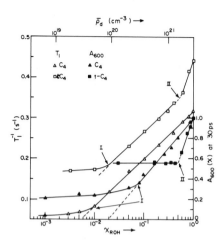

$\bar{\rho}_d$ (cm^{-3}) →

Fig.3 Comparison of NMR (T_1) and optical (A) data

$U_{ij}(r)$, V_0 and μ . The observation of a critical $\bar{\rho}_d$ value leads us further to suggest that a minimum amplitude in the configurational and density fluctuations is required to localize e_{qf}^- during the electron-medium scattering interactions [9] . This would predict that in high $\bar{\rho}_d$ (> 10^{22} cm^{-3}) pure fluids, the frequency of such useful fluctuations would be accordingly higher and thus electron localization extremely fast and efficient. Since most high $\bar{\rho}_d$ liquids contain molecules with more than one OH dipole/molecule, these liquids also support a significant degree of three dimensional H-bonding networks, which provide large clusters and cages for which rotational reorganization about e_t^- would be minimal. Induced molecular rotation would play a lesser role and we would predict ultrafast electron localization and solvation times (τ_s), in H_2O, diols and ammonia, NH_3 . Recent data support these predictions where in water, ρ_d = 3.3 x 10^{22} cm^{-3}, τ_s <5 x 10^{-13} s [10] : in ethylene glycol (EG) $\bar{\rho}_d$ = 2.2 x 10^{22} cm^{-3}, τ_s < 5 ps [5] : and in liquid ammonia $\bar{\rho}_d$ = 2.5 x 10^{22} cm^{-3}, τ_s < 5 ps [11] . Of pertinence too is a recent stydy [12] of carrier mobility in ammonia vapour which indicated a critical density $\rho \sim 10^{20}$ cm^{-3}, above which μ dropped a further 10^3 by ρ = 3.6 x 10^{21} cm^{-3}, indicating as predicted theoretically [6] electron localization had occurred in preexisting clusters in the ammonia vapour.

2. Laser-Induced Electron Transfer

We wish to report here a novel, laser-induced electron transfer mechanism proceeding through vibrationally selective, photodissociation of the acceptor molecule to give a stable negative ion fragment and a neutral atom. The donor species were optically excited electrons in alcohol clusters.   The technique involves "hole-burning" experiments to obtain picosecond relaxation times. First reported by us for e_{aq}^- in 1971, the subsequent theory and experiment have built on this approach and is reviewed in ref. 2.

In our current experiments, e_s^- are first generated in a two-photon photoionization of pyrene in ROH (Fig.1a), then subjected to a laser saturation pulse at a time delay τ_d, while the e_s^- absorption recovery is probed at times (τ_d + t) with a third laser pulse or a cw laser probe beam. By tuning the frequency (ω) of the second photon from just below (S_n) to well above the continuum, we are also attempting to observe directly the influence of the escape kinetic energy of e_{qf}^- on the electron solva-

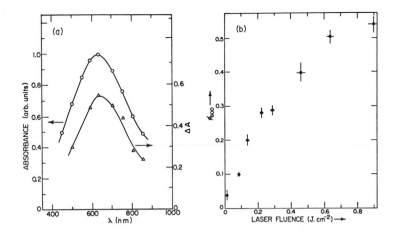

Fig.4   Laser photobleaching of e_s^- in ROH [9].

tion dynamics and on electron capture rates of e_{qf}^- by acceptor molecules. (We recently have observed field-induced ionization from high lying Rydberg-like states (S_n), where the lifetime of the state could possibly prolong the actual solvation process [13].) The intrinsic ps time resolution of this experiment is calibrated by hole-burning experiments in dyes such as DDI and cryptocyanine in methanol and has been described in detail elsewhere [9]. Figure 4a shows the first study reported of the full spectrum seen after hole-burning experiments in e_s^- in methanol [9]. Following ruby laser saturation at 694 nm, the absorption band was uniformly photobleached, within a time resolution of 10 ps, but at high laser fluxes (> 50 mJ·cm^{-2}) full recovery of the absorption band did not occur. The higher the laser flux, the larger the fractional e_s^- loss (ϕ) as Fig.4b shows, "fluence" being used to infer the possibility of multiphoton processes. Evidence led us to conclude that at high laser fluxes, the electronically excited e_s^{-*} had access to a highly reactive photochemical channel, which competed effectively with vibronic relaxation (~10 ps) that normally followed low laser intensity excitation. Preliminary isotope studies indicated that the OH vibration in the ROH molecules was probably involved. We describe now the latest isotope studies which reveal new aspects of this photochemical channel or laser-induced electron transfer (LIET).

Laser saturation studies were carried out at 694 nm (ruby) and 530 nm (Nd:YAG) on a series of e_s^- formed in deuterated alcohols and ROH/ROD mixtures. The permanent fractional loss of e_s^- (ϕ) follows the absorption cross-section of e_s^- and thus we will report data for the loss monitored at 633 nm (He-Ne) as representative of the whole band. Table 1 shows the isotope effects on ϕ (quoted to ± 0.05) for a given laser fluence for e_s^- in methanol at 0.35 J·cm^{-2}, in ethanol and EG at 0.44 J·cm^{-2} and in water at intensities up to 1 J·cm^{-2}. Since the photon:electron ratio was higher at 530 nm and the e_s^- absorption ~16% lower, the value of ϕ for a comparable incident laser fluence is still less at 530 nm than at 694 nm. A multipass cell was used in the Nd: YAG experiments in order to increase signal-to-noise, so there is a greater uncertainty on the effective laser fluence.

Table 1 Isotope Effects in LIET of e_s^- in ROH

Cluster	CH_3OH	CH_3OD	CD_3OH	CH_3CH_2OH	CH_3CH_2OD	EG	H_2O
ϕ_{694}	0.30	0.04	0.32	0.6	0.2	0.35	0
ϕ_{530}	0.17	0.06	0.16				0

The dramatic reduction in ϕ when the OH moiety of the molecule was deuterated clearly pinpoints the LIET mechanism as one involving the OH vibration of the alcohol molecules. In other experiments we had ruled out picosecond processes such as geminate ion recombination, associative attachment to pyrene, capture by impurity electron acceptors and photodetachment of the electron from the cluster and resolvation, an event governed by rotational relaxation in methanol, where $\tau_s = 11$ ps. Although the "empty" cluster might survive for $\sim 10^{-11}$ s (until rotational relaxation destroyed the molecular correlations) and thus act as an attractive, configurationally prepared, preexisting trap for e_{qf}^-, experiments with thermal electron acceptor molecules ruled this possibility out.

The only remaining mechanism was dissociative attachment through vibration to form the RO^- stable negative ion and H atoms. If so, which ROH molecules were involved? We prepared a series of mixed ROH/ROD clusters and followed ϕ as a function of the degree of deuteration in the cluster, assuming that the intermolecular forces between ROH and ROD were not too dissimilar and thus $\chi_{ROD} = 0.5$ implied an equimolecular mixture of ROH and ROD in the cluster hosting the excess electron. Figure 5a shows the results of laser photobleaching in the ethanol system. Only for $\chi_{ROD} \geq 0.5$ does ϕ show a significant dependence on deuteration. Similar evidence for selective solvation was seen in the ϕ dependence for $\chi_{ROH} \geq 0.05$ in ethanol: water mixtures.

These results demonstrate unequivocally that LIET is indeed an intracluster process during which, under high laser (ω) intensity excitation of e_s^-, vibrationally selective photodissociation and electron transfer occurs from e_s^{-*} in competition with vibronic relaxation via multiphonon processes within the cluster, to e_s^-.

$$e_s^- + nh\nu \rightarrow e_s^{-*} \rightarrow RO^- + H + (n-1)ROH$$

This is illustrated schematically in Fig.5b. We speculate that the symmetric stretching vibration of OH, at 3667 cm^{-1} in multimers [14] may be more likely to undergo rapid configurational deformation coupled to e_s^{-*} field than the lower frequency bending mode at 1335 cm^{-1}. One particular tantalising thought is that overtones of these OH stretching vibrations must be in the visible, and we predict the 5th overtone to be close to 694 nm, and the 6th to 530 nm. Could this be resonantly enhanced, vibrational photodissociation in the ROH? We are now pursuing this aspect of the problem in photoacoustic spectroscopy in order to elucidate more details of the electron–phonon coupling.

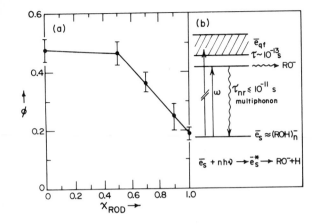

Fig. 5   (a) Isotope effects on ϕ in ROH/ROD clusters; (b) model for LIET.

Finally, we may also speculate on the apparent saturation in the fluence dependence.
If RO^- is indeed formed, then electron photodetachment of RO^- could be a source of
new e_s^- which would only be observed as <u>net</u> absorption recovery. Laser photodetachment experiments of RO^- indicate that the gas phase threshold is ~1.6 eV [15]. The
polarization energy of an ion in methanol, calculated via the Born approximation, increases this to ~3.9 eV. Perhaps coincidently, two ruby photons would provide $2\hbar\omega$ =
3.56 eV. Thus it is plausible, but still speculative, that the reason for saturation in
the ϕ dependence is that a new channel for the photogeneration of electrons opens up
at laser intensities, where two photon excitation of RO^- becomes significant. We are
now studying the profile of ϕ at other laser saturation frequencies and other alkoxide
ions with varying electron detachment thresholds, to test these ideas, employing subpicosecond, tunable dye laser spectroscopy.

3.   <u>Picosecond Pulse Diagnostics with CW Autocorrelator</u>

It is essential to know the temporal and spatial intensity profiles of the laser
pump and probe pulses, particularly when the relaxation dynamics to be inferred are
those of a molecular state prepared as a consequence of one or more nonlinear processes through virtual or real states. While autocorrelation measurements have
long been the mainstay of ps pulse diagnostics [1], only recently have "real time"
techniques emerged [16] to permit continuous display of laser pulses. We report
here an extension of our earlier work [17] on CW background free picosecond autocorrelation measurements, and report for the first time a "stereo autocorrelator"
device, operating at audiofrequencies and based on a double-speaker configuration in
lieu of the more conventional and slowly translating interferometer arms.

The CW "push-pull" autocorrelator employs two identical (and, in this case, disco!)
audio speakers as illustrated in Fig.6. The speakers are driven exactly 180° out-
of-phase from a common sine-wave source (SWG) and thus produce a net variation in
the relative difference between the interferometer arms which is <u>double</u> the total
displacement or "stroke" of one of the speakers. Essentially the speakers are util-

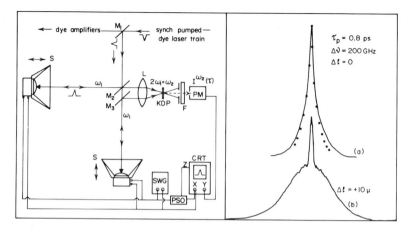

Fig.6   "Stereo" autocorrelator and autocorrelation traces of laser pulses.

ized as inexpensive transducers and permit this autocorrelation measurement to be
made rapidly and at audio frequencies (15-30 Hz) to allow a very convenient, CW display of the ps dye laser pulses. The sensitivity of the technique using one speaker,
and the significance of precise cavity matching in synchronous pumping, is illustrated
by comparison of the subpicosecond laser ($\tau = 0.8$ ps) pulse in Fig.6a with the partially
mode-locked pulse in (b) when the cavity length l is 10 μ too long. The dots in (a)
refer to an $\exp(-|\tau|/T)$ fit to a single-sided exp pulse with $\tau_p = 0.8$ ps. In general,
an asymmetric, two-sided exp pulse gives closer fits [17]. This autocorrelator in
particular offers a number of valuable and practical features. First, it has a large
dynamic range (≤ 300 ps) not accessible by other means and permitting CW autocorrelation measurements of mode-locked Ar ion laser pulses, and optical mixing of
unknown pulses with known reference pulses. Secondly, it has a continuously variable
range (by simply varying the sine-wave driving voltage) in effect allowing the user to
readily vary the "dispersion" of the displayed autocorrelation trace. Finally, this
technique offers a high degree of linearity. Any asymmetric nonlinearity in the speaker
displacement vs. instantaneous voltage behaviour is only noticeable when the speaker
is driven at near maximum stroke. However, because of the "push-pull" opposing
nature of the speaker motions, these asymmetric nonlinearities are very nearly exactly cancelled in the double-speaker autocorrelator with the result that the device
is surprisingly linear, even as maximum dynamic range is approached.

We gratefully acknowledge the contributions of Dr. G. E. Hall and Dr. C. D.
Jonah to this work and the financial support of the Connaught Foundation.

References

1 C.V. Shank, E.P. Ippen and S.L. Shapiro, eds., <u>Picosecond Phenomena</u>, Springer Series in Chemical Physics, Vol. 4 (Springer, Berlin, Heidelberg, New York, 1978)

2 G. A. Kenney-Wallace, Adv. Chem. Phys. $\underset{\sim}{47}$, 000 (1980).

3 L. F. Mollenauer, D. Bloom and A. del Gaudio, Optics Lett. $\underset{\sim}{3}$, 48 (1978).

4 D. Auston, P. Lavallard, N. Sol and D. Kaplan, Appl. Phys. Lett. $\underset{\sim}{36}$, 66 (1980).

5 G. A. Kenney-Wallace and C. D. Jonah, Chem. Phys. Lett. $\underset{\sim}{47}$, 362 (1977); ref. 1, p.208; and in press.

6 A. Gaathon and J. Jortner, Can. J. Chem. $\underset{\sim}{55}$, 1812 (1977).

7 V. Gibb, P. Dais, G. A. Kenney-Wallace and W. F. Reynolds, Chem. Phys. $\underset{\sim}{47}$, 407 (1980).

8 J. H. Baxendale, Can. J. Chem. $\underset{\sim}{55}$, 1996 (1977), and references therein.

9 G. A. Kenney-Wallace, L. A. Hunt, G. E. Hall and K. Sarantidis, J. Phys. Chem. $\underset{\sim}{84}$, 1145 (1980), and references therein.

10 J. Wiesenfeld and E. Ippen, Chem. Phys. Lett. in press.

11 J. Belloni, M. Clerc, P. Goujon and E. Saito, J. Phys. Chem. $\underset{\sim}{79}$, 2848 (1975); D. Huppert, P. M. Rentzepis and W. Struve, ibid., $\underset{\sim}{79}$, 2850 (1975).

12 P. Krebs and M. Wantschik, J. Phys. Chem. $\underset{\sim}{84}$, 1155 (1980).

13 S. C. Wallace, G. E. Hall and G. A. Kenney-Wallace, Chem. Phys. $\underset{\sim}{48}$, 000 (1980).

14 A. J. Barnes and H. E. Hallam, Trans. Faraday Soc. $\underset{\sim}{66}$, 1920 (1970), and references therein.

15 P. Engelking, G. E. Ellisen and W. C. Lineberger, J. Chem. Phys. $\underset{\sim}{69}$, 1826 (1978).

16 R. L. Fork and F. A. Beisser, Appl. Opt. $\underset{\sim}{17}$, 3534 (1978).

17 K. L. Sala, G. A. Kenney-Wallace and G. E. Hall, IEEE J. Quantum Electron. in press.

Picosecond Laser Photoionization in Polar Liquids and the Study of the Electron Solvation Process

J.C. Mialocq, J. Sutton, and P. Goujon

CEN-Saclay, Département de Physico-Chimie, B.P. n°2,
F-91190 Gif-sur-Yvette, France

Summary

The study of electrons in liquids and electron-solvent interactions has long been an important subject of theoretical and experimental work [1] . Recently the use of ultra rapid techniques for injecting electrons into liquids (relativistic electron beams in pulse radiolysis and laser photoionization of molecules and negative ions) coupled with rapid absorption spectroscopy has led to exciting new developments in the picosecond time range [2] .

Photoionization of halide anions which possess charge transfer to solvent spectra (CTTS) must be very fast and the solvated electron ($\bar{e}_s$) formation kinetics must be controlled by the solvation process itself. Using their picosecond echelon technique, RENTZEPIS et al. [3] have discussed such a mechanism in the laser photoionization of the ferrocyanide anion in aqueous solution at 265 nm. Electron solvation in liquid alcohols has been investigated by picosecond pulse radiolysis and discussed in terms of the Stokes-Einstein - Debye representation [2] , but temperature increases are to be expected in the spur after the deposition of radiation and the solvation process might be made faster. Picosecond laser photoionization seems convenient to avoid such effects but nothing has been done in alcohol solutions, as far as we know. Because in alcohol solution, the ferrocyanide anion is insoluble and the absorption spectra of the halide ions are blue-shifted with respect to the available 265 nm actinic light, we decided to study the photoionization of aromatic molecules and ions, for which a detailed description of the mechanism does not exist, even though the formation of a "semi-ionized", intermediate CTTS state or a "solvated Rydberg state" has long been discussed [4] .

In this paper we report photoionization of phenolate and phenol in alcohol solution, using a single exciting pulse of 27 ps duration at 265 nm. Solvated electron absorption was probed either with an analysing pulse at 630 nm obtained from the residual 530 nm pulse by stimulated Raman scattering (SRS) in a 2 cm cell containing dimethyl-sulfoxide [5] or with a continuum pulse generated by the 1064 nm fundamental in a 4 cm water cell (WC in Fig. 1).

I.   Electron Ejection Mechanisms in Aqueous Solution

Picosecond laser-induced photoionization of ferrocyanide, phenolate and phenol in aqueous solution has been discussed in great detail elsewhere [5] . The kinetics of the formation of the hydrated electron ($\bar{e}_{aq}$) absorption were measured at 630 nm where its molar extinction coefficient is ε_{630} = 15700 $M^{-1}cm^{-1}$. In order to verify that the absorption at 630 nm was due only to $\bar{e}_{aq}$, specific scavengers of this species were added to the solutions in known concentrations and the kinetics of its disappearance were measured after the end of the exciting pulse [5] The rate constants of the reactions ($\bar{e}_{aq}$ + S → S⁻) have been discussed in terms of the ionic strength effect, the relaxation time of the ionic atmosphere of $\bar{e}_{aq}$ and the time dependence of rate constants.

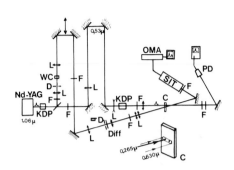

Fig. 1   Double beam picosecond
          spectrometer.

The kinetics of formation of $\bar{e}_{aq}$ are the same in the case of the ferrocyanide and phenolate anions [5] . The absorption spectrum of the ferrocyanide ion shows the presence of a CTTS band near 265 nm but that of phenolate involves a Π-Π* transition. The plateaux-values of the $\bar{e}_{aq}$ absorption obtained are in the ratio of the quantum yields of the $\bar{e}_{aq}$ formation as measured in steady—state photolysis with the N_2O scavenging technique. We think that in this laser experiment the photoionization process is mainly monophotonic.

In solutions of the phenol molecule, it must be emphasized that the electron ejection process is much shorter than the lifetime of the phenol excited state S_1, indicating that electron ejection arises from a non-relaxed excited state [5] . Moreover the $\bar{e}_{aq}$ formation is one order of magnitude higher than in steady state photolysis and a delay is observed in the appearance of the $\bar{e}_{aq}$ absorption. We have concluded that a consecutive two—photon process occurs where the second photon is absorbed by the excited singlet S_1, present in sufficient concentration shortly after inception of the exciting flash to compete for the UV absorption.

II.   Photoionization and Electron Solvation Process in Alcohols

The rate of appearance of the $\bar{e}_s$ absorption at 630 nm was compared in three different solvents : water, n-butanol and n-decanol (Fig. 2). The results show no delay in the more viscous solvents and a single component in the increase of the absorption signal with time. There is no slow component as observed in picosecond pulse radiolysis experiments [6, 7] , attributed to the solvation of initially "dry electrons" and correlated to the time for solvent monomer rotation. Moreover, similar kinetics (Fig. 3) were obtained in n-decanol at the two wavelengths 542 and 750 nm where the molar extinction coefficients of $\bar{e}_s$ are identical [8] . The presence of "dry electrons" would induce a change at the longer wavelength. We are unable at the pre-

sent time to follow the decay kinetics of the initial spectrum,
if any, in the near infrared, which might disappear as the visible $\bar{e}_s$ spectrum grows in.

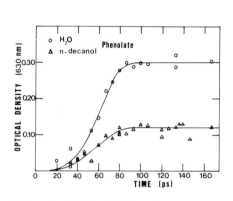

Fig. 2   Optical density, as a function of delay time.

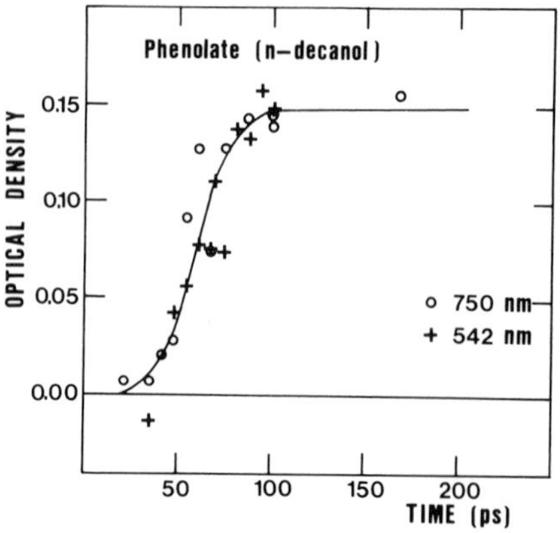

Fig. 3   Optical density, as a function of delay time.

From the plateaux-values of the $\bar{e}_s$ optical density at 630
nm, the already known extinction coefficients [9] and the absorbed laser energy, we deduce that the relative quantum yields
of $\bar{e}_s$ formation (normalized to the value measured in water)
are respectively 1.0, 0.67 and 0.43 in water, n-butanol and
n-decanol.

We suggest that in these photoionization experiments an electron is ejected from a charge transfer to solvent excited state
of the phenolate ion into a preexisting trapping site. However, we plan further experiments to investigate the presence
of initially localized electrons absorbing in the infrared.

References

1. L. Kevan, B.C. Webster (eds.): *Electron Solvent, Anion-Sol-
    vent Interactions*, Elsevier, Amsterdam (1976)
2. G.A. Kenney-Wallace, *Picosecond Phenomena*, ed. by C.V. Shank,
    E.P. Ippen, S.L. Shapiro, Springer Series in Chemical Physics,
    Vol. *4* (Springer, Berlin, Heidelberg, New York, 1978) 208
3. P.M. Rentzepis, R.P. Jones, J. Jortner, Chem. Phys. Lett.
    15, 480 (1972); and J. Chem. Phys. *59*, 766 (1973)
4. M. Ottolenghi, Che. Phys. Lett. *12*, 339 (1971)
5. J.C. Mialocq, J. Sutton, P. Goujon, J. Chem. Phys. (in press)
6. W.J. Chase, J.W. Hunt, J. Phys. Chem. *79*, 2835 (1975)
7. G.A. Kenney-Wallace, C.D. Jonah, Chem. Phys. Lett. *39*, 596
    (1976)
8. R.R. Hentz, G.A. Kenney-Wallace, J. Phys. Chem. *78*, 514
    (1974)
9. G.E. Hall, G.A. Kenney-Wallace, Chem. Phys. *32*, 313 (1978)

Dynamical Evidence for Preferential Structure in Electron Photoejection from Arylaminonaphthalene Sulfonates*

R.A. Auerbach[1], J.A. Synowiec, and G.W. Robinson

Department of Chemistry, Texas Tech University,
Lubbock, TA 79409, USA

1. Introduction

On variation of the solvent, arylaminonaphthalene sulfonates such as 2,6 toluidinyl naphthalene sulfonate (TNS) may exhibit enormous changes in their fluorescence properties while showing little or no change in their absorption spectra [1-3]. In particular, for TNS an extreme increase of two orders of magnitude in the nonradiative rate (k_{nr}) and a redshift of the emission maximum of several thousand wavenumbers is observed with water as a solvent as compared to other common solvents, such as simple alcohols. Previous work has shown that this large increase in nonradiative rate constant is accompanied by the appearance of solvated electron absorption [4]. Linear behavior of electron absorption with incident intensity, arguing against solvated electron production as an artifact of a biphotonic process, leads to the conclusion that electron photoejection is the primary process responsible for the high nonradiative rate in water [5]. Other workers [6] have shown that electron solvation is highly dependent on local solvent configuration. Electrons are thought to be solvated by initially being trapped in already existent molecular solvent clusters, which further relax to produce the typical solvated electron absorption [7]. Data in this paper suggest that, in the excited state, TNS in water has a preexisting local solvent structure that is nearly ideal for the electron transfer. Evidence based on the dynamics of solvent exchange in mixed solvent systems indicates that this structure consists of approximately twelve water molecules.

2. Experimental

Quantum yields and fluorescence spectra were measured with a Perkin Elmer MPF-44B spectrofluorimeter with a DCSU-2 correction attachment. Time-resolved fluorescence emission was measured using the 4th harmonic (263.4nm) of a mode locked neodymium/phosphate-glass laser for excitation, together with an Electo-Photonics Photochron II streak camera. The output of the streak camera is coupled to a PAR optical multichannel analyzer through an EMI image intensifier. Coaxial detection geometry was used, and polarization bias was eliminated by using an analyzing polarizer set at 54.7° to the polarization direction of the exciting beam. Data were analyzed on a PDP 11/34 computer with a nonlinear least squares program that deconvolutes the instrumental response of a laser pulse. A more complete description of the apparatus has been pre-

---

*This work was supported in part by the Robert A. Welch Foundation [K-099(e)] and the National Institutes of Health [GM-23765].

[1] Present address: Department of Chemistry, Tulane University, New Orleans, LA 70118, USA

sented elsewhere [8,9]. The TNS samples were obtained from Eastman Kodak Company and concentrations used were approximately 10^{-4}M.

3. Results

A comparison of properties of TNS in water and ethanol is presented in Table 1. In mixtures of these two solvents a nonlinear behavior occurs for k_{nr} (see Fig.1). Clearly, a great deal of water is necessary in the mixture for a major increase in nonradiative rate to occur. Fig. 2 illustrates the dynamical decay in a given mixed solvent as a function of emission wavelength. The overall emission has a roughly exponential decay. However, the blue edge of the band decays faster, and the far red-edge emission does not immediately

Table 1 Emission properties of TNS in water and ethanol at 18°C

	Emission Maximum [cm^{-1}]	Quantum Yield	Fluorescence Lifetime [ps]
Ethanol	23310	0.51*	8700*
Water	18980	0.0004	60

*Deoxygenated solutions. Quantum yield standard = quinine sulfate (H_2SO_4).

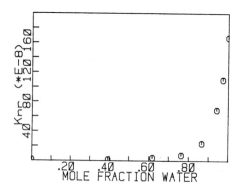

Fig.1 Nonradiative rate constant of TNS *vs* water mole fraction. Lifetimes were determined from observation of the entire emission band, assuming exponential decay. Quantum yields were determined from integrated emission intensity and compared with TNS in ethanol as the standard.

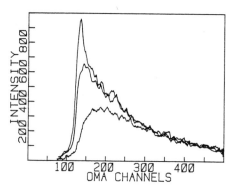

Fig.2 Emission decay of TNS in ethanol/water as a function of emission wavelength. Water mole fraction = 0.62. Top curve: 400-460nm. Middle curve: entire band. Lower curve: >570nm.

occur but shows initial growth before the decay. In such mixtures, there is an overall shift of the emission maximum to the red with time. This is indicative of solvent exchange occurring in the excited state, with water molecules replacing ethanol in the near neighbor environment. For extreme red-edge wavelengths (>570nm), where TNS in ethanol has no observable emission, a growth of emission before decay is observed for water mole fractions <0.9. This emission may be analyzed as a difference of exponentials, and the risetimes and falltimes vs water mole fraction are depicted in Figs. 3 and 4.

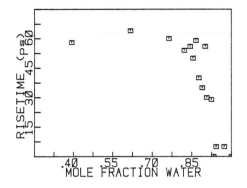

Fig.3 Risetime of TNS vs water mole fraction for wavelengths >570nm assuming a difference of exponential behavior.

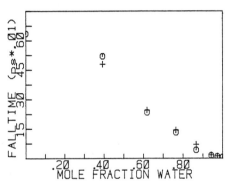

Fig.4 Falltime (crosses) of TNS vs water mole fraction for wavelengths >570nm assuming a difference of exponential behavior. Decay time (circles) for the entire band. These data are for air equilibrated solutions (cf. Table 1). The presence of dissolved oxygen affects only solutions with low water content and thus is not relevant to the discussions in this paper.

The behavior observed for all water mole fractions <0.8 is a characteristic risetime of 60ps. This is also the *decay time* of TNS in pure water. The falltimes for extreme red-edge wavelengths (>570nm) for all mole fractions closely correspond to the decay times for the entire emission band (>410nm) for the given mixed solvent. On increasing the mole fraction of water beyond 0.8, the risetime approaches zero. In contrast, dynamics for the pure solvents show no measurable risetime and are simple exponential decays for all wavelengths.

4. Interpretation and Discussion

There are many possible configurations of solvent around the ground electronic state of TNS in mixed solvents. Following excitation, the TNS, because of its charge transfer character [10], tends to become preferentially conjoined with water molecules through relative diffusion of all molecules. If a sufficiently high fraction of water molecules is present, a "complete shell" of water is formed. Such a complete shell is defined as having a local structure with enough congruence to structures present in a pure water solvent that the nonradiative rate is the same as in pure water. This shell need not totally surround the TNS molecule. As described in another paper presented at this conference [11], TNS in a complete water shell may be considered an "acceptor" species, while all other configurations may be considered "donor" species. In the limit of *high diffusion rates*, and ignoring for the moment the decay of TNS, a *steady state* gradient in water/ethanol concentration is formed. When this happens, the only dynamics remaining are the lifetimes of the species involved. "Acceptor" dynamics become a difference of exponentials of these lifetimes [11,12]. Our experiments on the red-edge emission were designed to follow the dynamics of this "acceptor" species. The results show that the red-edge risetime in the mixed solvent system is indeed equal to the 60ps falltime of TNS in pure water for all water mole fractions $0.4 < x(H_2O) < 0.8$. Thus, steady state has been reached in a time <60ps, implying that the relative diffusion of water and ethanol in the vicinity of the excited TNS molecule is sufficiently fast to maintain a constant concentration gradient as ethanol-rich shells transform into "complete shells" of water. Furthermore, the falltimes at the various mole fractions are the weighted averages of the "donor" species present. This then is the explanation of the behavior occurring in Figs. 3 and 4 for water mole fraction <0.8. For mole fractions of water <0.4, red-edge emission is very weak, and no "acceptor" experiments in this range were performed. An important contribution to the "donor" falltimes is the interconversion to species with higher water content.

When water concentrations are sufficiently high (x>0.8) statistically some "acceptor" species are already present in the ground state. These emit at the red-edge with a 60ps falltime and have no risetime. Results for water mole fractions >0.8 are therefore an admixture of these ground state *preformed* configurations and those configurations that are *postformed* in the excited state by virtue of the fast relative diffusion process. The observed "acceptor" dynamics can therefore be expressed as a weighted sum of these two types of terms,

$$S(t) = A[\exp(t/\tau_{pre})] + B[\exp(-t/\tau_{post}) - \exp(-t/\tau_{pre})] \qquad (1)$$

where $\tau_{pre} = 60ps$, while τ_{post} varies from 60ps at $x(H_2O) = 1.0$ to ~2000ps at $x(H_2O) = 0.8$. The fraction $A/A+B$ is a measure of the relative concentration of TNS having a "complete shell" of water in the ground state. Data analyzed in this way are presented in Fig. 5. Plotted on the same curve are scaled values of k_{nr}. The implication of this is that *only those TNS molecules with a complete shell of H_2O molecules have the large nonradiative (electron photoejection) rate.* There is therefore something exceptional about the local water structure around excited state TNS that promotes the electron transfer to the solvent. The onset of the effect with an increasing number of water molecules in this shell is surprisingly abrupt, and has important interpretational inferences in the use of such molecules as biological fluorescence probes [13].

A rough estimate of the number of water molecules necessary for the enhanced electron transfer may be determined as follows. If it is assumed that, for the ground state of TNS, solvent placement in the mixed solvent is statistical, then the probability of finding a water molecule at a required location near TNS is given by x, the water mole fraction. The probability F of having a "complete shell" of water is then,

$$F = x^n , \qquad (2)$$

where n is the number of water molecules necessary for the efficient electron transfer process. A fit of (2) to the data from the analysis of (1) in Fig. 5

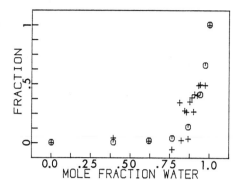

Fig.5 Fraction of TNS molecules emitting promptly at wavelengths >570nm *vs* water mole fraction (crosses). For comparison k_{nr} is scaled and also plotted as a function of water mole fraction (circles).

gives $n = 11 \pm 2$. A similar, but totally independent, fit of the k_{nr} data yields $n = 14 \pm 1$. Clearly, traditional exploitations of arylaminonaphthalene sulfonates as biological fluorescence probes for "polarity" [13] must be completely reexamined in light of these findings.

References

1. G. W. Robinson, R. J. Robbins, G. R. Fleming, J. M. Morris, A. E. W. Knight and R. J. S. Morrison, J. Am. Chem. Soc. 100, 7145 (1978).
2. W. O. McClure and G. M. Edelman, Biochem. 5, 1908 (1966).
3. E. M. Kosower, H. Dodiuk and H. Kanety, J. Am. Chem. Soc. 100, 4179 (1978).
4. G. R. Fleming, G. Porter, R. J. Robbins and J. A. Synowiec, Chem. Phys. Lett. 52, 228 (1977).
5. J. A. Synowiec, Ph.D. Thesis, London (1978).
6. J. P. Hansen and J. R. McDonald, *Theory of Simple Liquids* (Academic Press, New York 1976).
7. G. A. Kenney-Wallace, Acc. Chem. Res. 11, 433 (1978).
8. G. W. Robinson, T. A. Caughey and R. A. Auerbach, *Advances in Laser Chemistry*, A. H. Zewail, Ed., Springer Series in Chemical Physics (Springer Berlin, Heidelberg, New York 1978) pp. 108-125.
9. G. W. Robinson, T. A. Caughey, R. A. Auerbach and P. J. Harman, *Multichannel Image Detectors*, Y. Talmi, Ed., ACS Symposium Series #102 (American Chemical Society, Washington 1979) pp. 199-213.
10. C. J. Seliskar and L. Brand, J. Am. Chem. Soc. 93, 5405 (1971).
11. G. W. Robinson and R. A. Auerbach, paper WB1, this conference.
12. R. A. Auerbach, G. W. Robinson and R. W. Zwanzig, J. Chem. Phys. 72, 3528 (1980).
13. A. Azzi, Quart. Rev. Biophys. 8, 237 (1975).

Photodissociation, Short-Lived Intermediates, and Charge Transfer Phenomena in Liquids

K.B. Eisenthal, M.K. Crawford, C. Dupuy, W. Hetherington,
G. Korenowski, M.J. McAuliffe, and Y. Wang

Department of Chemistry, Columbia University,
New York, NY 10027, USA

It is the competition between a variety of chemical and physical channels for energy decay which determine the evolution of a molecular system.

In many molecules of the diazo family the dominant channel for energy degradation following ultraviolet photoexcitation is bond rupture, generating nitrogen and an important intermediate called a carbene. We can write the photodissociation as:

Diazo compound   Carbene

where R_1 and R_2 represent the groups attached, e.g. H atom, phenyl group - C_6H_5, etc. The class of carbenes, whether generated from diazo or other precursor molecules, has engaged the interest of chemists for many years {1-13}. This interest derives from their novel reactions, their unusual electronic structures in that most of them have two low-lying electronic states, a singlet and triplet, each of which has different reaction pathways, and for the simpler ones, the use of quantum mechanical calculations to examine their structures and reactivities. Although much is known about the chemical reactions of various carbenes, there is very little known about the dynamics of the photodissociation or the dynamics of formation and interconversion between the low-lying singlet and triplet states.

In the work reported here, the parent molecule was diphenyl-diazomethane. Irradiation of this precursor in the solvent 3-methyl pentane or acetonitrile at 25°C with a frequency quad-rupled Nd:YAG (266 nm) or Nd:phosphate (264 nm) pulse generated the diphenyl carbene. Writing the phenyl group ⌬ as φ we represent the relaxation pathways as follows:

Excited
singlet

The ground state of the diphenyl carbene molecule is a triplet. In a pioneering flash spectroscopic study of diphenyl carbene, DPC, it was established that there is a rapid equilibration between the low-lying singlet and triplet spin states{10}. The absolute rate constants for interconversion of the spin states of DPC were not measured due to the ultrafast

$$^1DPC \xrightarrow{k_{ST}} {}^3DPC$$

process, which was estimated to have a limiting value of the order of 10^{10} sec^{-1}.

Using a Nd:phosphate excitation pulse (FWHM∼6ps) with a streak camera-optical multichannel analyzer fluorescence detection system we find the rise time of the emission, $^3DPC^* \rightarrow {}^3DPC + h\nu^1$ to be roughly 15 ps. This means that the photodissociation step and the carbene excited singlet to excited triplet step, k^*_{ST} must occur in 15 ps or less. However, the very low quantum yield for production of $^3DPC^*$ which we find indicates that the principal source of the ground state ^3DPC is not the excited triplet, $^3DPC^*$. By establishing that the triplet fluorescence from $^3DPC^*$ varied linearly with the excitation intensity, we could conclude that the photodissociation results from the absorption of one photon at 264 nm. Knowing the energy of the exciting photon and the energy of the emitted photon, from our measurement of the fluorescence spectrum Fig.1, we conclude that the energy of the carbon-nitrogen bond is less than or equal to 2 eV.

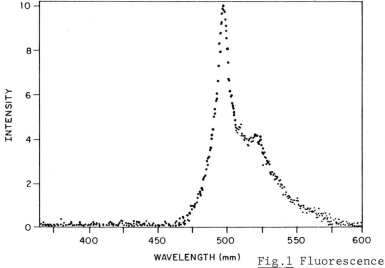

Fig.1 Fluorescence spectrum of $^3\overline{D}PC^*$ in 3-methylpentane at 25°C

The magnitude of k_{ST} was determined by a laser induced fluorescence method using a frequency quadrupled (266 nm) Nd:YAG laser pulse (FWHM 25<ps) to photodissociate the parent diphenyldiazomethane compound. A weak picosecond probe pulse at 266 nm was used to monitor the buildup of the ground state (^3DPC) by

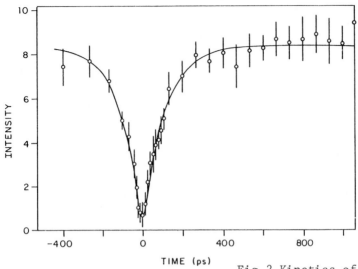

<u>Fig.2</u> Kinetics of ground state
³DPC formation in 3-methylpen-
tane at 25°C obtained by laser-
induced fluorescence method

measuring the triplet fluorescence intensity as a function of
the time separation between the excitation and probe pulses,
Fig.2.

The fluorescences resulting from the excitation pulse alone and
the dissociation and fluorescence resulting from the probe pulse
above were readily accounted for. The rate of formation of ^{3}DPC
was found to be $(9.1\pm1)\times10^9\,\mathrm{s}^{-1}$.

To place the origin of the ground state triplet ^{3}DPC on a
firmer basis than our quantum yield estimates of ^{3}DPC* production permits, we have determined the lifetime of ^{3}DPC*. From
measurements of the ^{3}DPC* fluorescence decay we find the lifetime to be 4 ns. Since the laser induced fluorescence measurements established that the ground state triplet, ^{3}DPC, appears
in 110 ps we conclude that the primary source of ^{3}DPC cannot be
from the more slowly decaying ^{3}DPC*. The dominant pathway for
^{3}DPC production is from the nearby ^{1}DPC state with a rate constant k_{ST} of $9.1\times10^9\,\mathrm{s}^{-1}$. Combining this value with the value of
$k_{ST}=1.7\times10^6\,\mathrm{s}^{-1}$ obtained in the laboratory of Prof.N.J.Turro, we
find that the value of the equilibrium constant, $K=k_{ST}/k_{TS}$ is
5.4×10^3 and the associated free energy difference at 25°C is
5.2 kcal/mole, to be compared with an upper limit estimate of
3.6 kcal/mole {10}. Correcting for the entropy difference arising
from the degeneracy of the triplet, we obtain a singlet-triplet
splitting in dephenylcarbene of 4.5 kcal/mole.

Another important energy dissipating pathway by which an
electronically excited molecule can relax is through charge
transfer (CT) interactions with surrounding ground state mole-

cules. The stabilization due to the CT forces between an excited and ground state molecule can produce new, albeit transient chemical species, an excited charge transfer complex called an exciplex {14-16}.

A key issue is the nature of any orientational requirements between the acceptor and donor molecules on the formation of the exciplex. This problem has been studied by many workers {17-21}, typically by hooking together the electron donor and acceptor molecules with a number of methylene (CH_2) groups, yielding the molecule A-$(CH_2)_n$-D. The relative orientation between the interacting A and D molecules (one of them being in an excited electronic state) is thereby restricted.

In the present study the molecule studied was 9-anthracene-$(CH_2)_3$-N,N-dimethylaniline.

This work complements our earlier transient absorption studies {19} and gives new information on the dynamic processes by following the fluorescence decay of A* and the fluorescence rise of the exciplex $(A^- - (CH_2)_3 - D^+)^*$

A single pulse extracted from a mode-locked Nd-phosphate laser, FWHM∿6 ps, was frequency tripled to 351 nm and used to excite the anthracene part of the molecule. The fluorescence of the anthracene was monitored at 410 nm and the exciplex at several wavelengths including 480 nm and 520 nm using a picosecond streak camera-optical multichannel detection system. The decays of excited anthracene in the nonpolar solvents isopentane, hexane, and decane, which cover a range of viscosities, were found to be exponential in all cases with values of 1.5 ns, 1.9 ns, and 2.9 ns respectively. Fig.3.

The fluorescence rise time of the exciplex was found to be the same as the anthracene decay times in each of the solvents. If there were no geometric requirements for exciplex formation, then we know from our earlier work {22} that the exciplex should have a rise time of less than 10 ps. The much slower formation times we find in the A-$(CH_2)_3$-D molecule confirm our earlier work and those of others {18,19} which indicated that rotational motions about the methylene groups are required to bring the two moieties into a favorable conformation to form the exciplex. The observation that the rate of anthracene decay and exciplex rise are equal indicates that there is no intermediate non-fluorescent state present before the formation of the fluorescent exciplex as had been suggested and, in turn, disputed by many workers. The kinetic curves and their viscosity dependence yields information on the dynamics of short chain systems and will be discussed elsewhere.

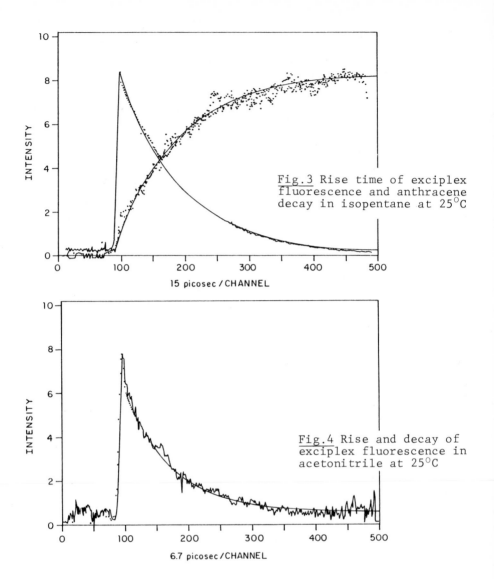

Fig.3 Rise time of exciplex
fluorescence and anthracene
decay in isopentane at 25°C

Fig.4 Rise and decay of
exciplex fluorescence in
acetonitrile at 25°C

A most interesting aspect of this study is the effect of solvent polarity on the charge transfer dynamics. In the polar solvent acetonitrile ($\varepsilon=37$) a very weak emission from the excited CT
complex was observed within the pulse width of our laser (6 ps)
and decayed in 580 ps. Fig.4.

This is in marked contrast to the observations in nonpolar solvents, Fig.3, where the exciplex is formed in nanoseconds and
has a decay time of more than one hundred nanoseconds. Of equal
interest is the rapid decay of the anthracene fluorescence ($\sim$10
ps) in contrast to the nanosecond decay times in the nonpolar

224

solvents. These results indicate that the dominant route for the quenching of the excited state anthracene does not lead to the formation of the excited CT complex. The very rapid (<6 ps) formation of the excited CT complex we observe results from the small population of ground state molecules that are, at the time of laser excitation, already in the required conformation for exciplex formation. For those molecules in all of the various other conformations unfavorable for exciplex formation, we believe that a direct electron transfer process from the donor moiety to the excited state anthracene occurs. This direct transfer produces a non-fluorescent ground state ion pair. The rotational motions about the methylene groups that are required in the pathway leading to the formation of the exciplex are too slow to compete with this new and direct ultrafast pathway and are thus effectively shut off, except for the small population of molecules already in the proper configuration for exciplex formation. The opening of this new electron transfer channel for energy dissipation in the polar solvent is made possible by the lowering of the energy of the ion pair due to the large solvation energy in the polar solvent {16,23,24}. The energy of the ion pair is below that of $A^*-(CH_2)_3-D$, as well as the exciplex in the polar solvent and therefore, the new fast decay channel becomes energetically available. In the nonpolar solvent, the ion pair is higher in energy than $A^*-(CH_2)_3-D$ and thus the exciplex decay channel, but not the ion pair decay channel, is available as an energy decay pathway.

Acknowledgements: This research was supported in part by the National Science Foundation, the Air Force Office of Scientific Research and the Joint Services Electronic Program (Contract # DAAG29-79-C-0079)

References:

1. R.A. Moss and M. Jones Jr., Carbenes, Vol.II, (J.Wiley, New York 1975)
2. W. Kirmse, Carbene Chemistry, Vol.I, 2nd Edition, (Academic Press, New York 1971)
3. G. Herzberg and J.W.C. Johns, Proc. Roy. Soc.London, Ser. A295, 107 (1966)
4. G. Closs, C.A. Hutchison, Jr., and B.E. Kohler, J. Chem. Phys. 44, 413 (1966)
5. I. Moritani, S.I.Murahashi, M. Nishino, K. Kimura, and H. Tsubomura, Tetrahedron Lett. 373, (1966)
6. A.M. Trozzolo, Accts. Chem. Res. 329 (1968)
7. W. Ware and P.J. Sullivan, J. Chem. Phys. 49, 1445 (1968)
8. P.F. Zittel, G.B. Ellison, S.V.O'Neil, E. Herbst, W.C. Lineberger, and W.P. Reinhardt, JACS 98, 3731 (1976)
9. J. Metcalfe and E.A. Halevi, J.C.S. Perkin II, 634 (1976)
10. G.L. Closs and B.E. Rabinow, JACS 98, 8190 (1976)
11. L.B. Harding and W.A. Goddard III, J. Chem. Phys. 67, 1777 (1977)
12. H.D. Roth, Accts. Chem. Res. 10, 85 (1977)
13. R.K.Lengel and R.N.Zare, JACS 100, 7495 (1978)
14. H. Leonhardt and A. Weller, Ber. Bunsenges. Physik. Chem. 67, 781 (1963)
15. N. Mataga, T. Okada, and K. Ezumi, Mol. Phys. 10, 203 (1966)
16. M. Ottolenghi, Accts. Chem. Res. 6, 153 (1973)

17. E.A. Chandross and H.T. Thomas, Chem. Phys. Lett. 9, 393 (1971)
18. T. Okada, T. Fujita, M. Kubota, S. Masaki, M. Mataga, R. Ide, Y. Sakata, and S. Misumi, Chem. Phys. Lett. 14, 563 (1972)
19. T.J. Chuang, R.J. Cox, and K.B. Eisenthal, JACS 96, 6828 (1974)
20. A. Weller, private communication
21. R.S. Davidson and K.R. Tretheway, J. Chem. Soc. Chem. Comm. 827 (1976)
22. T.J. Chuang and K.B. Eisenthal, J.Chem. Phys. 59, 2140 (1973); 62, 2213 (1975)
23. D. Rehm and A. Weller, Ber. Bunsenges. Phys. Chem. 73, 834 (1969)
24. T. Hino, H. Akazawa, H. Masuhara, and N. Mataga, J. Phys. Chem. 80, 33 (1976)

Excited-State Proton-Transfer Kinetics in 1-Naphthol, 1-Naphthol Sulfonates, and Organometallic Complexes

S.L. Shapiro and K.R. Winn

University of California, Los Alamos Scientific Laboratory,
Los Alamos, NM 87545, USA

J.H. Clark

Materials and Molecular Research Division, Lawrence Berkeley Laboratory, and
Department of Chemistry, University of California,
Berkeley, CA 94720, USA

Introduction

Upon photoexcitation many aromatic compounds are known to undergo ultrafast proton-transfer reactions in aqueous solution [1]. Picosecond spectroscopy offers the opportunity for modeling and understanding proton-transfer processes in solution by allowing direct measurements of proton-transfer kinetics to be carried out for broad classes of compounds. Given the important role of proton transfer in vast numbers of important chemical and biological processes, such studies can be expected to provide new, fundamental insight into the chemistry of the proton.

Aromatic molecules can become considerably more acidic upon absorption of light quanta due to the alteration of electronic structure, often changing dissociation constants for proton ejection by many orders of magnitude [1]. For naphthols and substituted naphthols in aqueous solution, the proton transfer to neighboring water molecules may be represented schematically as follows:

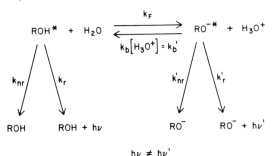

$$hv \neq hv'$$

Fig.1 Simplified schematic of excited-state proton transfer in naphthol compounds. Also possible are nonradiative deactivations induced by interactions with protons (vide infra).

where k_r and k_{nr} represent the radiative and nonradiative decay rates for the excited naphthol molecule (ROH*), k_f is the excited-state deprotonation rate, k_b represents the protonation rate from the excited naphtholate to the excited naphthol, and v and v' are the emission frequencies from the excited protonated and deprotonated species, respectively. The classic work of FORSTER [2] and WELLER [3] established the theoretical framework for understanding proton-transfer phenomena in the naphthols. Because the ground-state pK_a values for the naphthols are in the range 9-11, the only ground-state species present in solutions with pH < 7 is ROH. Following excitation of ROH with a picosecond pulse of the appropriate frequency, the protonation kinetics can be directly determined by observing the decay of emission from ROH* at frequency ν, or the risetime of the emission from RO⁻* at the red-shifted frequency ν'.

The dissociation of 1-naphthol in aqueous solution occurs so rapidly that the fluorescence from the neutral form, ROH*, has been previously described as "completely extinguished" [3] and as "hardly noticeable" [4]. Apparently nearly all of the fluorescence orginates from the naptholate ion. Here we report on the proton-transfer characteristics of a series of 1-naphthol compounds. We also report preliminary data on excited-state proton transfer in an organometallic complex of ruthenium.

Experimental Arrangement and Sample Preparation

Crucial to the success of these experiments and to their interpretation are sample preparation and purification. All samples were obtained from LC Laboratories (Newton, Massachusetts) and were determined to be greater than 99.5% pure by analytical liquid chromatography using a fluorescence detector. Samples were prepared in a nitrogen filled glove box, and all solvents were thoroughly degassed by undergoing successive freeze-pump-thaw cycles.

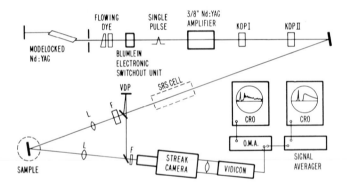

Fig.2 Experimental Arrangement for Measuring Rapid Proton Transfer.

The experimental arrangement is shown in Fig.2. A single 30-ps pulse is selected from the pulse train of a Nd:YAG modelocked laser, is amplified to a level of 10 mJ, and can be frequency doubled, tripled or quadrupled with the appropriate KDP crystals. The naphthols are excited with 266-nm radiation, and the organometallic complexes are excited with 355-nm radiation. Light emitted by the samples is collected onto the slit of a Hadland Photonics Photochron II streak camera. Streaks are imaged onto an OMA, and the individual shots can be accumulated on a Nicolet 1074 signal averager. A relative time reference provided by means of a precursor marker pulse ensures accurate signal averaging. Fall times of the protonated species are obtained by observing the emission through a Corning 7-54 filter, and risetimes of the deprotonated species with a Corning 2-62 filter.

1-Naphthol and 1-Naphthol Sulfonates: Results and Discussion

The experimental results for the relaxation of the excited protonated species in 1-naphthol and two of its sulfonated derivatives are summarized in Fig.3. Note that 1-naphthol and the two 1-naphthol

		τ_D
	l-naphthol	25 ± 10 ps
	l-naphthol-2-sulfonate	55 ± 15 ps
	l-naphthol-5-sulfonate	≤ 20 ps

Fig.3 Relaxation time for emission from the protonated form of 1-naphthol, 1-naphthol-2-sulfonate, and 1-naphthol-5-sulfonate.

derivatives dissociate extremely rapidly. Kinetics of the rapid proton transfer indicate that the excited-state pK value (pK*) for all three derivatives is in the vicinity of zero. This estimate is confirmed by a FORSTER-cycle calculation [2] based on the absorption and emission spectra of the pure compounds. Notice also that the dissociation rate for 1-naphthol-2-sulfonate is slower than for the other two derivatives. Evidently, the intramolecular hydrogen bond formed between the adjacent hydroxy and sulfonate groups leads to a slowdown in the proton ejection rate in this derivative. Such an effect has previously been postulated [5].

Our result in 1-naphthol-2-sulfonate differs strikingly from that of ZAITZEV et al. [6] who, using nanosecond techniques, reported that the dissociation rate for this compound is 5.4×10^8 sec^{-1} (1.85 ns). Our experiments indicate that this long decay time may be due to impurities. For example, a sample of 1-naphthol-2-sulfonate that was obtained from Eastman Kodak, for which no attempt was made at further purification or to prevent exposure to air during sample preparation, yields only a long-lived component, also a few ns in duration.

Further confirmation for the 55 ps lifetime reported here is obtained from the emission spectrum of purified, oxygen-free 1-naphthol-2-sulfonate excited at 313 nm, which is completely dominated by the emission from the deprotonated species. This is in contrast to the spectrum obtained by ZAITZEV, et al. [6], in which the ratio of the emission intensities of the deprotonated species to that of the protonated form is only 2 to 1.

On the other hand, we have found that the proton-transfer kinetics of 2-naphthol-6-sulfonate samples reported upon previously [7], do not appear to change when carefully purified samples are used. This may be due to a fortuitous coincidence between the proton-transfer rates of

2-naphthol-6-sulfonate and the fall time of the impurity fluorescence.
Sample analysis by high-pressure liquid chromatography is currently being
carried out to establish in more detail the role of impurities both in
the measurements reported here and in our previous measurements on
2-naphthol-6-sulfonate [7].

For the case of 1-naphthol, MARTYNOV, et al. [8] estimated a
dissociation rate of $> 3 \times 10^9$ sec^{-1}. Our measurement of 25 ± 10 ps
is in accord with this estimate. We have also observed a deactivation of
the excited 1-naphtholate due to interaction with hydronium ions. The
fluorescence lifetime of the deprotonated species varies from nanoseconds
to picoseconds with increasing solution pH. The values obtained from
these direct measurements are in accord with the quantum efficiencies
reported by WELLER [3].

Previously, we suggested that by using intense laser pulses, the pH of
a solution could be changed in a manner analogous to the temperature
change in a T-jump experiment [7]. This laser pH jump technique [7,9]
might allow the study of rapid acid-base reactions. The rapid
deprotonation of the 1-naphthol compounds demonstrates that the pH of a
solution may be manipulated on a time scale of less than 20 ps. This
rapid rate, the fastest intermolecular proton transfer process observed
to date, offers the possibility of producing very large and rapid pH
jumps for studying rapid chemical reactions.

Proton Transfer in an Organometallic Complex of Ruthenium

Organometallic complexes are known to display a rich and varied
photochemistry [10]. There is an increasing awareness that it may be
possible to exploit this chemistry for the efficient conversion of solar
energy to easily transportable fuels. Recently, excited-state proton
transfer was observed in an organometallic complex [11]. We report here
preliminary measurements on the kinetics of this process. Figure 4 shows

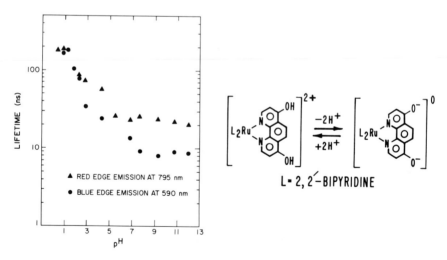

Fig.4 Lifetime of deprotonated (triangles) and protonated (circles)
forms of the organometallic ruthenium complex.

230

the schematic for the excited-state proton-transfer reaction of the
organometallic complex (2,2'-bipyridine)$_2$Ru(4,7-dihydroxy-1,10
phenanthroline). Also shown in Fig.4 are the decay times for the
fluorescence from the protonated and deprotonated forms of this complex
in nondegassed solutions, plotted as a function of pH. The emission from
the protonated form decays in about 200 ns at pH 1. Previous studies by
GIORDANO, et al. [11] indicate that between pH 5 and pH 2.5, the
protonated form will be the main species present in the ground state.
Upon excitation between these pH values, we could then expect to observe
a risetime for the 795-nm band corresponding to the formation of the
deprotonated species. Because of the insensitivity of the streak camera
S-20 photocathode at a wavelength of 800 nm, we have estimated this
risetime using a photomultiplier and a fast oscilloscope to temporally
resolve the emission. The risetime thus obtained is limited by the 5-ns
response time of the photomultipler tube. We are presently studying
organometallic proton-transfer processes from complexes that emit in a
region of the visible spectrum more amenable to streak camera detection.

Acknowledgments

Helpful discussions with P. Driedger and M. A. Tolbert are gratefully
acknowledged. This work was supported by the Division of Chemical
Sciences, Office of Basic Energy Sciences, U.S. Department of Energy
under contract No. W-7405-Eng-48.

References

1. J. F. Ireland and P. A. H. Wyatt, Adv. Phys. Org. Chem. 12, 131
   (1976).

2. Th. Forster, Z. Electrochem. 54, 42 (1950).

3. A. Weller, Z. Elektrochem. 56, 662 (1952).

4. L. D. Derkacheva, Opt. Spectrosk. 9, 209 (1960) [Eng. Trans. p. 110].

5. R. M. C. Henson and P. A. H. Wyatt, JCS Faraday II 71, 669 (1975).

6. N. K. Zaitsev, A. B. Demyashkevich and M. G. Kuz'min, Khimiya
   Vysokikh Energii 12, 436 (1978) [Eng. Trans. p. 365].

7. J. H. Clark, S. L. Shapiro, A. J. Campillo and K. R. Winn, J. Am.
   Chem. Soc. 101, 746 (1979).

8. I. Yu. Martynov. N. K. Zaitsev, I. V. Soboleva, B. M. Uzhinov, and M.
   G. Kuz'min, Zhurnal Prikladnoi Spektroskopii 28, 1075 (1978) [Eng.
   Trans. p. 732].

9. K. K. Smith, K. J. Kaufmann, D. Huppert, and M. Gutman, Chem. Phys.
   Lett. 64, 522 (1979).

10. P. D. Fleischauer, A. W. Adamson, and G. Sartori, Prog. Inorg. Chem.
    17, 1 (1972).

11. P. J. Giordano, C. R. Bock, and M. S. Wrighton, J. Am. Chem. Soc.
    100, 6960 (1978).

Excited-State Absorption Spectra and Decay Mechanisms in Organic Photostabilizers [1]

A.L. Huston, C.D. Merritt, and G.W. Scott

Department of Chemistry, University of California,
Riverside, CA 92521, USA

A. Gupta

Energy and Materials Research Section, Jet Propulsion Laboratory,
California Institute of Technology,
Pasadena, CA 91103, USA

1. Introduction

This paper discusses picosecond spectroscopic studies of excited-state decay
mechanisms in widely-used, ultraviolet-absorbing, polymer photostabilizers—
2-hydroxybenzophenone (I) and 2-(2'-hydroxy-5'-methylphenyl)-benzotriazole
(II)—the structures of which are shown below:

(I)   (II)

An intramolecular hydrogen bond in these molecules is thought to promote rapid
excited-state internal proton transfer and subsequent excited-state decay.
However, the detailed mechanism and kinetics of these processes have not been
determined. Related studies on organic photostabilizers have been reported
for salicylates [1,2], 2-(2'-hydroxyphenyl)-s-triazines [3], and 2-hydroxy-
benzophenones [4-6].

2. Experimental

Excitation of (I), (II), and a derivative of (II) was accomplished with a
third-harmonic pulse ($\Delta t \approx 8$ ps, $\lambda = 355$ nm) from a modelocked Nd^{+3}:glass laser
system. Excited state absorption spectra were obtained as previously de-
scribed [5] by probing the excited sample with a time-delayed continuum pulse
and recording a single-shot, double-beam-corrected spectrum (400-700 nm) with
a spectrograph and an optical multichannel analyzer (EG&G PARC, OMA 2). Excited-state absorption kinetics were obtained using a delayed probe pulse
from a short-cavity dye laser as previously described [7]. Absorption bleaching-recovery kinetics at 355 nm were also obtained with attenuated third-harmonic probe pulses.

---

[1]This research was supported by the ASTT project of Solar Thermal Power Systems of the Jet Propulsion Laboratory. Partial support by the California
Institute of Technology President's Fund and the Committee on Research at
the University of California, Riverside, is also acknowledged.

3. Results and Discussion

3.1. 2-hydroxybenzophenone

Excited—state absorption spectra of 2-hydroxybenzophenone (I), obtained at three delay times in several solvents, are summarized in the Table. These spectra are quite broad and weak. The maximum O.D. change observed is ~0.13. Under identical experimental conditions, a T-T absorption spectrum of benzophenone in ethanol has a maximum O.D. of 0.6 at ~530 nm.

Table   Summary of observed transient absorption spectra

Molecule	Solvent	Delay Time [ps]	λ_{max} [nm]	O.D.@λ_{max}
2-hydroxybenzo-phenone (I)	CH_2Cl_2	7	435	0.10
		20	450(v.broad)	<0.04
		485	---	<0.03
	isooctane	7	450	<0.04
		20	---	<0.03
		485	---	<0.03
	cis-1,3-pentadiene	7	450(v.broad)	0.05
		20	---	<0.03
	ethanol	7	435	0.08
		17	<400;475	>0.10;0.09
		50	420	0.13
		480	450	0.07
2-(2'-hydroxy-5'-methylphenyl)benzotriazole (II)	CH_2Cl_2	7	440	0.11
		20	---	<0.03
		485	460	0.10
	ethanol	7	435;500	0.07;0.07
		20	440	<0.04
		485	460	0.10
2-(2'-acetoxy-5'-methylphenyl)benzotriazole (acetoxy-II)	CH_2Cl_2	7	470	0.37
		20	475	0.42
		485	465	0.21

There has been some controversy regarding the decay mechanism of (I). KLÖPFFER [8] has proposed that the main decay route occurs via the triplet mainfold following proton transfer in the singlet state. In nonhydrogen-bonding solvents, however, no long-time absorption (i.e., $T_n \leftarrow T_1$) is observed in our experiments indicating that few of the excited singlet state molecules of (I) intersystem cross to the triplet. (See the Table; the experiment in cis-1,3-pentadiene confirms that the short-time absorption is probably due to a singlet.) In ethanol, some visible absorption remains at "long" time (480 ps) and thus some of the molecules of (I) may intersystem cross in hydrogen-bonding solvents. This interpretation is consistent with other recent studies [4] and with previously reported [10] quantum yields of quenchable triplets of (I) in nonhydrogen-bonding and hydrogen-bonding solvents ($\Phi_T \approx 0.03$ in cyclohexane and $\Phi_T \approx 0.15$ in ethanol).

3.2. 2-(2'-hydroxy-5'-methylphenyl)benzotriazole

Excited—state absorption spectra of 2-(2'-hydroxy-5'-methylphenyl)benzotriazole (II) and 2-(2'-acetoxy-5'-methylphenyl)benzotriazole (acetoxy-II) in methylene chloride are shown in Figs. 1A and 1B, respectively. These and similar spectra of (II) in ethanol are summarized in the Table above.

For (II) in both solvents, absorption at 7 ps in the blue decays to within the noise level by 20 ps. At longer time (485 ps), a different absorption in the blue appears. Our tentative interpretation of these spectra follows the decay mechanism suggested by WERNER [10];

$$S_0 \xrightarrow{h\nu} S_1 \xrightarrow[k_1]{H^+ \text{ trans.}} S_1' \xrightarrow[k_2]{i.c.} S_0' \xrightarrow[k_3]{H^+ \text{ trans.}} S_0 \quad , \qquad (1)$$

in which a prime designates proton transfer from oxygen to a nitrogen. Thus at 7 ps, we assign the observed spectrum of (II) to $S_n \leftarrow S_1$ based on similar spectra of (acetoxy-II) shown above. By 20 ps, rapid proton transfer has occured, yielding a weak, essentially featureless $S_n' \leftarrow S_1'$ spectrum in the visible. By 485 ps, internal conversion to the ground state of the proton-transferred species has occurred, yielding the $S_1' \leftarrow S_0'$ spectrum. In support of this last assignment, the absorption spectrum observed at 485 ps is the approximate mirror image of previously reported $S_1' \rightarrow S_0'$ fluorescence [10,11]. This interpretation suggests that there may be a barrier (of unknown height) to proton back-transfer in the ground state. It has been estimated [10] that the proton is favored to be on oxygen by $\Delta H \approx 16$ kcal/mole in the ground state.

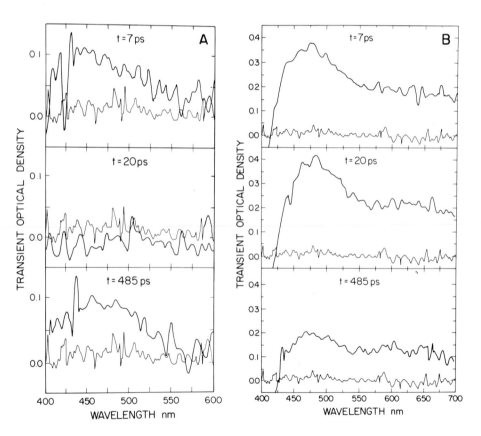

Fig. 1   Transient absorption spectra of (II) in CH_2Cl_2 (A) and of (acetoxy-II) in CH_2Cl_2 (B) at selected delay times

In (acetoxy-II), chemical modification precludes proton transfer. The observed spectra (Fig. 1B) are assigned to mostly $S_n \leftarrow S_1$ absorption. The absorption decay in (acetoxy-II) at 455 nm following excitation at 355 nm (Fig. 2) is consistent with exponential decay to a nonzero asymptote. The least squares fit rate constant is $(3.9\pm0.5) \times 10^9$ sec^{-1} (τ_{S_1}=0.26 ns). The long-

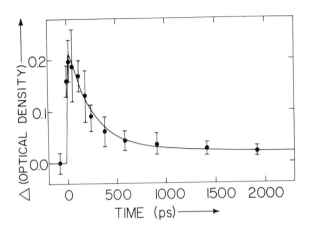

Fig. 2 Kinetics of optical density changes of (acetoxy-II) in CH_2Cl_2 at 455 nm

time, asymptotic absorption is assigned to $T_n \leftarrow T_1$ absorption. In a nanosecond flash photolysis experiment, $T_n \leftarrow T_1$ absorption peaking at ~425 nm decays with an ~200 ns lifetime [12]. We determined the triplet quantum yield of (acetoxy-II) to be $\Phi_T > 0.3$. Thus the S_1 lifetime of (acetoxy-II) is at least partially controlled by intersystem crossing.

Absorption bleaching-recovery kinetics at 355 nm were obtained for (II) in both ethanol and methylene chloride. The result in the methylene chloride solvent is shown in Fig. 3. In this experiment the ground state optical density of the sample was arranged to be ~0.6 at 355 nm, and the excitation and probe pulses had parallel linear polarization. The changes in optical density plotted in Fig. 3 correspond to decreases in optical density at 355 nm. In ethanol, a qualitatively similar result is obtained. The unusual behavior between 0 and 100 ps is in qualitative agreement with (1). That is, at 355 nm, S_1 absorbs less strongly than S_0; S_1 rapidly decays by proton transfer to S_1' (in <50 ps), a state which absorbs more strongly than the ground state; S_1' internally converts to S_0' (in ~100 ps after excitation), a state which absorbs less strongly than S_1'; and S_0' then decays more slowly to S_0. Although the data do not uniquely determine the rates of these processes, the smooth curve in Fig. 3 was obtained using k_1=15ns^{-1}, k_2=50ns^{-1}, and k_3=5ns^{-1} by convoluting rate expressions derived from (1) with the laser pulse durations.

Eq. (1) is not the only mechanism which can qualitatively reproduce the observed kinetics in Fig. 3. However, if one is restricted to a sequential decay mechanism like (1), then to obtain even qualitative agreement with the data, requires that $k_2 > k_1 \approx 10$-20 ns^{-1} > k_3. Furthermore, the extinction coefficients at 355 nm of the states involved must be ordered $\varepsilon(S_1') > \varepsilon(S_0) > \varepsilon(S_1)$ > $\varepsilon(S_0')$, and at least three intermediates are involved. The values of the rate constants and extinction coefficients are not uniquely determined by the

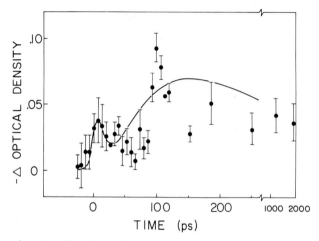

<u>Fig. 3</u>   Kinetics of optical density changes of (II) in CH_2Cl_2 at 355 nm.

data. Reorientation of the transition dipoles of the various states was not included in the kinetics model used to obtain the curve in Fig. 3.

We have also investigated the emission kinetics of solutions of (II) in ethanol and methylene chloride at room temperature, exciting with a 7-ns uv-pulse from a nitrogen laser. In both cases, prompt emission which followed the laser excitation pulse within the time resolution of the detection system (~0.5 ns) was observed. Thus, this emission, centered at ~420-450 nm, neces-sarily has a lifetime of <1 ns. Based on these observations, this emission is assigned to $S_1 \rightarrow S_0$ and is consistent with our other observations on (II).

References

[1] K.K. Smith and K.J. Kaufmann, J. Phys. Chem. <u>82</u>, 2286 (1978); and in A.H. Zewail (ed.), *Advances in Laser Chemistry*, Springer Series in Chemical Physics, Vol. 3 (Springer, Berlin, Heidelberg, New York 1978) 163
[2] P.J. Thistlethwaite and G.J. Woolfe, Chem. Phys. Letts. <u>63</u>, 401 (1979).
[3] H. Shizuka, K. Matsui, Y. Hirata, and I. Tanaka, J. Phys. Chem. <u>80</u>, 2070 (1976); <u>81</u>, 2243 (1977).
[4] S.-Y. Hou, W.M. Hetherington III, G.M. Korenowski, and K. B. Eisenthal, Chem. Phys. Letts. <u>68</u>, 282 (1979).
[5] C. Merritt, G.W. Scott, A. Gupta, and A. Yavrouian, Chem. Phys. Letts. <u>69</u>, 169 (1980).
[6] M.R. Topp, unpublished results.
[7] R.W. Anderson, Jr., D.E. Damschen, G.W. Scott, and L.D. Talley, J. Chem. Phys. <u>71</u>, 1134 (1979).
[8] W. Klöpffer, J. Polym. Sci: Symp. <u>57</u>, 205 (1976); Adv. in Photochem. <u>10</u>, 311 (1977).
[9] A.A. Lamola and L.J. Sharp, J. Phys. Chem. <u>70</u>, 2634 (1966).
[10] T. Werner, J. Phys. Chem. <u>83</u>, 320 (1979).
[11] J.-E.A. Otterstedt, J. Chem. Phys. <u>58</u>, 5716 (1973).
[12] A. Gupta, D.S. Kliger, and R. Liang, unpublished results.

Picosecond Studies of Temperature and Solvent Effects on the Fluorescence from Coumarin 102 and Acridine[1]

S.L. Shapiro and K.R. Winn

University of California, Los Alamos Scientific Laboratory,
Los Alamos, NM 87545, USA

1. Introduction

Picosecond kinetics often provides the only way of detecting and unravelling complex molecular interactions in liquids. Here we report about the kinetics of the fluorescence from two molecules using picosecond streak camera techniques. Upon excitation both coumarin 102 and acridine exhibit interesting spectral dynamical behavior depending sensitively on the temperature and solvent. For understanding energy dissipation in polyatomic molecules, the dependence of the relaxation times on emission and excitation wavelengths and upon temperature can be crucial. The behavior of the coumarin 102 emission with temperature and solvent allows us to interpret the results in terms of formation at an excited state [1,2] hydrogen bonded complex between the coumarin 102 and solvent molecules. The dependence of the fluorescence lifetime of acridine with temperature for several solvents demonstrates that an activation energy is present and that current models are inadequate.

We have previously shown that the complete protonation kinetics of coumarin 102 (2,3,5,6-1H,4H-tetrahydro-8-methyl-quinolazino [9,9a,1-gh] coumarin), in water can be established by varying the pH of the solution and detecting the fluorescence emission about appropriate wavelengths with a streak camera.[3] Here we report a solute-solvent interaction for coumarin 102 in different solvents regardless of the pH. We also report the temperature dependence of the fluorescence lifetime of acridine in ethanol, hexane and glycerol.

2. Experimental

Purified samples (> 99.5% purity) of both coumarin 102 and acridine were prepared by LC Labs under oxygen free conditions. Concentrations of purified samples ranged from 10^{-3} M to 10^{-6} M. Samples were prepared immediately prior to use in oxygen free, spectral quality solvents. These samples were contained in 2 mm quartz cuvettes designed to fit in a quartz dewar with high optical quality windows. The temperature could be lowered to $-80°C$ by means of a methanol dry ice bath that surrounded the cuvettes, and could be raised to higher values by heating a mineral bath. A thermocouple was placed inside the cuvette near the excitation area, and the output was plotted on a Brown recorder.

A 1064 nm pulse was selected from the pulse train of a Nd:YAG oscillator, was amplified in a Nd:YAG amplifier and was frequency tripled by

[1]Work performed under the auspices of the U. S. Department of Energy.

237

means of phase matched KDP crystals. Typical energies of the single 20 ps, 355 nm excitation pulse ranged from 20-30 µJ. Fluorescence emission from the samples was imaged onto the slit of a Hadland Photonics 675/II streak camera with a uv window and S-20 spectral response. Interference filters at 420, 435, or 450 nm (± 8 nm to 10% transmission points) were used to isolate the blue edge emission at coumarin 102, whereas a Corning 2-60 filter was used to isolate the red edge emission. A 450 nm filter was used for acridine samples.

3.   Results

The temporal emission characteristics of spectral bands of coumarin 102 in ethanol are shown in Fig. 1. The strong dependence of the temporal emission with temperature is displayed in Fig. 1a through 1c. As the temperature

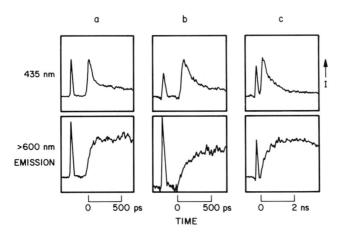

Fig. 1 Temporal display of fluorescence of coumarin 102 in ethanol
a) + 1°C b) -32°C c) -72°C

is lowered, the decay time of the selected blue edge spectral band increases. An increase in the corresponding risetime for the red edge emission is observed as well. The formation of the red edge emission is consistent with the falltime of the blue spectral band. The fluorescence decay of the blue spectral band can be fit with an exponential, except for a small background component with a lifetime of several nanoseconds.

The measured decay time for the blue edge emission for three normal alcohols is plotted versus the parameter η/T in Fig. 2, where η is the viscosity. For these normal alcohols, the relationship between τ and η/T is nearly linear with the same proportionality constant.

Our measurements of the temperature dependences of the fluorescence lifetime of acridine in ethanol, glycerol, and hexane are plotted in Fig. 3.

On a semi-logarithmic plot, the decay rate as a function of the inverse of the temperature is clearly a straight line for all three solvents over the region most affected by thermal processes. Moreover, all three lines are

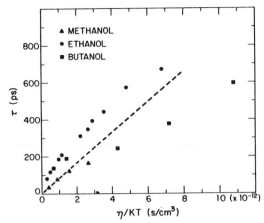

Fig. 2 Plot of measured lifetime vs viscosity divided by kT. For all these normal alcohols, the plot is nearly linear with the same proportionality constant, $\sim 10^{-22}$ cm^3, as indicated by dashed line

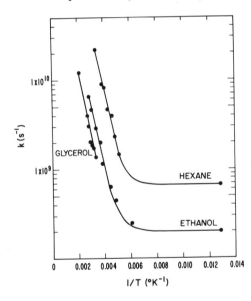

Fig. 3 Variation of decay rate with temperature for acridine in ethanol, hexane, and glycerol. Wavelength of observation is 450 nm. Solid line fit described in text

conspicuously parallel, their slopes being the same within experimental error. Our temperature data can be fit with curves of the form $k = k_0(77K) + k'\exp(-E/kT)$, where for ethanol $k_0 = 2 \times 10^8$ sec^{-1}, $k' = 0.97 \times 10^{12}$ sec^{-1} and $E = 1205$ cm^{-1}, and for hexane $k_0 = 6.7 \times 10^8$ sec^{-1}, $k' = 1.2 \times 10^{13}$ sec^{-1} and $E = 1291$ cm^{-1}, and for glycerol $k' = \cong 0.55 \times 10^{12}$ sec^{-1} and $E \cong 1225$ cm^{-1}.

4. Discussion

Results in coumarin 102 are consistent with the formation of a hydrogen bonded complex in the excited state. Upon excitation a dipolar species is known to be created,[4] and consequently hydrogen bonding of the hydroxyl group of a solvent molecule with the negatively charged carboxylic oxygen group of the solute molecule is likely. A simple kinetic analysis predicts the essential temporal features.[2] If we identify the falltime of the blue edge emission and the risetime of the red edge emission with the formation time for an excited state complex, then it follows from Fig. 2 that the formation time must be proportional to η/kT. According to a simple hydrodynamic model of molecular reorientation, the orientational relaxation time is given by $\eta V/kT$ where V is the molecular volume.[5,6] Because the formation time is proportional to η/kT and the proportionality constant of 10^{-22} cm^3 is close to the hydrodynamic volume of coumarin 102, our data suggest that the formation time can be identified with the orientational relaxation time. Thus the formation of such a complex is limited by the orientational response. Related measurements on time-resolved emission and absorption bands have also been reported in the picosecond regime.[7-10]

Our results for acridine in both protic and aprotic liquids yield about the same activation energy. Although our measurements are consistent with previous picosecond measurements at room temperature [11-14] and are consistent with an important role for hydrogen bonding,[11,12] our data as a function of temperature provides additional information that appears to be inconsistent with a model[15] used to describe earlier results[11,12] involving only a variation in the splitting of the $n\pi^*$ and $\pi\pi^*$ states.

According to this model,[15] part of the spectral behavior revolves about the role of $n\pi^*$ and $\pi\pi^*$ states which can be in close proximity in nitrogen heterocyclics and can interact through vibronic coupling.[16] The gap between the $n\pi^*$ and $\pi\pi^*$ states should depend upon the solvent, and therefore affect the electronic vibronic coupling.[15] Because the activation energies appear to be the same in all three solvents in spite of their hydrogen bonding characteristics, a new mechanism must be invoked. A possibility includes quantum mechanical tunneling due to the presence of distortion of the upper electronic state cause by the electronic-vibronic interaction of closely separated states. Franck-Condon overlaps may become poorer leading to the necessity of tunneling through a barrier prior to deactivation. A second possibility, however, is the presence of a higher lying level through which the deactivation proceeds, in a similar manner as proposed for isoquinoline.[17]

References

1.	C. V. Shank, A. Dienes, A. M. Trozzolo, and J. A. Myer, Appl. Phys. Letters 16, 405 (1970).

2.	S. L. Shapiro and K. R. Winn, Chem. Phys. Letters, to be published.

3.	A. J. Campillo, J. H. Clark, S. L. Shapiro, K. R. Winn, and P. K. Woodbridge, Chem. Phys. Letters, 67, 218 (1979).

4.	K.H. Dexhage: *Dye Lasers*, ed. by F.P. Schäfer, Topics in Applied Physics, Vol. 1 (Springer, Berlin, Heidelberg, New York 1977) 144

5.	P. Debye, Polar Molecules (Dover Publications, London, 1929) p. 84.

6.	T. J. Chuang and K. B. Eisenthal, Chem. Phys. Letters 11, 368 (1971).

7.	M. M. Malley and G. Mourou, Optics Communications 10, 323 (1974).

8. W. S. Struve and P. M. Rentzepis, Chem. Phys. Letters 29, 23 (1974).

9. H. E. Lessing and M. Reichert, Chem. Phys. Letters 46, 111 (1977).

10. L. A. Hallidy and M. R. Topp, J. Phys. Chem. 82, 2415 (1978).

11. V. Sundstrom, P. M. Rentzepis and E. C. Lim, J. Chem. Phys. 66, 4287 (1977).

12. L. J. Noe, E O. Degenkolb, and P. M. Rentzepis, J. Chem. Phys. 68, 4435 (1978).

13. H. B. Lin and M. Topp, Chem. Phys. 36, 365 (1979).

14. P. F. Barbara, L. E. Brus and P. M. Rentzepis, Chem. Phys. Letters 69, 447 (1980).

15. E. C. Lim, in Excited States, edited by E. C. Lim (Academic Press, New York, 1977), Vol. 3, p. 305-337.

16. R. M. Hochstrasser and C. A. Marzzacco, in Molecular Luminescence, edited by E. C. Lim (W. A. Benjamin, Inc., New York, 1969), p. 631-656.

17. J. R. Huber, M. Mahaney, and J. V. Morris, Chem. Phys. 16, 329 (1976).

Picosecond Proton Transfer

D. Huppert, E. Kolodny, and M. Gutman[1]

Department of Chemistry, Tel-Aviv University, Ramat Aviv, Israel

[1] Department of Biochemistry, Tel-Aviv University, Ramat Aviv, Israel

Introduction

Proton transfer is one of the simplest and most common of all chemical reactions. It is encountered in organic and inorganic chemistry, even more so in biochemistry where all reactions are carried out in aqueous solutions. In ATP synthesis [1], proton translocation across a biological membrane serves as the driving force for the reaction. The reactivity of the proton, especially in its hydrated form, is so high that most of its reactions are diffusion controlled. A rapid change in the proton concentration can be used to study chemical kinetics, in analogy to the T jump method. A system that can be used to generate such a pH jump has been known since Forster [2] interpreted the large Stokes shift seen in the fluorescence spectrum of certain aromatic molecules. The shifts were explained as changes in the acid-base properties of the molecules caused by electronic excitation. This is now known to be a general phenomenon. Aromatic alcohols or protonated amines become more acidic in the excited singlet state than in the ground state. From a large number of studies [3] it has become apparent that pK changes of -7 can be readily attained for molecules which become more acidic in the excited state. Using a high-intensity nanosecond laser pulse, 2-naphthol-3,6-disulfonate, or 8-hydroxy 1,3,6 pyrene trisulfonate in water at pH 7 can be excited to produce a final pH of $\sim$ 4 [4].

Recently, two groups, Campillo, Clark, Shapiro and Winn [5] and Smith, Kaufmann, Huppert and Gutman [6] have measured the rates of proton ejection from hydroxy-aromatic compounds by utilizing picosecond emission spectroscopy techniques. The rates from 2-naphthol, 6-sulfonate (pK*$\approx$1.66) [5], 2-naphthol-3,6-disulfonate (pK* $\approx$ 0.5) [6] and 8-hydroxy 1,3,6 pyrene trisulfonate (pK* $\approx$ 0.5) [6] have been found to be $1.0 \times 10^9 \mathrm{sec}^{-1}$, 1.5×10^{10} sec^{-1} and $1.2 \times 10^{10}\mathrm{sec}^{-1}$, respectively. A plot of log k vs. pK* shows a linear dependence with a slope close to unity [6]. This implies that the proton ejection rate is inversely proportional to the excited state pK*.

The property of ejecting a proton from excited hydroxy aromatic compounds with pK* greater than 0.5 is very sensitive to the solvent. Water was found to be the only solvent where proton transfer to the solvent occurs from these excited molecules. In order to elucidate the role of the solvent in the proton transfer mechanism, we have measured the rates of proton ejection from 8-hydroxy 1,3,6 pyrene trisulfonate in water-alcohol mixtures as a function of the binary solvent composition.

Experimental

8-hydroxy, 1,3,6 pyrene trisulfonate was purchased from Eastman. All samples were dissolved in solutions prepared from distilled water and absolute ethanol. The pH was adjusted by addition of perchloric acid or NaOH and checked with a Beckman BU pH meter. The steady state fluorescence was measured with a Perkin Elmer-Hitachi MPF-4 spectro-fluorimeter.

The picosecond emission apparatus was similar to that described earlier [6]. The samples were excited with 6 picosecond pulse at 353 nm obtained by generating the third harmonic from a mode-locked neodimium phosphate glass laser (KORAD). The emission from either the excited anion or the neutral species was isolated with glass color filters and recorded with a streak camera (Hamamatsu Model C 979). The output of the streak camera was imaged onto a silicon-intensified Vidicon target detector, (PAR 1205D optical multichannel analyzer).

Results

The fluorescence spectra of protonated 8 hydroxy 1,3,6 pyrene trisulfonate (HPS) exhibits two spectral bands with maxima at 445 nm and 510 nm; corresponding respectively to the protonated and deprotonated molecule. The rate of proton ejection from the electronically excited HPS was obtained either by measuring the decay of the protonated-molecule fluorescence at 450 nm or the rise of the unprotonated molecule fluorescence at 510 nm. A semi-logarithmic plot of the rates of proton ejection from excited HPS in water ethanol-mixtures as a function of the mole fraction of ethanol is shown in Fig. 1. The proton ejection rate is sharply decreased upon dilution of water with ethanol.

In addition to the direct picosecond observations of the proton ejection kinetics, complementary steady state fluorescence measurements were carried out. As mentioned above the fluorescence spectra of protonated HPS consist

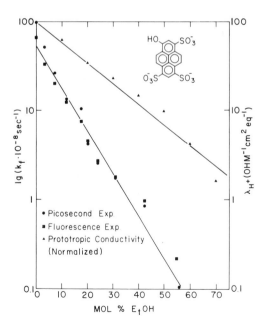

Fig.1 The dependence of the rate of proton ejection from the first electronically excited singlet of 8-hydroxy 1,3,6 pyrene trisulfonate on the mole fraction of ethanol in water-ethanol mixture. The rate of proton ejection was measured by picosecond emission technique (circles), or deduced from steady-state fluorescence (squares). The prototropic conductivity of the mixture (triangles) was taken from reference 7.

of two bands corresponding to the protonated and deprotonated form. The relative fluorescence intensity ratio ϕ'/ϕ, where ϕ' and ϕ are the fluorescence quantum yields of the deprotonated and the protonated form, respectively, provides indirect information on the dependence of the proton-ejection rate on ethanol concentration. The normalized relative fluorescence ratio ϕ'/ϕ is shown as a function of ethanol mole fraction in Fig. 1. It is evident from Fig. 1 that the ϕ'/ϕ dependence is the same as that of the direct picosecond measurements. As a comparison, the prototropic conductivity dependence on ethanol concentration taken from reference [7] is also plotted in Fig. 1. The proton ejection rate has a stronger dependence on solution composition than that of the bulk prototropic conductivity.

Discussion

We have correlated our results with the nonhydrodynamic fraction of the conductance of hydrogen ions in the mixed solvent. This non-Stokesian contribution to the ionic equivalent conductivity is commonly approximated [7] as

$$\lambda_{H^+} = \lambda^{\circ}_{H^+} - \lambda^{\circ}_{Na^+}$$

It is generally considered that the anomalous mobility of protons in aqueous solutions requires the presence of hydrogen-bonded ionic clusters (aggregates, chains, latticelike regions). The addition of nonelectrolytes (ethanol in this case) affects the structure of the solvent mainly by dissolving a fraction of the hydrogen bonds and changing the size of the clusters held together by these bonds. This structure-breaking effect decreases the clusters' size and hinders the prototropic conduction since monomer water molecules are unsuitable for this conduction. The conductivity depends on the bulk structure of the solvent and the average cluster order, <n> (which can be calculated for the various mole fractions [7].

The probability of the proton transfer process is dependent on the particular donor, i.e., the excited hydroxy aromatic compound, as well as the proton acceptor, liquid water. Theoretical [8] as well as experimental [9] studies have shown that the energy difference between H_3O^+ and $H_5O_2^+$ (n=2) is greater than 32 kcal/mole. Thus, from the energetical point of view, a water dimer is a much better acceptor than a monomeric water molecule. Higher degrees of water polymers might be even better proton acceptors than the dimer.

Subsequent to the primary proton transfer process, either diffusion of the proton out of the coulombic field of the excited anion, or ultrafast recombination of the proton with the donor takes place. The Debye radius, R_D for this molecule was calculated to be 28A [10]. The diffusion process of a proton within a cluster is an ultrafast process. Therefore, the probability of a proton to escape from this coulombic field is higher through large clusters than through small ones. Increasing ethanol concentration lowers the average cluster size and even more so, it decreases the number of large clusters. Thus the primary proton transfer process, as well as the diffusion out of the coulombic field, are strongly dependent on the ethanol concentration.

References

1. P. Mitchell, Chemiosmotic Coupling in Oxidative and Photosynthetic Phosphorylation (Glynn Research Ltd. Bodmin 1966).
2. T. Forster, Naturxiss. 36, 186 (1949).
3. J.F. Ireland and P.A.H. Wyatt, Adv. Phys. Org. Chem. 12, 131 (1976).
4. M. Gutman and D. Huppert, J. Biochem. Biophys. Methods. 1, 9, (1979).
5. A.J. Campillo, J.M. Clark, S.L. Shapiro and K.R. Winn, *Picosecond Phenomena*, ed. by C.V. Shank, E.P. Ippen and S.L. Shapiro, Springer Series in Chemical Physics, Vol. 4 (Springer, Berlin, Heidelberg, New York 1978) 319 - 326
6. K.K. Smith, K.J. Kaufmann, D. Huppert and M. Gutman, Chem. Phys. Lett. 65, 164 (1979)
7. T. Erdey-Gruz and S. Lengyel, *Modern Aspects of Electrochemistry No. 12* ed. by J. Bockris and B.E. Conway (Plenum Press, New York 1977) 1 - 40
8. W.P. Kaemer and G.H.F. Diereksen, Che. Phys. Lett. 5, 463 (1970)
9. P. Kebarle, S.K. Searles, A. Zolla, J. Scarbourough and M. Arshadi, J. Am. Chem. Soc. 89, 6393 (1967)
10. T. Forster and S. Volker, Chem. Phys. Lett. 34, 1 (1975)

Molecular Dynamics and Picosecond Vibrational Spectra

P.H. Berens and K.R. Wilson

Department of Chemistry, University of California, San Diego,
La Jolla, CA 92093, USA

The accessible time range for picosecond experimental measurements is now short enough, and that for molecular dynamics theoretical computation is now long enough, to significantly overlap. This overlap of theory and experiment can, at least in principle, be exploited to discover the trajectories of atoms during chemical processes in solution, for example rotational and vibrational relaxation [1], rotational and vibrational dephasing [2], and chemical reactions. This is significant in that our understanding of the microscopic basis of processes in liquids has fallen far behind our understanding of the gas phase. In fact, the atomic motions which form the microscopic mechanisms for solution processes are hardly understood at all, even though most processes of interest to most chemists occur in solution.

If we are efficiently to discover these microscopic atomic trajectories, i.e. the molecular dynamics of solution processes, we must be able to both theoretically compute and experimentally measure a macroscopic phenomena which depends closely upon the microscopic trajectories. We believe that transient vibrational spectra can serve this function. In this paper we will briefly discuss some of the theory for computing infrared and Raman spectra from molecular dynamics and will present an example of such a spectrum.

Two problems must be solved to compute such spectra. First, a theoretical approach must be developed which can be applied to the large number of atoms (~100-1000 including solvent molecules) involved in solution processes. This gives an incentive to develop ways to compute quantum reality largely from classical mechanics. Second, a computational approach must be developed capable of actually calculating spectra in a reasonable time. The latter problem is a significant one, as to compute 10^{-12} seconds of real time we find we may need to carry out 10^{12} floating point arithmetic operations, in order to sufficiently sample phase space.

To compute ir and Raman spectra we need three input functions. From $V(\{r_i\})$, the potential energy as a function of atomic positions, plus the initial positions $\{r_i(0)\}$ and velocities $\{\dot{r}_i(0)\}$ of the atoms, we compute, using classical mechanics, the set of atomic trajectories $\{r_i(t)\}$. From these trajectories and $\mu(\{r_i\})$, the dipole moment as a function of atomic positions, we compute $\mu(t)$, the dipole moment of the sample as a function of time. Using linear response theory [3-6], and suitable quantum corrections [6], we compute an ir spectrum. Finally, we average many such spectra over runs with different initial $\{r_i(0)\}$ and $\{\dot{r}_i(0)\}$ chosen from the ensemble which represents the experimental conditions of interest, usually a given temperature and pressure. In a similar manner, we use $\bar{\alpha}(\{r_i\})$, the polarizability matrix as a function of atomic positions, to compute the Raman spectrum. While the computations are mathematically relatively straightforward, the number of arithmetic operations is very large and we therefore carry them out on an array processor [7,8].

We have been refining this technique by comparing the computed spectra against known ir and Raman spectra of ordinary gas and liquid samples, i.e. time independent, equilibrium spectra. The results are encouraging. Fig.1 shows an example.

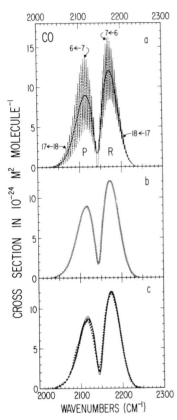

<u>Fig.1</u> The CO gas phase fundamental vibrational-rotational band at 298 K. The top panel shows the spectrum as computed from summing up all the applicable transitions calculated in the usual way from time dependent quantum perturbation theory. The dashed line shows the rotational transitions making up the P ($\Delta J=-1$) and R ($\Delta J=+1$) branches. The solid line shows the band contour produced by broadening out these rotational lines until they merge (which may be done experimentally by adding rare gas). The middle panel shows the quantum band contour (solid line) compared with the correspondence principle classical contour (open circles) calculated in the limit of quantum mechanics as Planck's constant $h \rightarrow 0$. Two simple and relatively small quantum corrections have been applied to the correspondence principle classical band contour. The rotational correction accounts for the different population ratios in classical and quantum mechanics of the rotational states connected by the transition. The vibrational correction is a frequency shift due to the fact that the classical diatomic vibrates at $\sim kT$ (~ 200 cm^{-1}), near the bottom of the well and thus in a largely harmonic region, while the quantum diatomic is sampling the potential in the $\sim \hbar\omega_o$ (~ 2000 cm^{-1}) energy range in which the effects of anharmonicity are more pronounced. As can be seen, after these quantum corrections the correspondence principle classical and the quantum band contours agree almost exactly, demonstrating that the band contour can be understood classically. The lower panel shows again the quantum band contour (solid line) and superimposed on it the quantum corrected Newtonian classical band contour (triangles) from classical molecular dynamics, classical linear response theory and classical statistical mechanical ensemble averaging. The filled circles show the shape of the experimental band contour for CO with added He as measured by ARMSTRONG and WELSH [9].

We see that the band contour results for all four of the following agree: i) quantum time-dependent perturbation theory, ii) quantum-corrected correspondence principle classical mechanics, iii) quantum-corrected Newtonian classical mechanics, and iv) experimental measurement. Thus, the vibrational-rotational band contour can be understood on the basis of classical trajectories of the atoms.

If we mix CO in Ar and keep increasing the Ar pressure, both by Newtonian classical mechanics [6] and by experimental measurement [10] the P and R branch peaks increasingly meld into one another until they fuse into the usual single-peaked liquid state vibrational band contour [11]. Therefore, in the liquid state as well, one can compute and understand the vibrational band contour in terms of classical atomic trajectories.

Finally, the techniques described here can be extended to treat the non-equilibrium domain, so that one can compute transient ir and Raman spectra during the course of picosecond processes, for example chemical reactions in solution. In this way, one can hope to combine the computation and the measurement of picosecond transient vibrational spectra to discover the microscopic mechanism, the atomic trajectories during chemical reactions in solution.

We wish to thank the National Science Foundation, Chemistry, the Office of Naval Research, Chemistry, and the National Institutes of Health, Division of Research Resources, for providing the support which has made this work possible.

References

1. D. W. Oxtoby, Adv. Chem. Phys., in press.
2. D. W. Oxtoby, Adv. Chem. Phys. 40, 1 (1979).
3. R. G. Gordon, Adv. Magn. Reson. 3, 1 (1968).
4. B. J. Berne in Physical Chemistry, An Advanced Treatise, Vol. VIIIB, edited by D. Henderson (Academic Press, New York, 1971).
5. D. A. McQuarrie, Statistical Mechanics (Harper and Row, New York, 1976) chapters 21 and 22.
6. P. H. Berens and K. R. Wilson, "Molecular Dynamics and Spectra: Diatomic Rotation and Vibration", submitted to J. Chem. Phys.
7. K. R. Wilson in Computer Networking and Chemistry, edited by P. Lykos (American Chemical Society, Washington, D. C., 1975).
8. K. R. Wilson in Minicomputers and Large-Scale Computation, edited by P. Lykos (American Chemical Society, Washington, D. C., 1977).
9. R. L. Armstrong and H. L. Welsh, Can. J. Phys. 43, 547 (1965).
10. R. Coulon, L. Galatry, B. Oksengorn, S. Robin and B. Vodar, J. Phys. Radium 15, 641 (1954).
11. U. Buonotempo, S. Cunsolo and G. Jacucci, J. Chem. Phys. 59, 3750 (1973).

Viscosity Dependence of Iodine Recombination

C.A. Langhoff, B. Moore, and W. Nugent

Chemistry Department, Illinois Institute of Technology,
Chicago, IL 60616, USA

Iodine photodissociation provided one of the original examples of the cage
effect in liquids. Since then, the concept of a cage has become of central importancè to chemists in guiding and understanding their choice of
solvents for particular chemical reactions. Recently, HOFFMAN, et al. [1]
applied picosecond techniques to study the dissociation and subsequent recombination of I in CCl₄ and hexadecane. They showed that the recombination occurs in two time scales, a very short (around 100 psec) and a much
longer one (greater than 1 nsec.). This was interpreted as a direct demonstration of the cage effect. The short time decay is the recombination of caged radicals (geminate) and the long time decay is due to recombination of randomly diffusing I atoms (nongeminate). This experiment
provides a good opportunity to test various diffusion theories, brownian
motion theories, and molecular dynamics descriptions of translational motion in liquids. Thus, we have extended these experiments to study the
recombination as a function of solvent viscosity and structure.

The experimental apparatus is shown in Fig. 1.

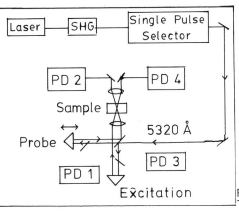

Fig. 1   Schematic of the laser system

The four photodiode signals, measuring incident and transmitted intensity
for both the probe and excitation pulses, are electronically integrated,
sampled and held, and subsequently digitized. The four signals are stored
in a computer (Cromemco Z2-D) where subsequent analysis takes place. The
raw data is treated as follows. For each delay line setting, two sets of
measurements are collected. The first set is typically 30 to 50 laser

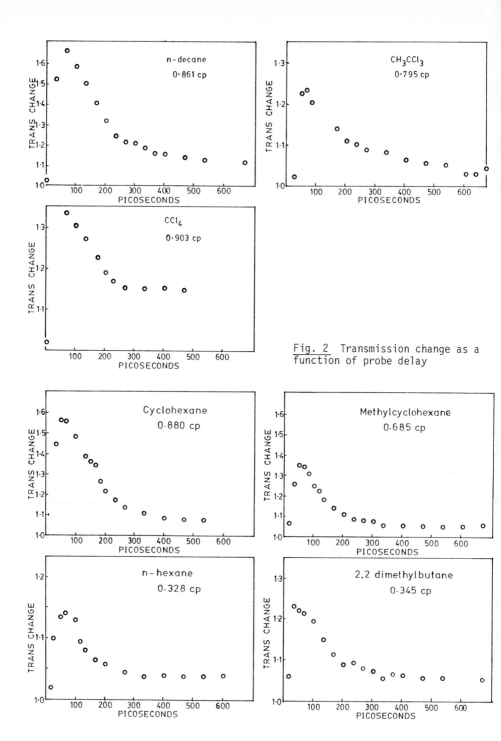

Fig. 2 Transmission change as a function of probe delay

shots and has the excitation beam blocked. This measures the zero intensity transmission and allows us to correct for changes in the geometry of the optics, or for long-time, permanent changes in the transmission of the sample. Both effects would bias the measured excitation induced transmissions. The second set is the actual experiment. Probe transmissions are measured in the presence of excitation. Again around 50 shots are collected for each delay line setting. The calculated probe transmissions are linear least squares fitted to the initial intensity of excitation. The linear fits are quite good with correlation coefficients generally exceeding 0.75. The transmission is then calculated for a selected intensity (which remains constant for the entire experiment) from the parameters of the least squares line. This means that transmissions for all decay line settings are evaluated at the same intensity removing the effect of fluctuating laser intensities. The calculated transmission is then divided by the no excitation transmission as mentioned above. A complete decay curve can be collected in 2 to 3 hours. The data evaluation scheme described above has been found to be essential in obtaining accurate and reproducible data. Statistical accuracy is found to be around 1%.

The data for seven solvents is shown in Fig. 2. These solvents cover a range of 3 in viscosity, from 0.328 cp for n-hexane to 0.903 cp for CCl_4. All solvents are hydrocarbons and show very little interaction with iodine molecules or atoms [2]. We at present have on order several solvents which will extend the range of viscosities up to 30 cp.

There are two general points to be noted about the data. First, the dependence of the initial decay (geminate recombination) on viscosity is weak, if it is at all dependent on viscosity. The time from the peak of the bleaching to the beginning of the long time decay is about 125 to 150 psec for all solvents (with perhaps the exception of CH_3-CCl_3). It remains to be seen if this constancy holds for a wider range of viscosities. Secondly, the ratio of the peak bleaching to the bleaching at long times varies quite a bit. N-hexane has a very small peak bleaching, whereas the peak for n-decane is quite large. At present we can offer no firm explanation for these effects without having more data available. We are in the process of measuring these decays for more solvents and, in addition, to investigating the concentration and temperature dependence of the recombination.

A possible, although as yet unproven, conjecture about these results can be made. Assuming the constancy of the geminate decay time is verified, this may mean that the time measured is limited by a kinetic process rather than the presumed diffusional mechanism. It is known, and has been the concern of several workers [3], that the recombination of I atoms may not be simply a recombination on the ground state potential surface, but may involve some low-lying excited electronic states. The geminate recombination time may be just this chemical process. The variation in peak to long-time transmission could be interpreted as a measure of the quantum yield of the caging process. This ratio would then be a measure of the diffusional forces in the liquid. We stress that the conjectures above are only suggestions and await further investigation, both experimental and theoretical.

References

1. T.J. Chuang, G.W. Hoffman, and K.B. Eisenthal, Chem. Phys. Lett. 25, 201 (1974).
2. C.A. Langhoff, K. Guadig, and K.B. Eisenthal, Chem. Phys. 46, 117 (1980).
3. J.T. Hynes, R. Kapral and G.M. Torrie, J. Chem. Phys. 72, 177 (1980).

High-Power Picosecond Molecular Fragmentation and Rapid C-X Radical Formation[1]

B.B. Craig[2], W.L. Faust, L.S. Goldberg, P.E. Schoen, and R.G. Weiss[2]

Naval Research Laboratory, Washington, D.C. 20375, USA

[2] Department of Chemistry, Georgetown University, Washington D.C. 20057, USA

1. Introduction

We report the first high-power picosecond uv photofragmentation experiments on simple organic molecular gases. These studies extend into the picosecond realm the capability to resolve primary dissociative processes and rapid fragment formation, even at gas pressures approaching one atmosphere. We have examined the time dependence of characteristic visible emission from product species C_2, CH, CN, and H upon irradiation of ketene, methane, nitromethane and carbon monoxide at pressures 10-500 torr. The excitation mechanism invokes multiphoton absorption and dielectric breakdown processes, a consequence of the high uv flux densities ($>10^{11}$W/cm^2) and pressures of our experiment. We find that the fragment distributions produced by this excitation are remarkably similar to those found by Ambartzumian et al.[1], who employed CO_2 TEA laser multiphoton excitation at gas pressures of one torr or less. Rotationally-hot C_2 diradicals, emitting the characteristic $d^3\pi_g \rightarrow a^3\pi_u$ Swan bands, are produced from ketene and from methane more rapidly than would be anticipated from thermalized collisions. For CO there is a different formation process yielding C_2 Swan emission, with very different features of vibration-rotation distribution and of slow temporal development.

2. Experiment

The Nd:YAG laser system (Fig.1) operated at a flashlamp repetition rate of 1 Hz. The oscillator employed hybrid modelocking (active acousto-optic loss modulator and passive saturable absorbing dye) to enhance the reproducibility of pulse trains. A single 1064 nm pulse switched from the train had ca. 0.4 mJ energy and typically 30 ps duration. Two stages of amplification, apodization, and spatial filtering of the beam provided an ir pulse of high spatial quality and of 30-40 mJ energy. Efficient doubling and redoubling in KD*P crystals generated a 266 nm photolyzing pulse of up to 10 mJ energy. The laser beam was focused to a 0.15 mm dia. spot within a static gas cell having LiF windows. Light emitted from a visible streak within the focal region was collected at right angles and focused into a 0.8 m spectrometer coupled to a Nuclear Data ND100 intensified vidicon recording system (spectral sensitivity 380-800 nm). For time-resolved emission, a Varian VPM-154 crossed-field photomultiplier was coupled to the exit slit. The transient signal was displayed on a Tektronix 7104 oscilloscope, giving a system risetime ca. 400 ps. Observations also were made with an Electrophotonics streak camera (S-20 photocathode) having time resolution of 10 ps.

---
[1] Research supported in part by the Office of Naval Research and by the Defense Advanced Research Project Agency

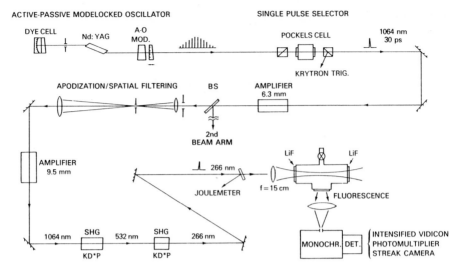

ACTIVE-PASSIVE MODELOCKED OSCILLATOR   SINGLE PULSE SELECTOR

DYE CELL   Nd: YAG   A-O MOD.   POCKELS CELL   1064 nm 30 ps

KRYTRON TRIG.

APODIZATION/SPATIAL FILTERING   BS   AMPLIFIER 6.3 mm

2nd BEAM ARM

AMPLIFIER 9.5 mm

LiF   LiF

266 nm

JOULEMETER

f = 15 cm

FLUORESCENCE

1064 nm   SHG   532 nm   SHG   266 nm

KD*P   KD*P

MONOCHR. DET.

INTENSIFIED VIDICON
PHOTOMULTIPLIER
STREAK CAMERA

Fig.1   Schematic of the laser system

3. Results & Discussion

Figure 2 shows low-resolution (2 nm) emission spectra obtained from irradiation of ketene (CH_2CO) and carbon monoxide, at 100 torr. Spectra (Figs.3,4) at higher resolution (0.1 nm) demonstrate that the predominant emission belongs to the C_2 diradical in its triplet $d^3\pi_g \rightarrow a^3\pi_u$ Swan transition. Figure 3 shows the C_2 $\Delta v = -1$ emission from ketene, methane, and carbon monoxide; other transitions from $\Delta v = -3$ to $+2$ are found between ca. 670 and 435 nm. The spectra from ketene and from methane exhibit strong nonthermal rotational structure. In addition, there is a weak underlying continuum emission associated with a plasma (duration ca. 2 ns) that extends throughout the visible region. This plasma emission is most prominent for methane. In the $\Delta v = +2$ region (Fig.4), there are striking differences in the emission derived from ketene and from carbon monoxide. For ketene, weak CH emission bands are observed. Only for CO among the gases studied are the C_2 high-pressure (HP) bands observed, They are attributed to selective population of a specific upper vibrational level (near v'= 6) of the $d^3\pi_g$ state [2,3].

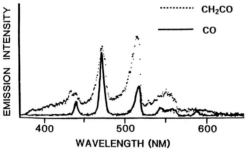

CH₂CO

CO

EMISSION INTENSITY

400   500   600

WAVELENGTH (NM)

Fig.2   Low-resolution (2 nm) emission spectra obtained when ketene and carbon monoxide, at 100 torr, are irradiated with individual laser pulses at 266 nm

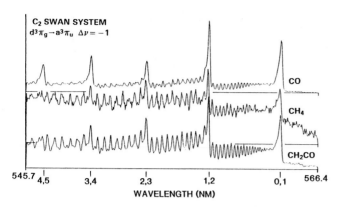

Fig.3 High-resolution (0.1 nm) spectra of the Swan system emission ($\Delta v = -1$) derived from 266 nm irradiation of carbon monoxide, methane, and ketene, at 100 torr. Data are accumulated from 30 laser shots

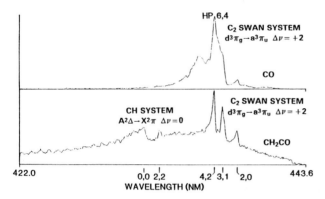

Fig.4 Spectra of the Swan system emission ($\Delta v = +2$). Conditions as in Fig.3. The spectra also show the high-pressure 6,4 band and CH emissions

Distinct formation mechanisms for C_2 from CO and from the other parent molecules are indicated by the time-dependent oscilloscope data (Fig.5), as well as by the spectra. Figure 5A (lower trace) shows the C_2 emission monitored at 516 nm for ketene at 100 torr. This risetime-limited signal incorporates a component of emission from the prompt plasma radiation, which is represented in the upper trace monitored at 616 nm where Swan emission is negligible. In initial measurements at 10 torr ketene, for which the plasma emission is minor, there is an indication of a C_2 formation time in excess of the instrumental risetime. In contrast, no prompt C_2 emission is seen from 100 torr CO at 516 nm (Fig.5C) in either the normal or the high pressure bands. The trace shows only the brief plasma emission. In Fig. 5D for 100 torr CO we observe slow formation and decay of C_2, consistent with a collisional process [2] involving long-lived intermediates. The C_2 emission from 100 torr ketene (Fig.5B) has a decay time of ca. 70 ns (cf. collision-free lifetime [4] ca. 120 ns). This indicates quenching by the parent molecule and/or dissociation fragments (e.g., we observe a strong Balmer H_α emission line of atomic hydrogen at 656.3 nm).

Strictly prompt C_2 formation would suggest a unimolecular mechanism. On the other hand, rapid formation may be explained if reacting fragments are created with substantial kinetic energy or if there is ground-state association of parent molecules. Initial attempts to observe the formation with a streak camera have proved inconclusive; the spectral resolution required to discriminate against the prompt (rise ca. 20 ps) plasma background has not yet allowed sufficient C_2 signal to be detected. For methane, of course, C_2 formation must be collisional. The plasma radiation, however, dominates the photomultiplier signal for ca. 2 ns, by which time the C_2 signal is already fully developed.

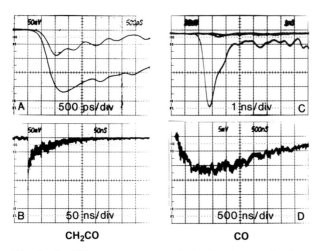

Fig.5 Oscilloscope traces of the C_2 Swan emission at 516 nm (0,0 transition). A ketene, 100 torr, 500 ps/div. The upper trace shows the plasma radiation at 616 nm. B ketene 100 torr, 50 ns/div. C CO, 100 torr, 1 ns/div. D CO, 100 torr, 1 µs/div

The power dependence of C_2 is displayed in Fig. 6 for ketene and for carbon monoxide at 100 torr. At high-input pulse energies both gases exhibit a limiting linear slope indicative of a saturation regime. The high order of the excitation process is clearly evident from the steepening of the curves towards lower input energies. Carbon monoxide and methane show no emission at pressures below 10 torr. Ketene, however, possesses a single-photon transition at 266 nm and exhibits luminescence even at pressures below 1 torr.

For nitromethane (Fig. 7) the C_2 Swan emission is substantially weaker than for other parent molecules. The emission shows excess rotation as seen for ketene and methane. The predominant emitting fragment is CN, in its $B^2\Sigma^+ \rightarrow X^2\Sigma^+$ transition, with band heads at 421.6 and 388.3 nm. Studies of the CN time development are in progress to determine whether the fragment is produced unimolecularly or by collision.

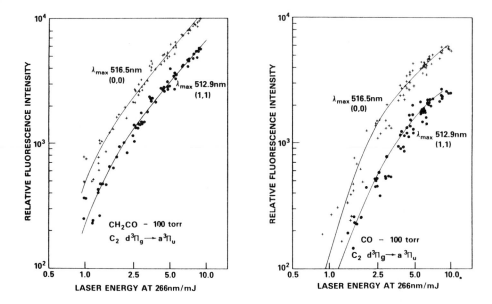

Fig.6 Dependence of the C$_2$ Swan emission intensity on input laser pulse energy at 266 nm, from ketene (left) and carbon monoxide (right), at 100 torr

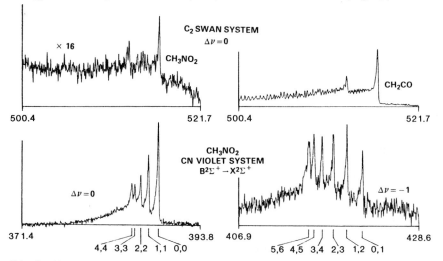

Fig.7 Above Spectra of the C$_2$ Swan emission ($\Delta v = 0$) derived from ketene at 10 torr and nitromethane at 15 torr. Note reduced intensity of the signal from nitromethane. Below Spectra of the CN violet system emission ($\Delta v = 0$, $\Delta v = -1$) derived from nitromethane at 15 torr. Data are accumulated from 30 laser shots

References

1 R.V. Ambartzumian, N.V. Chekalin, V.S. Letokhov, E.A. Ryabov, Chem. Phys. Lett. 36, 301 (1975).
2 C. Kung, P. Harteck, and S. Dondes, J. Chem. Phys. 46, 4157 (1967).
3 S.M. Read, and J.T. Vanderslice, J. Chem. Phys. 36, 2366 (1962).
4 J.R. McDonald, A.P. Baronavski, and V.M. Donnelly, Chem. Phys. 33, 161 (1978).

Photoinduced Isomerism of Stilbenes: Picosecond Timescale Results [1]

F.E. Doany, B.I. Greene[2], Y. Liang[3], D.K. Negus, and R.M. Hochstrasser
Department of Chemistry,
Laboratory for Research on the Structure of Matter, University of Pennsylvania,
Philadelphia, PA 19104, USA

Introductory Review:

It is well known that substituted ethylene can undergo cis-trans isomerism following optical excitation. A large literature already exists on this topic and the particular case of stilbenes (1,2-diphenylethylenes) was recently reviewed by Saltiel and coworkers [1]. Phenyl substituted ethylenes have received considerable attention from photochemists because their optical spectra are in the near ultraviolet, readily accessible to traditional light sources. Although the ground work for understanding the photoisomerism process in solutions was laid on the basis of conventional (pre-laser) experiments [1,2] the intrinsic molecular steps involved are now known from vapor phase studies to be subnanosecond [3,4]. The basic model for the isomerism is that a twisted, less conjugated form of stilbene is formed following optical excitation [5]. This twisted form may then undergo relaxation processes that result in both cis and trans structures [6]. The effect of the solvent is to slow down the isomerization and at sufficiently low temperature or high viscosity the rate is found to be even slower than radiative processes [7]. The fluorescent properties of the system therefore can be used to monitor the isomerization. Saltiel and coworkers [8,9] have shown that the principal photochemistry of stilbene arises from the singlet excited surface although the triplet state can be involved when sensitizers are present [9,10]. There is also evidence that isomerization is not the exclusive photoprocess since small yields of rearranged species have been reported [11], but such processes will not be considered further in this article.

The electronic states of stilbenes were first characterized by their low temperature spectroscopy in mixed crystals [12]. These measurements by Dyck and McClure refer to essentially planar molecules - so constrained by the host crystal lattice. Under such circumstances the optical absorption starting at 325 nm exposes a single electronic transition with the bulk of the intensity appearing in two modes at 212 cm^{-1} and 1600 cm^{-1}. The lower frequency mode corresponds to an in-plane symmetric bend of the ethylenic

---

[1]
This research was supported by a grant from the National Science Foundation (CHE 800016) and in part by the NSF/MRL Program under Grant DMR76-80994.
[2]
Present Address: Bell Laboratories, Holmdel, N. J. 07733, USA.
[3]
IBM Predoctoral Fellow.

portion of the molecule while the other progression forming mode is a car-
bon-carbon stretch involving the conjugated backbone of the molecule. No
progressions of low frequency torsional modes are seen in the spectrum indicating no substantial alteration in planarity of the ethylene system in
the excited state. On the contrary the electronic spectrum of ethylene [13]
displays a long Franck-Condon progression of the out-of-plane bending mode.
These experiments show that the relevant states of ethylene and stilbene
are quite different, and indeed this is confirmed by theoretical calculations as discussed below. Warshel [14] has calculated an equilibrium geometry for stilbene subject to no anisotropic constraints and finds that the
phenyl rings are significantly rotated from the planar configuration in
both the trans and cis isomers. This conclusion is consistent with electron
diffraction studies [15]. The molecular spectroscopy of the nonplanar ground
states of the stilbenes is not structurally informative: Both the vapor [16]
and the solution spectra are sufficiently diffuse that the details of the
vibrational structure are lost. Presumably this situation will be reme-
died as a result of future spectroscopic studies in supersonic jets. The
calculations of the optical spectrum based on Warshel's structure [14] indicate that significant changes in the phenyl ring orientations might occur
in the electronically excited state. Recent studies of the variation with
exciting wavelength of fluorescence from trans-stilbene excited in the region of the 0-0 band [3,4] have indicated that the absorption spectrum of
cold molecules should reveal valuable dynamical information. The existence
of strong two photon absorption [17] (λ_{max} = 2 x 250 nm) has indicated the
presence of a gerade state in stilbene that may be important for the photochemistry.

Theoretical studies of the electronic structure of stilbenes have advanced far beyond π-electron calculations of the spectra [18]. The more
recent work has attempted to compute the potential surfaces for ground and
excited states corresponding to twisting about the ethylenic double bond
and internal rotations of the phenyl groups [19-22]. A significant con-
ceptual advance was made when Orlandi and Siebrand suggested that the singlet
isomerism might be mediated by an avoided crossing of the delocalized $\pi\pi^*$
(first singlet) surfaces with a 'gerade' type state involving mainly two-
electron excitation [19]. This suggestion is analogous to current views
on the isomerism of polyenes [23,24]. The avoided crossings of the $\pi\pi^*$
and other types of states having lower energy in the twisted configuration
have also been invoked to explain the photoproperties [25]. These theoretical calculations suggest that the photoisomerization is an adiabatic process that may involve a small barrier in passing from the trans to the
twisted conformation. Although the theories lead to qualitative agreement
with earlier experiments, there is as yet little theoretical information
on the effects of coordinates other than twisting about the double bond,
nor do the calculations lead to strong predictions regarding the depth of
the potential minimum of the twisted conformation.

Picosecond laser experiments on the stilbene system have resulted in
the confirmation of some of the earlier ideas by allowing direct measurements of subnanosecond fluorescence and singlet state absorption spectra
of solutions at room temperature. The fluorescence lifetime was measured
as a function of temperature and solvent using streak camera techniques
[26-28]. Transient absorption of trans-stilbene was discovered at around
550 nm after pumping with the Nd: glass fourth harmonic [29]. The decay
of this absorption occurred with a lifetime of 90 ps leading to products
that were transparent in the region 400-700 nm [29]. These measurements

probably signal the primary step in the process of isomerism. The cis iso-
mer disappears too quickly to have yet been exposed by short pulse experiments. Estimates based on the absence of transient absorption signals place
its lifetime at less than 1 ps in hexane at 300 K [29]. At lower temperature in rigid media the cis - singlet lifetime is much longer and both fluorescence and transient absorption have been observed [30]. In a recent experiment the trans form was observed to appear in a time less than 20 ps
after pumping the cis [31]. Further experiments along these lines are
needed to elucidate the isomerization dynamics.

<u>Vibrational Relaxation</u>: Optical excitation at 265 nm introduced in addi-
tion to thermal energy about $6450\ cm^{-1}$ vibrational energy excess onto the
excited singlet state surface. The excited state absorption spectrum
changed in shape during the first 50 ps following excitation [29]. This
was interpreted as being due to the presence of thermally unrelaxed molecules. During this same period the excited state population decays exponentially with a time constant of 68 ps (n-hexane, 300K). The decay constant is not very dependent on the vibrational energy excess. This is most
readily seen from the fluorescence decay profile (Figure 1) which shows only

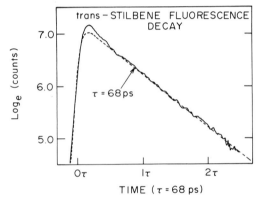

TIME (τ = 68 ps)

<u>FIGURE 1</u>   Fluorescence decay of trans-stilbene in hexane. Experimental data
(full line) average of seven traces using 265 nm single pulse excitation and
GEAR-Pico-V streak camera. The dashed line represents the convolution of a
68 ps. decay with the average single pulse shape.

a slight deviation from a single exponential near t = o. These important
results imply that the hot molecules that exist prior to 50 ps (as observed
with ca. 10 ps pulses) do not isomerize much more efficiently than those
that are thermalized. The vibrational relaxation is therefore faster than
than the vibrational energy redistribution leading to isomerism for most
levels reached in the cascade from the optically prepared state.

     Evidence in favor of the hot molecule interpretation of the spectral
shape changes was recently obtained in experiments where the sample was
excited with a 306 nm pulse [32]. The transient absorption spectrum showed
essentially no time evolution under these circumstances when the system was
pumped and probed with ca. 10 ps pulses. These 306 nm pulses generate much
cooler molecules having ca. $1300\ cm^{-1}$ vibrational energy excess in addition
to thermal energy. It is hard to eliminate the possibility that these differences at 265 nm and 306 nm excitation are a result of hole-burning an

inhomogeneous distribution of molecules since the transient spectrum is too complex to yield structural information. One piece of evidence favoring the present interpretation is that the fluorescence quantum yield of the solution is essentially independent of excitation wavelength [33,34].

Isomerization Dynamics: Figure 2 shows a sketch of the singlet surface on which the various configurational changes are occurring. Excitation at the trans configuration results first in relaxation at that configuration. The

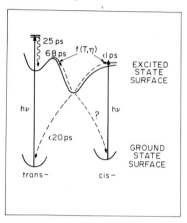

FIGURE 2   Schematic diagram of the relevant portion of the potential surface of stilbene. The pathways of the various picosecond processes that have been explored in solutions are indicated. $f(T,\eta)$ indicates a process known to depend on temperature and viscosity.

isomerization then presumably occurs by the transmission of appropriately activated molecules across the barrier towards the twisted configuration, with the rate being controlled by the hydrodynamic drag modified by the inertia introduced through the molecular potential. An important issue concerns the recurrence of excited trans molecules. If twisted molecules were thermally reformed into excited trans states that could radiate, then the fluorescence decay should have a nanosecond component.

Non-exponential decays of trans fluorescence have been reported. Heisel et al. [28] suggested that the deviation from exponential was caused by an impurity, but Taylor et al. [26,35] have insisted that their high repetition rate streak camera has such unique dynamic range that their slow (ca. 1.5 ns) fluorescence component would not be seen with the conventional streak cameras used by Heisel et al. We have used a GEAR streak camera to investigate the fluorescence of trans-stilbene out to ca. 5 lifetimes. In this experiment the dynamic range of the system is a few thousand (not the 'typical' value of 30 quoted in refs. 26 and 35). The results are given in Figure 3 on a semilog plot of the fluorescence decay. These data (OMA output) fit an exponential decay having the time constant of 68 ps, whereas, the sum of the two exponentials reported by Taylor et al. [26,35] (shown as a curve on Figure 3) does not fit the measurement. It is apparent that the conclusions from the previous work are incorrect, and that there is no significant recurrence of excited trans molecules.

A Dynamical Model: Our results indicating that the isomerization initiates from thermalized molecules is suggestive that the changes in molecular shape are a result of solvent motion, though mediated by the molecular potential function. McCaskill and Gilbert[36] have discussed at length the application of the Fokker-Planck equation to conformational relaxation in solution.

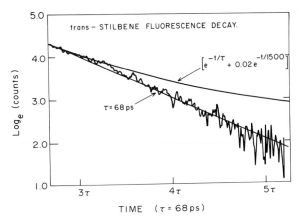

FIGURE 3   Fluorescence of trans stilbene in hexane.   Data are an average of traces from 13 laser shots (265 nm).

They found that the Kramers equation (below) was able to account for the picosecond twisting dynamics of binaphthyl.   The Kramers equation relates the

$$1/\tau = (1/\tau_v) \; (\nu/4\pi \; \nu') \; [(1 + \{\pi\nu'/\tau_v\}^2)^{\frac{1}{2}} - 1] \; e^{-Q/kT} \qquad (1)$$

time constant (τ) for passing over a hill in the molecular potential to the barrier height (Q), the absolute temperature (T), the correlation time (τ_v) for the angular velocity of twisting, and to characteristic frequencies ν' and ν of the piecewise parabolic potentials that compose the hill and the valley from which the system evolves.   The velocity correlation frequency $1/\tau_v$ is the ratio of the angular drag coefficient to the moment of inertia about the axis of rotation.   The drag coefficient for spherical objects having hydrodynamic radius d and radius of gyration r is $4\pi\eta \; dr^2$, with η being the viscosity of the solvent.   When $\nu'\tau_v \ll 1$ equation (1) reduces to a Smoluchowski result for passage over the barrier, in which τ is proportional to the viscosity.

Taylor et al. [26] have measured the decay time of the fluorescence of trans-stilbene as a function of solvent composition in ethanol-glycerol mixtures.   We have evaluated the variation of the faster of the two decays reported by these workers as a function of the solvent viscosity, and find that ($d\tau/d\eta$) is approximately constant at 87 ps/poise for viscosities in the range 0.1 to 1.0 poise.   These results are consistent with equation (1) for $\nu\nu'/c^2 = 6.6 \times 10^5$ cm^{-2} (we use $I = 9.2 \times 10^{-38}$ gm cm^2; d = 1.7 Å; r = 2.42 Å; Q = 3.2 kcal mole^{-1}).   A value for ν/c of 450 cm^{-1} is reasonable for the frequency of twisting about the double bond according to Warshel, so that ν'/c is ca. 1400 cm^{-1}.   These frequencies are significantly less than $1/\tau_v$ for the chosen range of viscosities.   The results are consistent with the isomerism being an essentially classical statistical processes understood in terms of a simple separation of stilbene and solvent motions.   In this case, the isomerizing stilbene molecule is subject to large numbers of binary uncorrelated collisions with the solvent at each angular increment along the twisting coordinate.   At much higher viscosities the values of ($d\tau/d\eta$) are significantly smaller and the present simple model appears to be inapplicable.

263

The solution phase measurements [29,32] suggest that in hexane at 300K during the period 0 to 200 ps after pumping, there should exist a wide range of molecular conformations ranging from trans through the twisted form. Perhaps it will be possible to detect these by means of transient Raman effects.

Properties of Isolated Molecules vs. those in Solution

The vibrational energy redistribution leading to isomerization of trans-stilbene with 265 nm excitation in a low pressure gas occurs in 15 ps [3,4]. This indicates that an intrinsically fast molecular process exists. The overall effect of the solvent seems to slow down the conformational change. However, the excess energy dependence of the twisting rate shows that for small energy excesses over thermal, the lifetime is lengthened to ca. 1 ns in the isolated molecule [4]. Thus the isomerization is initiated in 1000 ps (indicated by the disappearance of the optically prepared state) for the free molecule, whereas the thermally assisted passage over the barrier in hexane solution occurs with a characteristic time of 68 ps - a time that decreases with decreasing viscosity [1]. Presumably the solution process is accelerated because the thermal distribution is dynamically maintained by exchange of energy with the medium. In the isolated molecule the isomerization is limited by the rate of vibrational energy redistribution and only those molecules that have sufficient vibrational energy may isomerize. It is apparent from this discussion that the vapor phase fluorescence decay could be non-exponential for optical excitations causing low energy excess. Experiments to observe this were not yet carried out though we assumed in a previous study of the quantum yields of fluorescence [4] that the decay would be exponential.

References

1. J. Saltiel, J. D'Agostino, E. D. Megarity, L. Metts, K. R. Neuberger, M. Wrighton and O. C. Zafiriou, Org. Photochem., 3, 1 (1973).
2. E. Fischer, M. Frankel and R. Wolovsky, J. Chem. Phys., 23, 1365 (1955).
3. B. I. Greene, R. B. Weisman and R. M. Hochstrasser, J. Chem. Phys., 71, 544 (1979).
4. B. I. Greene, R. B. Weisman and R. M. Hochstrasser, Chem. Phys., in press.
5. G. S. Hammond and J. Saltiel, J. Am. Chem. Soc., 85, 2516 (1963).
6. J. Saltiel, J. Am. Chem. Soc., 89, 1036 (1967).
7. G. Fischer, G. Seger, K. Muszkat and E. Fischer, J. Chem. Soc., Perkin II, 1569 (1975).
8. J. Saltiel, E. D. Megarity and K. Kneipp, J. Am. Chem. Soc., 88, 2336 (1966).
9. J. Saltiel, J. Am. Chem. Soc., 90, 6394 (1968).
10. G. S. Hammond, J. Saltiel, A. A. Lamola, N. J. Turro, J. S. Bradshaw, D. O. Cowan, R. C. Counsell, V. Vogt and C. Dalton, J. Am. Chem. Soc., 86, 3197 (1964).
11. R. Srinivasan and J. C. Powers, Jr., J. Chem. Phys., 39, 580 (1963).
12. R. H. Dyck and D. S. McClure, J. Chem. Phys., 36, 2326 (1962).
13. G. Herzberg, Electronic Spectra of Polyatomic Molecules (New York: Van Nostrand Reinhold Company, 1966) pp. 533-5.
14. A. Warshel, J. Chem. Phys., 62, 214 (1975).
15. M. Traetteberg, E. B. Frantsen, F. C. Mijlhoff and A. Hoekstra, J. Mol. Struct., 26, 57 (1975).
16. O. P. Kharitonova, Opt. Spectry., 10, 394 (1960).
17. T. M. Stachelek, T. A. Pazoha, W. M. McClain and R. P. Drucker, J. Chem. Phys., 66, 4540 (1977).
18. H. Suzuki, Bull. Chem. Soc. Jpn., 33, 379 (1960).

19. G. Orlandi and W. Siebrand, Chem. Phys. Lett., $\underline{30}$, 352 (1975).
20. G. Orlandi, P. Palmieri and G. Poggi, J. Am. Chem. Soc., $\underline{101}$, 3492, (1979).
21. A. Wolf, H.-H. Schmidtke and J. V. Knop, Theoret. Chim. Acta, $\underline{48}$, 37 (1978).
22. P. Tavan and K. Schulten, Chem. Phys. Lett., $\underline{56}$, 200 (1978).
23. B. S. Hudson and B. E. Kohler, Chem. Phys. Lett., $\underline{14}$, 299 (1972).
24. K. Schulten, I. Ohmine and M. Karplus, J. Chem. Phys., $\underline{64}$, 4422 (1976).
25. M. C. Bruni, F. Momicchioli, I. Baraldi and J. Langlet, Chem. Phys. Lett., $\underline{36}$, 484 (1975).
26. J. R. Taylor, M. C. Adams and W. Sibbett, J. Photochem., $\underline{12}$, 127 (1980).
27. M. Sumitani, N. Nakashima, K. Yoshihara and S. Nagakura, Chem. Phys. Lett., $\underline{51}$, 183 (1977).
28. F. Heisel, J. A. Miehe and B. Sipp, Chem. Phys. Lett., $\underline{61}$, 115 (1979).
29. B. I. Greene, R. M. Hochstrasser, and R. B. Weisman, Chem. Phys. Lett., $\underline{62}$, 427 (1979).
30. K. Yoshihara, A. Namiki, M. Sumitani and N. Nakashima, J. Chem. Phys., $\underline{71}$, 2892 (1979).
31. M. Sumitani, N. Nakashima and K. Yoshihara, Chem. Phys. Lett., $\underline{68}$, 255 (1979).
32. F. E. Doany, B. I. Greene and R. M. Hochstrasser, Chem. Phys. Lett., in press.
33. R. M. Hochstrasser, Can J. Chem., $\underline{37}$, 1367 (1959).
34. G. Bartocci and U. Mazzucato, Chem. Phys. Lett., $\underline{47}$, 541 (1977).
35. J. R. Taylor, M. C. Adams and W. Sibbett, Appl. Phys. Lett., $\underline{35}$, 590 (1979).
36. J. S. McCaskill and R. G. Gilbert, Chem. Phys., $\underline{44}$, 389 (1979).

Study on Fast Processes in Radiation Chemistry by Means of Picosecond Pulse Radiolysis

Y. Tabata, Y. Katsumura, H. Kobayashi, M. Washio, and S. Tagawa

Nuclear Engineering Research Laboratory,
Faculty of Engineering, University of Tokyo,
Tokaimura, Ibaraki, Japan 319-11

1. Single Picosecond Electron Pulse for Study on Radiation Chemistry

In order to obtain picosecond electron pulses, our S-band linac has been specially designed [1]. Important factors affecting picosecond pulses obtained are mentioned and some of those are discussed.

A computer simulation for the bunching process of electrons has been done to make clear the important factors affecting the process, and to find criteria for obtaining a narrow, intense, single shot without satellite pulses. A following simple formula has been used for the calculation.

$$t_2 = t_1 + \frac{S}{\sqrt{\frac{2eV_0}{m}} \sqrt{1 + \frac{V_1}{V_0} \sin\omega t}}$$

where, t_2 is time arriving at a distance of S from the gap in a subharmonic buncher (SHB) which was designed to get a single picosecond pulse, t_1 is time travelling the position of the gap in SHB, V_0 is 90 KV which is incident electron energy at the entrance to SHB(energy of electrons emitted from the gun), V_1 is 13.5 KV which is electric field applied for the modu-

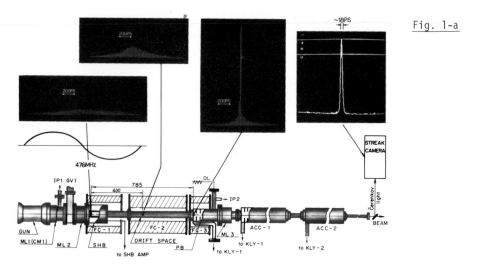

Fig. 1-a

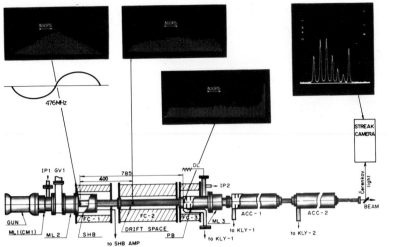

Fig. 1-b

lation, and ω is frequency of microwave energized in SHB cavity which is
1/6 of 2856 MHz. The shape and width of emission pulse from the gun is
expressed as a function of t_1. We assumed simply the shape of pulse as
a symmetric triangle. As the experimental results have shown that the
pulse shape can be represented well as a triangle, the computer simulation
could be done very accurately. The pulse shape can be estimated as a function of t_1 and S under other fixed conditions. The results are shown in
Figs. 1-a and b. If the matching of phase between the emission pulse and
476 MHz subharmonic microwave energized in the SHB cavity is well controlled
and the peak position of the emission pulse is adjusted to coincide with
the center of microwave, the bunching proceeds very effectively. The bunching proceeds very rapidly in the drift space and most electrons are bunched within 350 ps duration at the end of the drift space, and then bunched
further by accelerating tubes, as shown in Fig. 1-a. The shape of the
bunched pulse obtained by the simulation is very similar to that obtained by
experiments. If the phase of 476 MHz is shifted in degree of 180°(1 ns in
time), the situation is drastically changed, as shown in Fig. 1-b. In this
case, the debunching proceeds. The injected electrons are spreaded and can
not be bunched into one wavelength of subharmonics. The effect of the width
of emission pulse for the bunching process has been also examined.

All signals to the high speed pulse amplifier, the SHB prebuncher, the prebuncher and acceleration tubes, and to the triggering system for detection
with a certain time delay, if necessary, are given from the same origin; a
master oscillator. This is important and essential for obtaining highly
controlled picosecond pulses. In Fig. 2, a single shot pulse and its
accumulation by memorizing 100 shots repeatedly are shown. It is clear
from the results that the jitter of our system is fairly small. Total
jitter has been estimated to be ±11 ps. The characteristics of the single
pulse beam are summarized.

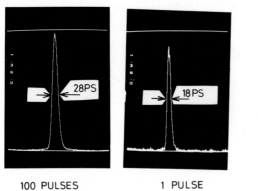

Table	CHARACTERISTICS OF THE SINGLE PULSE

PULSE WIDTH: <18 ps
CHARGE   :   1 nC
BEAM DIA. :   2 mmφ
ENERGY   :   36.6 MeV
STABILITY :   1.3 %/30 min
PULSE REPETITION
          :   MAX 200 pps
TIME JITTER: ± 11 ps
               (STREAK CAMERA)
               ± 6 ps
               (SAMPLING OSC.)

100 PULSES   1 PULSE

Fig. 2

2. Studies on Fast Formation Processes of Solute Excited States in Liquid Hydrocarbons by Means of Picosecond Pulse Radiolysis

A lot of work has been carried out on excited states in liquid hydrocarbons by both emission and absorption spectroscopies [2]. However, little work in a subnanosecond time region has been done. By means of picosecond pulse radiolysis, the formation process of solute excited state in a liquid hydro- carbon such as cyclohexane was observed by Beck and Thomas in 1972 [3]. Since they used pulse trains from a L-band linac, information from the ex- periment was limited from 60 to 770 ps. The pulse to pulse separation is 770 ps for the L-band linac. Recently, using single picosecond electron pulse from a S-band linac with about 30 ps time resolution, observation of the emission from liquid scintillation systems such as 2,5-diphenyloxazole in liquid hydrocarbons has been made by the present authors without time limitation after electron pulse irradiation [4]. The block diagram of the picosecond single pulse radiolysis system is shown in Fig. 3.

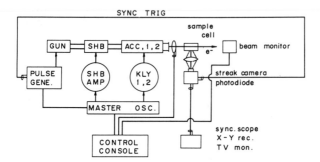

Fig. 3 The block diagram of picosecond single pulse radiolysis system. The present system is a combination of a fast detection system and the generation of a single picosecond electron pulse. The fast detection system is composed of a streak camera (C979 HTV), a SIT camera (C1000-12, HTV), an analyzer (C1098, HTV), PDP 11/34 and display systems

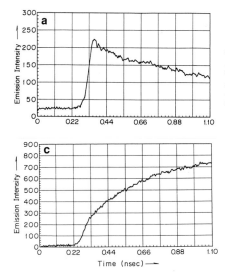
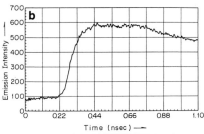

Fig. 4 Emissions from excited 2,5-
diphenyloxazole in ethylalcohol;
a, cyclohexane;b, and phenylcyclohexane;c. The concentrations
of the solute are 10 mM.

In Fig. 4, time profiles of emission from excited 2,5-diphenyloxazole in
ethlalcohol, cyclohexane and phenylcyclohexane are shown. In ethylalcohol,
growth of emission from the solute excited state was completed at the end of
Čerenkov light pulse induced by electron pulse and the emission decays gra-
dually with decay time of 1.6 ns. As a lifetime of excited 2,5-diphenyl-
oxazole in ethylalcohol obtained by photo-excitation was reported to be 1.6
ns [5], above experimental result suggests that most of the solute excited
states are already formed immediately after the pulse. This rapid growth of
emission has been called that faster formation of solute excited state. In
phenylcyclohexane, at a lower solute concentration, Čerenkov light is clearly
seen and at the end of Čerenkov light pulse, a certain fraction of excited
states are already formed and a slow growth is also seen over 1 nsec. This
slow growth has been called the slower formation process. At 10 mM 2,5-
diphenyloxazole solution in phenylcyclohexane, ratio of the slower formation
to the faster one is large and the faster one is hardly discriminated. On
the other hand, in cyclohexane, both the slower and the faster formation can
be seen, but the growth of the emission corresponding to the slower formation is faster as compared with those in phenylcyclohexane.

Normally above two formation processes can be seen in any kind of solvent.
However, contribution of each process is dependent on the kind of solvent
and concentration of solute molecules. Details of the mechanisms for the
formation of solute excited states in liquid hydrocarbons by electron
irradiation were discussed elsewhere [4].

3. Picosecond Time Resolved Fluorescence Studies of Poly(N-vinylcarbazole)

Much attention has been recently paid to excimers of polymers with pendant
aromatic chromophores. Normally fluorescence is emitted from a sandwich
type excimer in which π orbitals of aromatic chromophores overlap as in the
case of pyrene. In the case of poly(N-vinylcarbazole);PVCZ, however, two
distinct intramolecular excimers have been reported. One is a sandwich type
excimer [6], and the other is a so-called second excimer [7]. Preliminary

experiments about the formation processes of these excimers have already reported [8]. The present paper mainly describes the configurational effects on the time resolved fluorescence spectra of PVCZ, the formation processes of two distinct excimers of PVCZ, and the primary processes of radiation induced reactions in polymer solutions by using single picosecond pulse radiolysis.

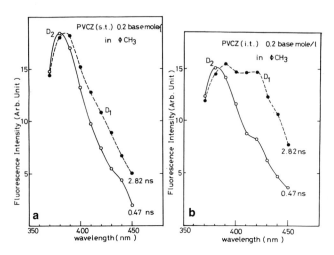

Fig. 5 Time resolved fluorescence spectra observed for irradiated 0.2 base mole/1 PVCZ solutions in toluene under the same condition. a) syndiotactic rich PVCZ; b) isotactic rich PVCZ. o: at 470 ps and ●: at 2.82 ns after the pulse

Both isotactic rith PVCZ(i.t.PVCZ) and syndiotactic rich PVCZ(s.t.PVCZ) were used. Figs. 5-a and b show the time resolved fluorescence spectra of 0.2 base mole/1 s.t.PVCZ and i.t.PVCZ solution in toluene, respectively.
In these spectra, the fluorescence around 375 nm is due to the so-called second excimer (D_2). Most of the so-called second excimer is already formed within 10 ps electron pulse. This fact indicates that the so-called second excimer should be formed in pre-existing excimer traps. However, the precise structure of the pre-existing excimer traps is not yet clear. The fluorescence around 420 nm is due to the sandwich type excimer (D_1), and is formed in several nanosecond. Formation rate of D_1 and decay rate of D_2 are different. Therefore, it can be said that the so-called second excimer is not the main precursor of the sandwich type excimer. In the case of s.t.PVCZ, the fluorescence spectrum at 0.47 ns after the pulse is composed of the so-called second excimer and a small portion of the sandwich type excimer. While, the formation yield of the sandwich type excimer in i.t.PVCZ system is larger than in s.t.PVCZ system. This fact suggests that π-orbitals of carbazolyl chromophores in i.t.PVCZ is much easier to overlap than in s.t.PVCZ. The fluorescence near 350 nm which is considered to be due to the excited singlet state of the monomer of PVCZ, starts to decrease at 2 ns after the pulse. The excited singlet state of the monomer is believed to be the precursor of the sandwich type excimer. The sandwich type excimer is mainly formed from the direct interaction between an excited carbazolyl and ground state carbazolyl chromophore after the formation of the excited carbazoly chromophore.

270

References

1. Y.Tabata, J.Tanaka, S.Tagawa, Y.Katsumura, T.Ueda and K.Hasegawa;
   J. Fac. Eng. of Univ. of Tokyo $\underline{34}$, 619 (1978)
   Y.Tabata, S.Tagawa, Y.Katsumura, T.Ueda, K.Hasegawa and J.Tanaka;
   J. Atomic. Energy Soc. Japan. $\underline{20}$, 473 (1978)
2. J.K.Thomas; Int. J. Radiat. Phys. Chem. $\underline{8}$, 1 (1976) and references cited
   therein.
3. G.Beck and J.K.Thomas; J. Phys. Chem. $\underline{76}$, 3856 (1972)
4. S.Tagawa, Y.Katsumura and Y.Tabata; Chem. Phys. Lett. $\underline{64}$, 258 (1979)
   S.Tagawa, Y.Katsumura, T.Ueda and Y.Tabata; Rad. Phys. Chem. $\underline{15}$, 287
   (1980)
   Y.Katsumura, T.Kanbayashi, S.Tagawa and Y.Tabata; Chem. Phys. Lett. $\underline{67}$,
   183 (1979)
   Y.Katsumura, S.Tagawa and Y.Tabata; J. Phys. Chem. 84, 833 (1980)
5. I.B. Berlman, Handbook of Fluorescence Spectra of Aromatic Molecules
   (Academic Press, New York 1971)
6. C.David, M.Pines and G.Geuskeus; European Polym. J. $\underline{8}$, 1291 (1972)
7. C.D. Johnson; J. Chem. Phys. $\underline{62}$, 4697 (1975)
8. S.Tagawa, M.Washio and Y.Tabata; Chem. Phys. Lett. $\underline{68}$, 276 (1979)

Part VI

Applications
in Solid State Physics

Picosecond Nonlinear Spectroscopy of Electrons and Phonons in Semiconductors

D. von der Linde, J. Kuhl, and R. Lambrich

Max-Planck-Institut für Festkörperforschung,
D-7000 Stuttgart 80, Fed. Rep. of Germany

1. Introduction

Electrons and holes with large excess energies can be photoexcited in semiconductors by absorption of light of suitable frequency. The electronic energy relaxes very rapidly, primarily via the interaction with phonons. A very
efficient energy loss mechanism in polar semiconductors is the emission of
long wavelength LO-phonons [1].

The carriers also interact strongly with each other; at high density collisions establish quasi-thermal equilibrium between electrons and holes before
energy relaxation and recombination of the carriers is completed [2]. The
electronic system can be characterized by some electronic temperature T,
different from the lattice temperature T_L. The rapid energy loss due to phonon emission causes a fast decrease of the temperature of the electron-hole
plasma.

The rate of phonon emission from the relaxing e-h plasma can be faster
than the phonon decay processes. As a result an excess phonon population is
expected to build up in certain lattice modes [3].

In this report we summarize recent results of our investigations of photoexcited e-h plasmas in GaAs. Picosecond nonlinear absorption spectroscopy
yields the variation of the plasma temperature with time and provides information on exciton formation and exciton scattering. A new technique of
time-resolved nonlinear Raman scattering is used to detect optical phonons
released by the cooling plasma, and the phonon relaxation time is measured.

2. Absorption and Gain in Photoexcited GaAs

GaAs platelets typically a few microns thick are photoexcited by two-photon
absorption of single light pulses of 25 ps duration (λ = 1.064 μm). Two-
photon excitation provides a uniform density of e-h pairs of about 10^{17} cm^{-3}.
The absorption induced by the e-h plasma is measured with probe pulses selected from the pulse train of a synchronously pumped optical parametric oscillator. Synchronous pumping ensures precise synchronism of the pump pulses
(1.064 μm) and the frequency tunable probe pulses [4].

Figure la depicts absorption spectra of a 2 μm thick GaAs crystal measured
at different times Δt after excitation. The curve at the very top of Fig. la
is measured with probe pulses arriving earlier than the pump pulses, i.e.,
this curve represents the absorption spectrum of GaAs in the absence of the
electron-hole plasma. The spectrum clearly shows the n = 1 exciton at $\hbar\omega$ =

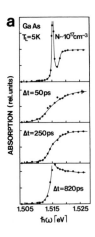

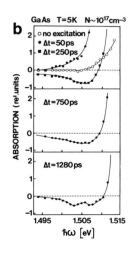

Fig. 1

Absorption coefficient versus
photon energy of the probe
pulse for various values of
the delay time Δt; a) 2 μm
thick GaAs, b) 4 μm thick
GaAs

1.515 eV and, at higher energies, the continuum absorption due to valence-to-
conduction band transitions. On the other hand, the exciton peak is absent
in the spectrum measured at Δt = 5o ps, and the absorption rises smoothly
over approximately 15 meV. At Δt = 25o ps we still observe a monotonic increase of the absorption with energy, but now the absorption edge is much
steeper.

The data for Δt = 82o ps give an example of the qualitative changes of
the spectra occuring at much longer delay times. An absorption maximum builds
up near the n = 1 exciton resonance, but the width is still much bigger than
that of the exciton peak in the absence of photoexcited electrons and holes.

Very useful information is also obtained from absorption measurements at
energies $\hbar\omega$ < 1.515 eV, i.e., below the exciton energy. While the absorption
coefficient at the exciton peak reaches a value of a few times 10^4 cm^{-1} the
absorption at lower energies is about two orders of magnitude smaller. Nevertheless, this weak absorption can be readily measured when thicker crystals
are used.

Figure 1b shows absorption spectra of a 4 μm thick crystal of GaAs. The
open circles in the top part represent the crystal absorption before excitation, i.e. we are measuring the low energy wing of the undisturbed exciton.
The spectrum at Δt = 5o ps shows that the e-h plasma induces absorption extending well beyond the exciton energy to lower energies. At Δt = 25o ps the
induced absorption is negative for energies $\hbar\omega$ < 1.51o eV, i.e. we observe
optical gain. The gain is found to persist for delay times well in excess of
1ns with only minor changes in the shape of the spectrum.

The principal physical phenomena showing up in the time-resolved absorption spectra are outlined in the following.
i) The transition from the spectrum at the top of Fig. 1a (Δt < 0) to that
for Δt = 5o ps indicates the rapid screening of the exciton resonance by
the photoexcited electron-hole plasma [5]. The self-interaction of the carriers
reduces the band gap [6], and absorption (and gain) [7] is observed at energies lower than the exciton resonance, which is the lowest-excited state of
the unperturbed semiconductor.

276

ii) A detailed analysis [8] shows that the spectra up to Δt = 25o ps represent the energy relaxation of the electron-hole plasma, which cools from
T = 65 K at 5o ps to somewhat less than 4o K at 25o ps (T_L = 5 K). The transition from absorption to optical gain, and, at higher energies, the steepening of the absorption edge are characteristic of a decrease of the electronic
temperature. The energy relaxation mechanism is identified to be the emission
of LO-phonons [8].
iii) The broadened exciton peak reappearing at Δt = 82o ps in Fig. 1a indicates that the density of carriers is too low for complete screening of the
exciton; obviously recombination has already substantially diluted the electron-hole plasma at 82o ps. On the other hand, the optical gain persists
essentially unchanged over more than 1 ns. Thus the gain cannot be attributed
to stimulated emission of the electron-hole plasma, because that process is
very sensitive to changes of the carrier density. We believe that the optical
gain at these long delay times is due to excitonic scattering processes [9].

3. Raman Scattering from Non-Equilibrium LO-Phonons

In a different experiment [1o] we use Raman scattering to detect the burst of
LO-phonons released during the relaxation of hot carriers in GaAs.

A plasma of 10^{17} e-h pairs per cubic centimeter with an excess energy of
typically 1o times the LO-phonon energy is expected to emit about 10^{18} phonons
per cubic centimeter. These phonons occupy modes with wavevectors q ranging
from 3 x 10^5 cm^{-1} to about 10^7 cm^{-1}, as determined from energy and momentum
conservation for interaction of the carriers with LO-modes. In GaAs the hot
carriers preferentially emit long wavelength LO-phonons, because the Fröhlich-
interaction is the dominant coupling mechanism. We estimate that about 1 to
2 phonons per mode are generated around q = 7.7 x 10^5 cm^{-1}, which corresponds
to the momentum transfer in backward Raman scattering (λ = 575 nm). Thus a
sizable non-equilibrium population is expected, provided the electronic relaxation is faster than the relaxation of the phonons.

In the Raman experiment [1o] the electron-hole plasma is generated via
single photon absorption using very short pulses from a synchronously pumped
Rhodamin 6 G dye laser (λ = 575 nm). The average power of the continuous
pulse train is 8o mW, and the pulse duration is 2.5 ps. The laser output is
divided into two equally intense portions, a variable time delay is introduced, and the two beams are recombined at the surface of the GaAs crystal.
Raman scattering of the dye laser pulses is detected in a backward scattering
configuration using a photon counting spectrometer.

Both Stokes scattering (S) and anti-Stokes scattering (A) from LO-phonons
($\hbar\omega_{LO}$ = 37 meV) is observed. The phonon occupation number N = $(S/A-1)^{-1}$ is
measured to be o.7. Lattice heating by the laser light can be ruled out [1o];
the temperature of the probed crystal volume does not rise over the ambient
temperature of ~ 8o K, so that anti-Stokes scattering from thermal phonons
can be neglected. Thus the observed anti-Stokes signal must be due to non-
equilibrium phonons generated by the relaxation of the photoexcited carriers
[3].

The dynamics of the non-equilibrium phonons can be unraveled by measuring
the anti-Stokes signal A as a function of Δt. If Δt is much larger than the
phonon relaxation time τ', one measures just the sum of the scattering of the
two pulses. However, when Δt is comparable with τ', the scattering of the
delayed pulse is enhanced due to the phonons left behind by the first pulse.

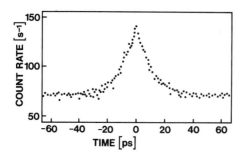

Fig. 2

Anti-Stokes intensity of Raman
scattering from LO-phonons
versus delay time. Two beams
with parallel polarization

In Fig. 2 the measured anti-Stokes signal is plotted versus Δt. A maximum is obtained at $\Delta t = 0$, and the curve falls off symmetrically to the background signal which is given by the sum of the anti-Stokes generated by one pulse train alone. The decay of the signal to the background signifies the decrease of the concentration of excess phonons.

An interesting variant of the experiment can be realized when the Raman light is polarized. By adjusting the polarization of the light beams perpendicular to each other, the Raman light due to one of the beams can be blocked by a suitably oriented analyser in front of the spectrometer. The other beam then simply probes the phonon population created by the first beam. In this experiment the anti-Stokes signal is given by the convolution of the probe pulse with the phonon population $N(t)$, and the complete rise and decay of the phonon burst is revealed.

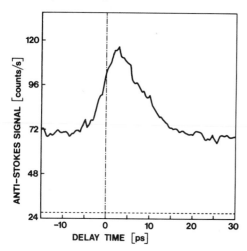

Fig. 3

Anti-Stokes intensity versus
delay time. The horizontal
dashed line indicates the background signal due to laser
light and detector dark current

Figure 3 shows the result of such an experiment [1o] for GaAs with (100) surface orientation, and with pump pulses and the probe pulses polarized along (010) and (001), respectively. The rise of the anti-Stokes signal signifies the growth of the phonon population caused by the emission of LO-phonons from the photo-excited e-h plasma. The decay is exponential with a time constant $\tau' = (7 \pm 1)$ ps, in agreement with time constant inferred from the wings of the symmetric curve of Fig. 2.

We estimate that in this experiment the hot carriers lose more than 9o % of the excess energy in about 5 ps [1o]. This estimate is in agreement with the observed rise time [11]. We note also that the zone center phonons cannot propagate out of the probed crystal volume. It is therefore concluded that the decay of the anti-Stokes signal represents the decay of the excess population, and that $\tau' = 7$ ps is the relaxation time of the non-equilibrium LO-phonons.

4. Discussion

From gain and absorption measurements we infer that the electron-hole plasma cools by ~ 25 K in 2oo ps via LO-phonon emission [8]. On the other hand, the Raman experiment [1o] shows that the phonon emission occurs within 5 ps. The apparent inconsistency of the two results is explained by the very strong temperature dependence of the energy relaxation rate [1]. In [8] we examine just the final stage of the relaxation at a rather low temperature, $T \sim 65$ K. The remaining small fraction of the total initial excess energy relaxes quite slowly in a time long compared with the relaxation of the phonons. In the Raman experiment [1o], on the other hand, ten times shorter excitation pulses (2.5 ps) are used, and a much higher plasma temperature is reached ($\sim 5oo$ K). In this case about 9o % of the LO-phonons are generated while the light pulses are on, e.g., in a time shorter than the phonon relaxation time.

References

1. See, e.g. E.M. Conwell, *High Field Transport in Semiconductors*, Suppl. 9 to *Solid State Physics*, ed. by F. Seitz, D. Turnbull, and H. Ehrenreich (Academic Press, New York, 1967)
2. R. Stratton, Proc. Roy. Soc. London, A 246, 406 (1958); J. Shah, Solid State Electron. 21, 43 (1978)
3. J. Shah, R.C.C. Leite, and J.F. Scott, Solid State Commun. 8, 1089 (1970)
4. D. von der Linde and R. Lambrich, *Picosecond Phenomena*, ed. by C.V. Shank, E.P. Ippen, and S.L. Shapiro, Springer Series in Chemical Physics, Vol. 4 (Springer, Berlin, Heidelberg, New York 1978)
5. C.V. Shank, R.L. Fork, R.F. Leheny, and J. Shah, Phys. Rev. Lett. 42, 112 (1979)
6. G. Beni and T.M. Rice, Phys. Rev. B 18, 768 (1978)
7. O. Hildebrand, B.O. Faltermeier, and M. Pilkuhn, Solid State Commun. 19, 841 (1976)
8. D. von der Linde and R. Lambrich, Phys. Rev. Lett. 42, 1o9o (1979)
9. E. Göbel, K.L. Shaklee, and E.R. Epworth, Solid State Commun. 17, 1185 (1975)
1o. D. von der Linde, J. Kuhl, and H. Klingenberg, Phys. Rev. Lett. 44, 15o5 (198o)
11. We are grateful to C.V. Shank for pointing out that coherent coupling could affect our data for $|\Delta t|$ less than the coherence time (~ 2.5 ps). This effect has not yet been taken into account in the interpretation of the rise time.
    See also: E.P. Ippen and C.V. Shank, in *Ultrashort Light Pulses*, ed. by S.L. Shapiro, *Topics in Applied Physics*, Vol. 18 (Springer Verlag, Berlin, Heidelberg, New York 1977)

Picosecond Spectroscopy of Semiconductor Microstructures

R.L. Fork, C.V. Shank, and B.I. Greene
Bell Laboratories, Holmdel, NJ 07733, USA

F.K. Reinhart and R.A. Logan
Bell Laboratories, Murray Hill, NJ 07092, USA

An understanding of carrier dynamics in semiconductor microstructures is both of fundamental interest and important to the design of new devices. This understanding, however, depends on our capacity to examine carrier dynamics in structures with dimensions of the order of, or less than, an optical wavelength with time resolution in the picosecond or subpicosecond time regime. We describe here the use of picosecond continuum spectroscopy [1] to obtain data in this difficult experimental regime. In particular, we characterize carrier transport [2] in thin (1μ-2μ) semiconductor layers and carrier relaxation and screening dynamics [3] in very thin (50Å-200Å) semiconductor layers, in both cases with a time resolution of 0.5 psec. Of special interest is the first direct observation of nonequilibrium transport including evidence for velocity overshoot behavior.

1. Nonequilibrium Transport

Theoretical analyses of carrier transport in semiconductors for short distances and short times predict nonequilibrium transport behavior [2] distinct from behavior at long times. For example in GaAs, which has a low mass central valley, carriers introduced into the central valley for moderately large electric fields are predicted to exhibit a large initial velocity (velocity overshoot) followed by rapid relaxation to the saturation velocity observed at long times. This nonequilibrium behavior occurs on a time scale of a few picoseconds and a spatial scale of approximately one micron. Direct experimental observation has therefore been difficult.

To examine nonequilibrium transport [2] we developed a novel optical technique in which the optical pulse width (0.5 psec), rather than electronic circuit time constants, determines the time resolution. The technique consists of optically exciting carriers in a thin semiconductor layer in a field normal to the plane of the layer and then using the change in the Franz-Keldysh effect [4] at the band edge to monitor transport as the carriers are swept out of the layer. The carriers create a weak screening field opposing the applied field as they propagate. The reduction in the Franz-Keldysh effect as that screening field develops provides a simple way of optically monitoring transport. By maintaining low carrier densities ($\sim$2x10[14] cm^{-3}) this screening field is kept small compared to the applied field and the transport closely approximates that in a static applied field. Plots of the optically observed change in the band edge transmission are shown in Fig. 1 for a time delay (20 psec) when sweepout was essentially complete. The sample for the experiment reported here was a reverse biased p-i-n diode grown by liquid phase epitaxy. It was formed by a double heterostructure configuration consisting of a 2μ thick layer of GaAs between two 1.3μ layers of Al$_{.2}$Ga$_{.8}$As. The light was incident from the p side and the lattice temperature was 77 K.

There are a number of unique advantages to this experimental approach. Most importantly, the time resolution is limited solely by the optical pulse width. In addition an independent check on the applied field value is obtained from the optical signal. As is evident in Fig. 1 the optical signal varies with field and serves both as a measure of the applied field and as proof that screening has not caused excessive reduction of the applied field. Another advantage occurs in that carriers can be injected at any of a range of initial energies by varying the wavelength of the exciting pulse. For example, in this work the pump pulse was at 805nm which given a band edge at 1.51eV implies initially injected carriers with energies comparable to those of carriers at room temperature. A final advantage is that errors due to baseline drift are avoided by taking the optical change Δα as the sum of areas of equal widths, one under the central positive peak and the other under a negative peak (see Fig. 1).

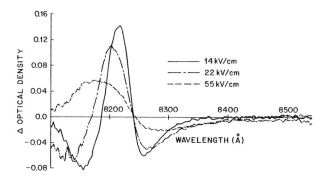

Fig. 1 Change in optical density due to carrier transport (t=20 psec).

The transport dynamics are obtained by measuring the time evolution of the optical change $\Delta\alpha(t)$ and comparing it with the calculated value for an electron velocity $v_e(t)$ and a hole velocity $v_h(t)$. The calculated value is

$$\Delta\alpha(t) = [d(1-\exp-(\alpha_p d))]^{-1}\left\{(\int v_e dt - \int v_h dt \cdot \exp(-\alpha_p d))\right.$$

$$\left. + \alpha_p^{-1}[1-\exp(-\alpha_p \int v_h dt) + \exp(-\alpha_p d) \cdot (1-\exp(+\alpha_p \int v_e dt))]\right\} \qquad (1)$$

Here $\Delta\alpha(t)$ is normalized to have unit value at infinite time. For values of t where the integrals $\int v_e dt$ and $\int v_h dt$ exceed d, they are to be set equal to d. Since α_p, the absorption coefficient at the pump wavelength, and d, the sample thickness, are known we obtain $v_e(t)$ for a given $v_h(t)$ and vice versa. (This equation was obtained by assuming a Beer's law absorption of the pump light and that the change in optical absorption was everywhere linearly proportional to the change in electric field.

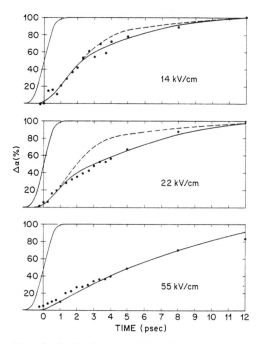

Fig. 2 Optical absorption change $\Delta\alpha$ vs. time delay. The individual points are experimental. For 14kV/cm the calculated curve (solid line) is a least squares fit of (1) using two constant velocity regimes. Here v_e = 3.0 x 10^7 cm/sec (t<2.4 psec) and v_e = 1.0 x 10^7 cm/sec (t>2.4 psec). The dashed segment shows the continuation of $\Delta\alpha$(t) if v_e remains constant for t > 2.4 psec. For 22kV/cm the solid curve is a two velocity least squares fit of (1) with v_e = 4.4 x 10^7 cm/sec (t<1.1 psec) and v_e = 1.2 x 10^7 cm/sec (t>1.1 psec). The dashed curve shows the continuation of $\Delta\alpha$(t) if v_e remains constant for t > 1.1 psec. For 55kV/cm the solid curve is a single velocity least squares fit with v_e = 1.3 x 10^7 cm/sec for all t. The curve at the far left of each trace shows the instrumental response.

In calculating $\Delta\alpha$(t) for Fig. 2 the holes were approximated as accelerating ballistically to a saturation velocity of 1.0 x 10^7 cm/sec. The sample was not fully depleted for the 14kV/cm and 22kV/cm cases, hence d was taken as the depletion width, or 1.45μ and 1.8μ respectively for those traces. The quoted fields are thus also average fields.

We interpret the behavior shown in Fig. 2 as direct evidence for the previously predicted nonequilibrium transport behavior [2]. In particular, the data imply large initial electron velocities followed by relaxation to velocities approximating the electron saturation velocity. Also the duration of the initial velocity transient decreases with increasing field. For example, at 14kV/cm the mean initial velocity obtained from a least squares fit of (1) is 3.0 x 10^7 cm/sec and persists for 2.4 psec. For 22kV/cm the mean initial velocity is larger (4.4 x 10^7 cm/sec) and persists for a shorter time (1.1 psec) as predicted. Finally for 55kV/cm the duration of the initial transient is too short to resolve hence the data has been fitted for a single velocity which, as expected, approximates the saturation velocity. The fact that the data for 55kV/cm at short times lies above the calculated curve is consistent with the expected initial velocity transient.

This technique appears applicable to a wide variety of semiconductors. The principal requirments are an optical pump pulse of short duration with an energy slightly above the band edge and a continuum probe pulse with a bandwidth of ∿ 30nm, or more, centered near the band edge. The thickness of the sample is a limitation only in that it must be thin enough to transmit an observable light level and thick enough to cause an appreciable change in transmission.

2. Carrier Dynamics in Superlattices

Carrier dynamics in semiconductor microstructures where the layer thickness is sufficiently small that quantum size effects become important [5] were also explored. Here the thickness is comparable to the carrier DeBroglie wavelength with the result that resolved quantum well states occur and the semiconductor band structure is altered. To explore carrier dynamics in this case we employed a pump-probe technique similar to one used previously on bulk material [3]. We find considerable information concerning the dynamics of carrier screening and relaxation in these superlattice structures.

The experimental technique consisted of exciting carriers in a super-lattice of alternating layers of GaAs (50Å-205Å thick) and AlGaAs (sufficiently thick to ensure carrier confinement) with a pump pulse at 745nm. Overall sample thickness was designed to give an optical density near unity. A continuum pulse spectrally overlapping the sample band edge was then used to probe the sample transmission at various time delays. We show in Fig. 3 the change in transmission for a sample with 205Å thick GaAs layers for several time delays. We see, e.g., induced bleaching of absorption peaks at each of the two-dimensional (n=1,2,3) band edges which we attribute to screening of the excitons associated with those edges, an overall bleaching at short times which we attribute in large part to band filling, and a change in the spectrum with time which implies carrier recombination and cooling to the n = 1 level. While more interpretive work remains it is clear that this technique is an effective probe of carrier dynamics in a situation where quantum size effects play a prominent role.

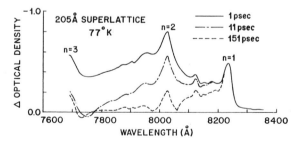

Fig. 3 Optical density change in an optically pumped superlattice.

We gratefully acknowledge collaboration on the superlattice study with R. Dingle, C. Weisbuch, A. C. Gossard and W. Wiegmann.

REFERENCES

1. For a description of the picosecond continuum spectroscopy technique see E. P. Ippen and C. V. Shank in "Picosecond Phenomena", C. V. Shank, E. P. Ippen and S. L. Shapiro (eds.) (Springer-Verlag, Berlin 1978, pp. 103-107).

2. For calculations of nonequilibrium transport and references to earlier work see, e.g., T. J. Maloney and J. Frey, Journal of Applied Physics 48, 781 (1977).

3. C. V. Shank, R. L. Fork, R. F. Leheny and J. Shah, Phys. Rev. Lett. 42, 112 (1979).

4. See, e.g., K. Seeger, Semiconductor Physics (Springer-Verlag, New York, 1973) p. 351.

5. For a review of superlattice structures see R. Dingle, Festkorperprobleme (Advances in Solid State Physics) edited by H. J. Queisser (Pergamon/ Vieweg, Braunschweig, 1975), Vol. XV, p. 21.

Picosecond Photoconductivity in Amorphous Silicon

A.M. Johnson
City College of New York, NY 10031, USA

D.H. Auston, P.R. Smith, J.C. Bean, and J.P. Harbison
Bell Laboratories, Murray Hill, NJ 07974, USA

D. Kaplan
Thomson CSF, F-91405 Orsay, France

In amorphous semiconductors, states lie in bands separated by energy gaps, as in the crystalline case. In an effort to explain some of the differences between the amorphous and crystalline states, Anderson [1] has proposed that the spatial randomness of the atomic potential, due to the intrinsic disorder of the lattice, gives rise to states localized to within a few lattice sites - localized states. The high-mobility valence and conduction-band states are called extended states. There is also extrinsic disorder within the lattice due to the presence of structural defects (divancies, microvoids, etc.), which depend heavily upon the method of preparation of the amorphous semiconductor. The question of electrical transport in amorphous semiconductors has evoked a vast compendium of new physics and many unresolved questions [2]. For example, there is the question of the relative importance of the various modes of electronic conduction such as extended state transport, hopping between localized states in a continuous random network, hopping between isolated defect sites, and small polaron motion [2].

The drift mobility of amorphous silicon was first measured by the time-of-flight technique, by LeComber and Spear [3]. They measured a trap-limited drift mobility of $10^{-1} cm^2 V^{-1} s^{-1}$. From the temperature dependence of the drift mobility and together with both estimates of the density of states at the bottom of the conduction band and estimates of the trap density, they extrapolated a microscopic extended state mobility of $1-10 cm^2 V^{-1} s^{-1}$. Photoconductivity measurements by Loveland, Spear, and Al-Sharbaty [4] in amorphous silicon yielded a photocurrent that was proportional to the product of the carrier mobility and the carrier relaxation time, $\mu\tau$. They found that the $\mu\tau$ product decreased with increasing defect density. With the use of picosecond time-resolved photoconductivity, we have been able to make direct measurements of the initial transient extended-state carrier mobility. We have measured the rate at which photoexcited carriers relax from extended states to localized states, due to the presence of structural defects, which act as effective trapping and recombination centers. Time-resolution is essential to separating the carrier mobility, μ, from the carrier relaxation time, τ.

A widely used indicator of the lower limit of the density of structural defects in amorphous silicon is the number of unpaired spins associated with dangling bands. The approximate spin densities noted in the literature [2], as determined by electron spin resonance (ESR) measurements, for the three forms of amorphous silicon chosen are: ultra-high vacuum-deposited a-Si (UHV a-Si), 10^{19}-10^{20} cm^{-3}; chemical vapor-deposited a-Si (CVD a-Si), 10^{18}-10^{19} cm^{-3}; and glow-discharge-deposited a-Si:H (GD a-Si:H), $<10^{16}$ cm^{-3}. Structural defects are associated with the extrinsic disorder of the lattice and the defect density depends heavily on the method of preparation. We find that the relaxation rate of the photoexcited carriers depends upon the degree of extrinsic disorder (defect density). We also find that the magnitude of the initial transient mobility ($\sim$1cm^{2}V^{-1}s^{-1}) is independent of sample preparation and thus depends upon the intrinsic disorder in the lattice.

The measurements were performed with a cw Rh6G dye laser that was synchronously pumped by a mode-locked argon ion laser. The optical pulse duration (FWHM) was estimated to be 3.5 ps by a second harmonic generation auto-correlation measurement in KDP. The wavelength of excitation was 575 nm (2.15 eV), approximately 0.5 eV above the absorption edge [4]. The laser had a peak power, pulse energy, and repetition rate of 350 W, 1.2 nJ (=3.5×10^9 photons) and 81.9 MHz, respectively.

In a recent paper [5], CVD a-Si was utilized to produce a fast linear gap cell-type photodetector whose response was limited by the rise time (25 ps) of the sampling oscilloscope. The scope-limited response had a risetime of 25 ps and a duration of 40 ps (FWHM). When the active area of the photodetector is a low-defect density material (i.e., GD a-Si:H), where the relaxation time of the photoexcited carriers is much longer than the optical pulse width, the photocurrent can be completely resolved on the sampling oscilloscope. The expression for the peak photocurrent (before the relaxation of the photoexcited carriers to localized states) in this time-resolved regime is

$$i_p = \eta(1-R)(1-e^{-\alpha d}) \frac{e}{\hbar\omega} \mu \frac{V_b}{\ell^2} \varepsilon_p; \quad \tau_p < \tau_r \qquad (1)$$

where η is the quantum efficiency for the generation of electron-hole pairs, R is the reflectivity, α is the absorption constant, d is the film thickness, $\hbar\omega/e$ is the photon energy in volts, μ is the mobility, V_b is the bias voltage, ℓ is the gap length, ε_p is the optical pulse energy, τ_p is the optical pulse duration, and τ_r is the carrier relaxation time. For improved signal-to-noise, the photoresponse is averaged in a multi-channel analyzer (MCA). The peak time-resolved photocurrent is proportional to the product of the quantum efficiency, η, and the mobility, μ. For GD a-Si:H, we measure $\eta\mu$=0.8 cm^{2}V^{-1}s^{-1}(see Fig. 1).

When the active area of the photodetector is a high-defect density material (i.e., CVD a-Si or UHV a-Si), where the relaxation time of the photoexcited carriers is comparable to or less than the response of the sampling oscilloscope, we measure the area under the photoresponse curve in the MCA. This is equivalent to measuring the total charge, Q, in the photocurrent pulse. If we assume a photocurrent that decays exponentially with time constant τ_r, the integration of the photocurrent yields a charge, Q, that is proportional to the product of the quantum efficiency, the mobility, and the carrier relaxation time, $\eta\mu\tau_r$.

$$Q = \eta(1-R)(1-e^{-\alpha d}) \frac{e}{\hbar\omega} \mu\tau_r \frac{V_b}{\ell^2} \varepsilon_p \qquad (2)$$

Improved time resolution is needed to obtain τ_r. The carrier relaxation time, τ_r, is obtained by the electronic correlation of two photoconductors [6]. This correlation method is very similar to the nonlinear optical autocorrelation used to measure the optical pulse widths. In the optical measurement, the temporal resolution is limited by the speed of the electronic nonlinearity responsible for SHG. The temporal resolution of the electronic correlation is limited by the material relaxation time (τ_r), the finite capacitance of the photoconducting circuit, and the duration of the optical pulses used to activate it. With an approximate deconvolution of the electronic autocorrelation function, we can obtain τ_r and use it in conjunction with the measured total charge, Q, to obtain $\eta\mu$.

The electronic autocorrelation of CVD a-Si [6] yields two exponentials with time constants of 16 ps and 30 ps. We model the two-component relaxation in CVD a-Si by the sum of two exponentials, weighting the faster decay heavier than the slower. The convolution of this response with itself also has two components. From an analysis of this type, one can use the observed decay constants of the autocorrelation function to obtain an effective relaxation time, τ_{eff}, which is a weighted sum of the observed decays. The total charge, Q, is proportional to $\eta\mu\tau_{eff}$. For CVD a-Si, τ_{eff} is approximately 21 ps (see Fig. 1).

Three different film-growth techniques were utilized to intentionally produce large differences in the electrical and optical properties of amorphous silicon. Furthermore, within the ultra-high vacuum-deposited a-Si (UHV a-Si), three samples with different deposition conditions were studied. A measure of the large differences in the types of amorphous silicon utilized is the large range of measured dark resistivities, as indicated in Fig. 1.

The carrier relaxation times decrease with increasing defect density. The $\eta\mu\tau$ product is a good indicator of the density of defects and the temporal resolution needed to resolve the carrier relaxation. A small value of $\eta\mu\tau$ corresponds to a high defect density and a need for high temporal resolution. In sample UHV-1, we were able to measure a carrier relaxation time of 4 ps. In samples UHV-3 and UHV-4, slightly improved temporal resolution is needed and is presently under investigation. We use 4 ps as an upper limit for the carrier relaxation in these two samples, thus slightly underestimating the mobility. The mobility of all the samples is on the order of 1 $cm^2V^{-1}s^{-1}$. We include the quantum efficiency, η, although it is generally assumed to be unity. For a non-unit quantum efficiency, the mobilities listed

SAMPLE	ESR [cm^{-3}]	T_d [°C]	ρ_d [Ω-cm]	$\eta\mu\tau_r$ [cm^2 v^{-1}]	τ_r	$\eta\mu$ [cm^{2}v^{-1}s^{-1}]
UHV-1	10^{19}-10^{20}	300	2800	4.2×10^{-12}	~4ps	~1.1
UHV-2	10^{19}-10^{20}	20	1300	1.5×10^{-12}	<4ps	>0.4
UHV-4*	10^{19}-10^{20}	300	4700	3.7×10^{12}	<4ps	>0.9
CVD-1	10^{18}-10^{19}	600	9.2×10^6	2.3×10^{-11}	~21ps	~1.1
GD-1	$<10^{16}$	250	10^8			0.8

*O_2 INTRODUCED ~1 AT%

T_d : DEPOSITION TEMPERATURE

ρ_d : DARK RESISTIVITY

are a lower limit. The magnitudes of the mobilities are consistent with the theoretical values expected for extended state conduction (1-10 cm^{2}V^{-1}s^{-1}) [2] associated with the intrinsic disorder of the lattice. Although the initial transient mobilities are roughly equal, the relaxation rates for transitions from extended to localized states varies greatly with sample preparation. Thus we believe that picosecond time resolution enables the delineation of intrinsic and extrinsic transport of photogenerated carriers.

Although these results are preliminary and much work remains to be done, we can nevertheless draw the following conclusions:
1. The relatively large initial transient mobility observed in all three forms of amorphous silicon provides strong evidence for the existence of a mode of conduction in extended states which is an intrinsic property of amorphous silicon (the mobilities due to hopping and small-polaron transport are expected to be smaller by 2 or 3 orders of magnitude [2]).
2. The strong dependence of the relaxation rate of the photocurrent on the defect density suggests the dominant mode of relaxation is trapping and/or recombination of carriers at isolated defect sites rather than Anderson [1]-type localized states that are due to the intrinsic disorder of the lattice.

One of us (AMJ) would like to acknowledge support by the Bell Laboratories Cooperative Research Fellowship Program.

REFERENCES

1. P. W. Anderson, Phys. Rev. 109, 1492 (1958).

2. See, for example, N. F. Mott and E. A. Davis, "Electronic Processes in Non-Crystalline Materials", 2nd edition, (Clarendon Press, Oxford 1979), and M.H. Brodsky (ed.): *Amorphous Semiconductors*, Topics in Applied Physics, Vol. 36 (Springer, Berlin, Heidelberg, New York 1979)

3. P. G. LeComber and W. E. Spear, Phys. Rev. Lett. 25, 509 (1970).

4. R. J. Loveland, W. E. Spear, and A. Al-Sharbaty, J. Non-Cryst. Solids 13, 55 (1973/74).

5. D. H. Auston, P. Lavallard, N. Sol, and D. Kaplan, Appl. Phys. Lett. 36, 66 (1980).

6. D. H. Auston, A. M. Johnson, and P. R. Smith, "Recent Advances in Picosecond Optoelectronics", Digest of Second Topical Meeting on Picosecond Phenomena (1980), Cape Cod, MA

Picosecond, 1.06 Micron Laser-Induced Amorphous Phases in Thin, Single Crystal Silicon Membranes

M.F. Becker, R.M. Walser, J.G. Ambrose, and D.Y. Sheng

The University of Texas at Austin, Department of Electrical Engineering, Austin, TX 78712, USA

Recent reports of laser-induced amorphous phases in crystalline silicon due to picosecond 532nm and 266nm radiation [1] and nanosecond 266nm radiation [2] raise questions regarding the dynamics of such a transition. These results may not be adequately described by equilibrium heat-flow analysis for several reasons.

1. The system is clearly not near equilibrium; in fact, the absorption length and the energy deposition length are of the order of a few phonon mean-free paths. At this point the assumptions of equilibrium thermodynamics break down. Energy transfer rates are affected and the resulting final state is not given by equilibrium heat-flow considerations.

2. The anomalous heating effects observed by HARRINGTON [3] in metals, semiconductors, and insulators have not been taken into account. This represents a physical manifestation of Item 1, above.

3. The possibility of alternative models, such as those incorporating the effects of excited state absorption.

These last two items both point out the possible necessity for considering non-equilibrium or non-normal processes, such as black body radiation, electric field induced instabilities, or surface electron or atom emission, for the equilibration of a solid excited far from equilibrium. These unanswered questions led us to an investigation of the laser interaction with thin silicon films. The transparency of such films facilitates the study of optical absorption and transient effects as well as TEM without intermediate thinning. We report the first observation of laser-induced amorphous phase with picosecond radiation at 1.06µm. Previous models based on heat-flow and fast quench rates have not predicted this behavior. We suggest that localization of the excitation is required to explain our results.

The thin silicon samples were prepared from low resistivity (111) Si wafers which had 1.5 or 2.5 µm thick epitaxial layers of un-doped Si 15-30 Ω-cm [4]. The wafers were then masked to expose an area of approximately 0.5 cm^2 to be electrochemically etched [5]. The selective etch left the epi layer intact. Further thinning of selected areas to thicknesses less than 300 nm was done by ion beam milling. Chemical etching and jet thinning were also used in the preparation of TEM samples and gave results equivalent to ion beam milling.

The laser pulses were supplied by a passively mode-locked Nd:YAG laser. Single pulses were selected, and they had an average FWHM duration of 38 psec. When needed, the second harmonic was generated in an angle tuned KDP crystal. The laser could also be operated in the Q-switched mode, giving pulses 22 nsec long. At the sample, the pulses were focused to intensities

from less than 0.3 GW/cm² to more than 10 GW/cm². The Si substrates were
free standing in air at room temperature and at elevated temperatures up to
∿ 500°C. In addition to intensity, other experimental parameters were the
number of laser pulses incident on a single location (from one to >1000),
laser polarization and wavelength, and sample thickness. Transient optical
absorption, optical microscopy, and SEM were used with the thicker samples
while the <300 nm samples were prepared specifically for the TEM.

The results of various irradiations are best described as a function of
laser intensity. Below the threshold for any permanent change for large
numbers of pulses, 0.5 GW/cm², picosecond optical absorption transients were
observed at elevated temperatures (≳ 200°C). No transients were observed for
room temperature samples. A typical transient is shown in Fig. 1. The fast
decay time constant is 50 to 55 psec and is independent of sample temperature.
This time constant is consistent with the estimate of free carrier lifetime
just before melt estimated by AUSTON et al [6]. A second, longer time constant is present and is bounded between 1 nsec and 10 μsec. Overall, an
upper limit on absorbed laser energy was established at 2%.

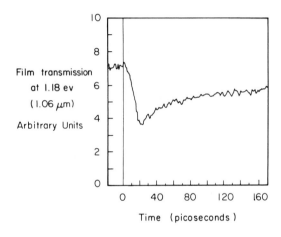

<u>Fig. 1</u> Film transmission at
1.06μm versus time at a
temperature of approxi-
mately 300°C. The rise
time constant is 50-55
psec

Just above the threshold for permanent change, small areas within the
laser beam profile changed from orange and transparent to black, opaque and
grainy. These regions were located at the beam entrance face of the sample.
The threshold for single shot effects was 1 GW/cm² compared to 0.5 GW/cm² for
multiple pulses. As shown in Fig. 2, TEM diffraction shows the characteristic α-Si rings (not polycrystalline). The actual amorphous region overlays
a region which is still single crystal.

For higher intensities the size of the amorphous area increased until it
filled the entire laser beam profile. This behavior was essentially the same
for multiple pulse or single pulse experiments; more pulses at lower intensity being equivalent to fewer higher—intensity pulses. Another effect
observed at higher intensities was a similar amorphous layer on the exit face
of the sample. In the 1.5 micron thick samples, a region of single crystal
silicon could be clearly observed by optical microscopy between the two
amorphous surface layers.

291

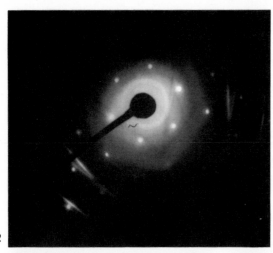

Fig. 2 TEM diffraction pattern of a typical amorphous region on a < 300 nm thick silicon sample

Fig. 3 SEM photo of a 1.5μm thick Si film irradiated at 533nm by multiple linealy polarized pulses. A close-up view of the smaller ridge structures at 10,000X, 30° tilt

2

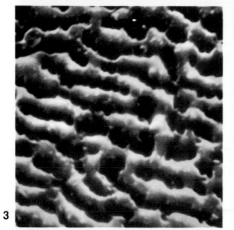

3

4

Fig. 4 SEM photo of a 1.5μm thick SI filme irradiated at 1.06μm by multiple circularly polarized pulses at 20,000X

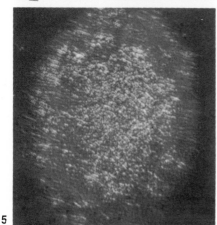

5

Fig. 5 A Normarski optical micrograph photo of a 1.5μm thick Si film irradiated at 533nm with multiple linearly polarized pulses. Note the small ridges at the periphery and the large orthogonal ridges at the center, 1280X

For higher intensities and larger numbers of pulses, interesting surface morphology was observed. At lower intensities small ridges are formed, as shown in Fig. 3. These are formed with irridation at both 1.06μm and 533nm. The ridge spacing scales proportional to the laser wavelength. The ridges always form normal to the optical electric field and do not form at all for circular polarization, as shown in Fig. 4. At higher intensities, generally in the center of the laser spot, larger ripples form. These are shown near the center in Fig. 5. These ridges are always parallel to the optical electric field and do not scale linearly with the incident wavelength.

At the highest intensities used ($\sim$ 10 GW/cm^2), the films repeatedly punctured and cracked. However, epi layers which were never etched from their 20 mil substrates could withstand much higher intensities. In fact, similar surface structures were observed on such samples but typically at intensities ten times greater than for the unsupported epi layers.

Our most important conclusion is that spatial energy concentration is required to achieve sufficient excitation to amorphize the surface silicon layer. For an inter-band absorption coefficient of 10 cm^{-1}, carrier densities will be in the range of 10^{18}-10^{19}/cm^{-3} for laser intensities in these experiments. Even for a maximum total film absorption at threshold of 2%, the carrier density should only reach 1.4×10^{19} cm^{-3}. We suggest that the damage observed in these experiments results from the transfer of energy from the laser electric field to these excited states rather than to the direct production of a large number of free carriers by the laser photons. The damage morphology in the center of Fig. 5 is consistent with melting following an avalanche multiplication process but the amorphous structure in Fig. 3, produced by slightly smaller intensities, is not.

Finally, we speculate that these laser-induced amorphous phases in silicon are an example of the classical nucleation and growth process. Evidence of the irregular amorphous regions at low intensities growing larger for higher intensities or more pulses is very suggestive of nucleation and growth from heterogeneous sites. [7].

References

1. P. L. Liu, R. Yen, and N. Bloembergen, Appl. Phys. Lett. 34, 864 (1979).
2. R. Tsu, R. T. Hodgson, T. Y. Tan, and J. E. Baglin, Phys. Rev. Lett. 42, 1356 (1979).
3. R. E. Harrington, J. of Appl. Phys. 37, 2028 (1966);
   ....., J. of Appl. Phys. 38, 3266 (1967);
   ....., J. of Appl. Phys. 39, 3699 (1968).
4. The authors would like to thank Dr. A. Syllaios and Dr. R. Sandfort of Monsanto for providing the silicon epi wafers.
5. I. Nashiyama, Phys. Rev. B 19, 101 (1979).
6. D. H. Auston et al. AIP Conference Proceedings: Laser Solid Interactions and Laser Processing 1978, P. 11, A.I.P. (1979).
7. This research was supported by the DoD Joint Services Electronics Program through the AFOSR contract F49620-77-C-0101.

Picosecond Time-Resolved Resonant Two-Photon Absorption in Cu_2O

A. Migus, J.L. Martin, R. Astier, and A. Antonetti

Laboratoire d'Optique Appliquée, Ecole Polytechnique - ENSTA,
F-91120 Palaiseau, France

D. Hulin and A. Mysyrowicz

Groupe de Physique des Solides, Ecole Normale Supérieure, Tour 23,
2, place Jussieu, F-75005 Paris, France

While there has been a considerable effort devoted to two-photon absorption
spectroscopy in solids, (in which process the intermediate state of the transition is virtual), relatively little has been done in the two-photon resonant
case, where one of the incident frequencies is tuned to an absorption level of
the system. This situation may provide informations upon the real intermediate
state, particularly its lifetime, and this aspect has already been exploited
in resonant Raman spectroscopy [1] . With available light sources of subpicosecond duration, it is possible to perform a time-resolved study of the transition process ; in this way, the pure two-photon signal can be discriminated
from the two-step process, and a measure of the intermediate state lifetime is
obtained directly. (By contrast, resonant Raman spectroscopy measures the variation of intermediate state lifetime with input frequency, but not its magnitude).

We present results concerning the study of resonant two-photon absorption
in Cu_2O, a semiconductor well-known for its excitonic properties [2] . The experimental set-up is shown schematically in fig. 1. Part of the amplified subpicosecond pulse of passively mode-locked Rh 6G laser is used as a pump beam
ω_1 . The probe beam ω_2 is given by the broad band subpicosecond emission generated by focusing the rest of beam ω_1 in a cell containing H_2O. The induced
absorption at frequency ω_2 is measured in function of delay Δt between the
pump and probe pulses. By changing the sample temperature it is possible to

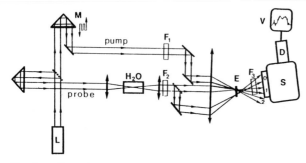

Fig.1   Experimental setup: L is the output of an amplifier of subpicosecond pulses (0.5 ps , 1 mJ , 10 Hz) ,M a step motor, F1,F2,F3 opti -
cal densities and coloured filters.The beam 2 ,the pump,is focused
on the sample E in spatial coincidence with the white probe beam 1 .
Beams 1 and 0 (reference white pulse) enter in the spectrometer S
and the OSA D.Induced absorption spectrum appears on display V.

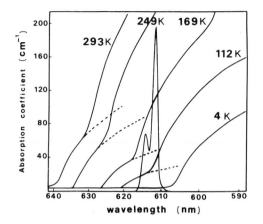

Fig.2   Absorption coefficient
         of Cu_2O at different tem-
peratures.   Also shown is the
spectrum of the pump beam.

pass continuously from a pure two-photon transition, for which both ω_1 and ω_2 fall in a transparency region (but with $\hbar(\omega_1 + \omega_2) > E_g$ the energy gap) to a situation where ω_1 lies inside an intrinsic absorption band of the crystal.

The fundamental absorption spectrum of Cu_2O near ω_1 is shown in fig. 2 for different temperatures, together with the position of pump beam ω_1. The absorption continuum corresponds to a phonon-assisted transition (fig. 2) from the ground state to the $n = 1$ term of the yellow excitonic series [3]. The lowest edge is due to the creation of an exciton with simultaneous absorption of an optical phonon of symmetry Γ_{12}^- and energy $\hbar\omega_0 = 13.5$ meV [3], and, it is consequently very sensitive to the crystal temperature. The second edge is due to the Γ_{12}^- phonon creation process. Since the pump pulse duration ($\Delta t \simeq 0.5$ ps) is of the order of one cycle of the Γ_{12}^- phonon vibration, we have verified, in a preliminary experiment, that the phonon assisted absorption band is similar if measured with CW or subpicosecond light.

The time evolution of the induced absorption at $\omega_2 = 1.87$ eV is shown in figure 3. At $T < 100$ K only a fast component is observed, with a time response corresponding to the input pulse duration. As T is increased, a long-lived component appears within the pulse duration and shows no significant decay in the time range explored in our experiment. We also observed that the amplitude of this component increases with the temperature, while at the same time the peak at $\Delta t = 0$ ps decreases.

Using time-dependent perturbation theory to second order, the two-photon transition rate may be expressed as :

$$W_{if} = c\, I_1 I_2 \; \left| \sum_{\ell} \frac{\langle f|\vec{\varepsilon}_1\vec{p}|\ell\rangle\langle\ell|\varepsilon_2\vec{p}|i\rangle}{E_\ell - E_i - \hbar\omega_2} + \sum_{\ell \neq \ell(R)} \frac{\langle f|\varepsilon_2\vec{p}|\ell\rangle\langle\ell|\vec{\varepsilon}_1\vec{p}|i\rangle}{E_\ell - E_i - \hbar\omega_1} + R \right|^2$$

w h e r e the resonant term R is also expressed in the framework of perturbation theory to the third order [4], because it involves the participation of a parity conserving phonon as stated above (see Fig. 4a and b).

From the experimental results, the respective contributions from the two-photon and the two-step processes may be readily sorted out: two-photon absorption is instantaneous and should be the only transition present below T = 100 K. This corresponds to the behaviour of the fast signal apparent for a time delay $\Delta t = 0$.

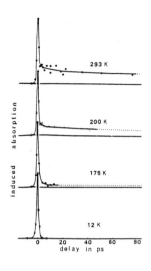

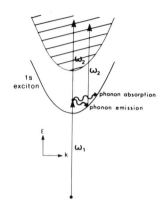

Fig.3 Induced absorption of the probe beam as a function of delay between pump and probe pulses. The peaks have been normalized.

Stepwise excitation involves the creation of real excitons as a first step. Because of the large spectral width of the pump beam (see Fig.2) a superposition of exciton states with different K vectors is initially produced. The second step of the excitation completes the transition towards the continuum of electron-hole pairs. Under the assumption of a smooth final density of states, the second step does not discriminate between the contributions involving different excitons states within their kinetic energy band; therefore, the signal amplitude versus time delay Δt probes the total exciton population decay in the $n = 1$ level rather than the lifetime of an exciton with a particular wave vector $\underline{K}$, as it is the case in resonant Raman scattering.

From the experimental results, we infer an exciton population lifetime in excess of 10^{-10} sec at 300 K. No measurement of exciton lifetime from luminescence decay is available beyond T = 100 K, but from the relative radiative efficiency of luminescence between 77 (where $\tau_x \simeq 1$ μsec) and 300 K, a lifetime τ_x (300 K) > 10^{-9} sec is estimated [5]; such a large value although unusual at high T, can be understood for Cu_2O : the exciton binding energy is very large in this material, with $E_x = 0.13$ eV, so that thermal dissociation is not important even at room temperature. Further, the direct radiative recombination of $n = 1$ excitons is forbidden in dipole approximation for this

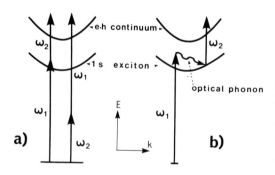

Fig. 4 Schematic representation of two-photon (left) and two-step (right) contributions to the induced absorption. The real creation of excitons (right) requires participation of negative parity phonons to conserve parity

296

direct gap material because of parity conservation law. The most efficient free exciton radiative decay requires the participation of optical phonons. Such processes are known to lead to long exciton lifetimes in indirect gap materials such as Ge or AgBr.

In conclusion, it is possible by time-resolved measurements on a fast time-scale, to distinguish between true two-photon absorption and two-step processes which are both present in the non-linear absorption of Cu_2O at high T. A lower limit of the exciton lifetime at room temperature is obtained.

References

1. P.Y. Yu, Y.R. Shen, Y. Petroff, L.M. Falicov, P. R. Lett. <u>30</u>, 283 (1973).

2. See for instance R.S. Knox, Theory of excitons, Acad. Press, N.Y. (1963).

3. S. Nikitine in "Optical properties of Solids" (Plenum, N.Y. 1969)

4. R. Loudon, Proc. Roy. Soc. <u>A 275</u>, 218 (1963).

5. A. Mysyrowicz, D. Hulin, A. Antonetti, P. R. Lett. <u>43</u>, 1123 (1979) E <u>43</u> 1275 (1979) and unpublished results.

Determination of Dispersion Curves of Excitonic Polaritons by Picosecond Time-of-Flight Method

Y. Aoyagi, Y. Segawa, T. Baba, and S. Namba

The Institute of Physical and Chemical Research
Wako-shi, Saitama 351, Japan

Dispersion curves of the excitonic polaritons in CuCl and CdS were precisely determined by a picosecond time-of-flight method in which the propagation time of a tunable picosecond light pulse in a crystal was measured. The experimental results and the availabilities of this technique were discussed.

Introduction

Excitonic polaritons (the mixed state of excitons and photons in a crystal) gives rise to a nonlinear dispersion in an energy region where the photon dispersion crosses to the exciton band. This dispersion curve is strongly dependent upon the oscillator strength and the effective mass of the exciton. Since the group velocity of the polaritons in the crystal corresponds to the derivative of the dispersion curve, we can precisely determine small changes of the dispersion curve, the effective mass, and the oscillator strength of excitons by observing the group velocity of a picosecond light pulse in the crystal as a function of the incident photon energy (a picosecond time-of-flight method).

Experimental Results and Discussion

1. Determination of Fine Structure in Dispersion Curves [1]

Figure 1 shows our experimental setup for the picosecond time-of-flight experiments. The output beam of the tunable picosecond laser was perpendicularly incidents on the surface of the crystal and the time needed for the transmitance of the light through the sample was analyzed by a Kerr shutter system.

Figure 2 shows various arrival times of the light propagated to the Kerr shutter in CuCl at various wavelengths. The highest curve shows the arrival time of the laser light without the sample. The arrival time was increased with the increase of the photon energy from the transparent energy region of the crystal to the Z_3 exciton absorption band (3.203 eV). At the incident photon energy of 3.192 eV, we observed the delay of more than 400 psec for the 1.6 mm thick sample.

The observed relative polariton group velocities to the light velocity in vacuum were plotted in Fig.3-a for various photon energies. As the incident photon energy sweeps the lower branch of the polariton from its lower energy region, we can see monotonous decrease of the relative group velocity. Above the Z_3 exciton band there is a maximum group velocity at 3.230 eV. In this energy region, there are two polariton modes. One mode is the upper branch polariton which has nearly the same momentum with the

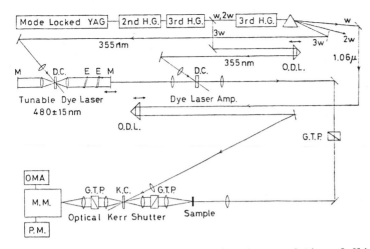

Fig.1   The experimental setup for the picosecond time-of-flight experiments

incident photon.   Another mode is the lower-branch polariton which gains
its energy from the kinetic energy of the Z_3 exciton.   In our experiment
only a signal from the upper-branch polariton was observed.   The monotonous
decrease of the group velocity of the lower-branch polariton was explained
by a one-oscillator model but this model could not explain the experimental
results of upper-branch polariton as shown in the dotted line.   The group
velocity of the upper-branch polariton observed in this experiment was
anomalously slow compared with the group velocity calculated from the one
oscillator model.   This discrepancy is explained using a two-oscillator
model of Z_3 and Z_{12} excitons as shown in the solid line in the figure.

The dotted curve and the solid curve
of Fig.3 shows the calculated dispersion curves by one-and two-oscillator
models, respectively.   The effect
of the Z_{12} exciton to the dispersion
curves is very small but is significantly shown in the difference between
the two solid and dashed curves of
the group velocity.   This feature
demonstrates that the measurement of
the group velocity of the polariton
is apparently a useful technique for
the precise determination of the
dispersion curves.

2.   Precise Determination of Oscillator Strength of Exciton

Figure 4 shows the transmittance and
reflectance spectrum for the light
with the electric vector E parallel
and perpendicular to the c-axis of
CdS.   Since the absorption of the
A exciton for the light of E∥c is

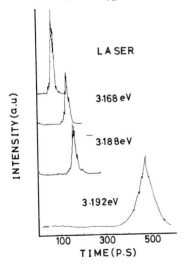

Fig.2   Various arrival time of
the light propagated to the
Kerr shutter in CuCl at various
wavelengths

forbidden, we can estimate the polariton dispersion of the A and B excitons independently by observing the group velocity for the light with the E vector parallel and perpendicular to the c-axis. The ratios of the observed polariton group velocity to the light velocity in vacuum were shown in Fig.5-a for the light E∥c and E⊥c. As the incident photon energy increases in the energy region from 2.50 to 2.55 eV for the lower branch of the dispersion curve we can see monotounous decrease of the relative group velocity. Fig. 5-b shows the relative group velocity near the absorption of the A exciton for E⊥c. In this energy region we can see the upper branch polariton in the energy higher than 2.556 eV. In Fig.5 the solid curves are calculated by the one and two oscillator models for E∥c and E⊥c, respectively. The experimental results were well fitted with the calculated ones by using the parameters W_1=2.5528 eV, W_1'=2.5548 eV, W_t=2.5686 eV, W_t'=2.5720 eV and ε_∞=8.0, where W_1, W_1', W_t, W_t' are the frequency of the longitudinal and transverse A and B excitons, respectively, and ε_∞ is a dielectric constant.

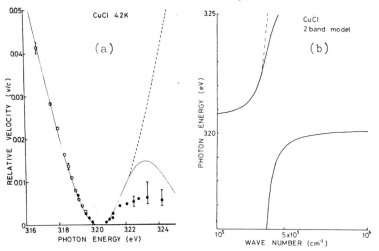

Fig.3 (a) Relative polariton group velocities at various photon energies.
: 1.6 mm thick crystal. : 30 m thick crystal. The dashed line: calculated by the one oscillator model. The solid line: calculated by the two oscillator model. (b) Polariton dispersion relation near the Z_3 exciton. Dashed line: one oscillator model. Solid line: Two oscillator model.

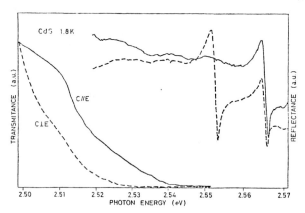

Fig.4 The transmittance and refelctance spectrum for the light with $E_{\parallel}c$ and E⊥c in CdS

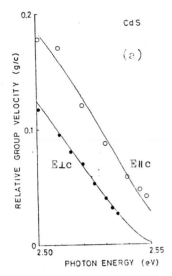

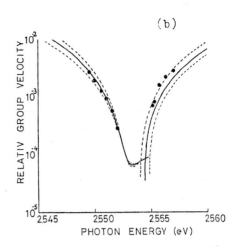

Fig.5　(a) The ratios of the observed polariton group velocity to the light velocity for the light E⊥c and E∥c in CdS.　(b) The ratios of the observed polariton group velocity to the light velocity for the light E⊥c near the A exciton absorption in CdS.

In Fig.5-b the dotted lines are calculated by using the parameters deviated by 0.0004 eV from the values given above.　In this deviated case the calculated curves could not explain the experimental results.　This result shows that the time-of-flight method gives precise determination of the oscillator strength of excitons (L-T splitting).

3.　Precise Determination of Effective Mass of Exciton

As shown in Fig.6 the dispersion curve strongly depends on the effective mass of the exciton in the energy region from 2.557 to 2.560 eV in the case of CdS.　By observing the group velocity through this region we can estimate the effective mass of the excitons precisely.

4.　Experimental Determination for Additional Boundary Condition Problem

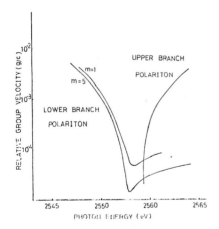

Fig.6　The relative group velocity of the excitonic polariton near the absorption of the A exciton as a function of the effective mass.

Figure 7 shows the dispersion curves of the upper and lower branch polaritons.　In the large wave-number region of the lower branch polariton,

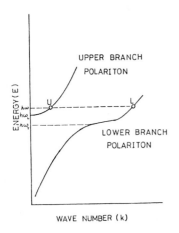

UPPER BRANCH
POLARITON

LOWER BRANCH
POLARITON

WAVE NUMBER (k)

Fig.7 The dispersion curves of the upper–branch and lower–branch polariton

the kinetic energy of the exciton influences the polariton dispersion curve to increase the energy. In the energy region larger than hW_1 the incident light can have two independent wave numbers U and L (two independent waves). In this case the conventional boundary condition to determine the electrical field in the dielectric material is not enough to determine the intensity of the two possible waves in the crystal and an additional boundary condition is needed. Since the group velocities for the U and L waves are differ each other, we can separate the two waves experimentally in our method. By measuring the intensity of the two waves as a function of the energy of the incident wave it will be possible to determine the additional boundary condition experimentally by comparing the experimental results with several model proposed theoretically [2-4].

Conclusion

The picosecond time-of-flight method is a powerful method to determine
1. fine structures of the dispersion curve of excitonic polaritons,
2. the precise ossilator strength of excitons,
3. the precise effective mass of excitons, and
4. the additional boundary condition, experimentally.
We can determine the fine structure of dispersion curves in CuCl and oscillator strengths precisely in CdS.

References

1. Y. Segawa, Y. Aoyagi and S. Namba: Solid State Commun. 32, 229 (1979).
2. J. J. Hopfield and D. G. Thomas: Phys. Rev. 132, 563 (1963).
3. C.S. Ting, M. J. Frankel and J. L. Birman: Solid State Commun. 17, 1285 (1975).
4. J. L. Birman and J. J. Sein: Phys. Rev. B6, 2482 (1972).

Picosecond Infrared Spectroscopy in Narrow-Gap Semiconductors

B.D. Schwartz and A.V. Nurmikko

Division of Engineering, Brown University,
Providence, RI 02912, USA

Presently, narrow-bandgap semiconductors occupy an increasingly important role in mid-infrared electro-optics. They are also interesting from a purely physical point of view because of characteristics arising from their bandstructure. One example of experimental interest to us concerns the properties of dense, nonequilibrium electron-hole gases in such materials as PbTe, $Hg_{1-x}Cd_xTe$ and InSb. In particular, we have applied picosecond infrared techniques at 10.6 μm and 5.3 μm to the study of the highly degenerate plasmas in these semiconductors. The emphasis in the work discussed below centers on the transient dynamics of the electron-hole system, specifically the recombination kinetics. Our measurements have provided a first indication of recombination rates at excess densities $n \geq 10^{19} cm^{-3}$.

In a typical experimental arrangement we employed a modelocked Nd-glass laser to operate a picosecond infrared switch thereby generating ultrashort pulses of 10.6 μm radiation [1]. Part of the 1.06 μm radiation was also used to excite a dense electron-hole gas in a sample under study through single photon interband absorption. Time resolved measurements of the transmission and reflectivity at 10.6 μm were made, complemented in selected instances by transmission measurements at 5.3 μm (radiation obtained by second harmonic generation in Te or $AgGaSe_2$).

An example of our experimental results is shown in Figure 1 which displays the transmitted intensity of short 10.6 μm pulses (t_p = 10 psec) through a film of epitaxial PbTe (t = .5μm, T = 300K)as influenced by a dense, transient electron-hole gas. The characteristic feature in all our temporal transmission data corresponding to high excitation shows an initial rapid decrease in the transmitted intensity followed by an apparently constant (but finite) region of transmission whose duration is dependent on the level of excitation. Subsequently, the transmission recovers with a time constant characteristic of each semiconductor. The origin of time in Figure 1 has been chosen not to signify the coincidence of the excitation and probe pulses but rather to identify the beginning of the recovery of transmission. Qualitatively similar features were observed at T = 300K for epitaxial $Hg_{.70}Cd_{.30}Te$ and bulk InSb. The recovery of transmission was, however, measured to be faster. In this article we limit the quantitative discussion mainly to PbTe.

Measurements of the time resolved reflectivity R at 10.6 μm yielded complementary information with temporal features which agreed well with the probe pulse transmission data through the relation R ≈ 1-T. Thus for the range of excitation used in these experiments the transmission appeared to be mainly affected by reflectivity changes at the first surface of a sample

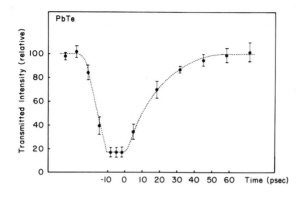

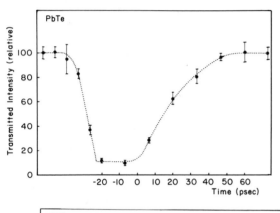

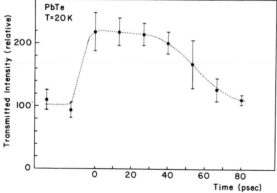

Figure 1 Transmission at 10.6 µm in PbTe as influenced by a dense free carrier plasma. The level of excitation in (b) is 5 times larger than in (a)

Figure 2 Transmission at 5.3 µm in PbTe at T = 20K

where the excitation of an electron-hole gas took place.

An illustration of the probe transmission measurements at 5.3 µm is shown in Figure 2 for PbTe (epitaxial) at T = 20K. Unlike at 10.6 µm where intraband single particle and collective (plasma) modes of the photoexcited electron-hole gas couple with the radiation field, at 5.3 μm interband transitions ($\hbar\omega > E_g$) in a degenerate system can lead to saturation of absorption (through bandfilling effects) and gain for an inverted population.

A significant challenge in quantitative interpretation of our experimental results concerns an accurate determination of the instantaneous excess carrier density and the dielectric modeling of a high density electron-hole gas in a narrow-gap semiconductor. This is a complicated problem since, for example, the (quasi) Fermi energy of the system may significantly exceed the value of the energy gap in our experimental conditions. The strong band nonparabolicities are formally equivalent to a relativistic description which complicates the interpretation of the highly energetic quasi-metallic plasma. As a first approximation we have identified the data such as in Figure 1 ($\hbar\omega < E_g$) with the contribution of a free electron-hole gas in a simple Drude like model, appropriately corrected for band nonparabolicities:

$$\varepsilon(\omega,t) = \varepsilon_\infty - [4\pi N(t)e^2/m_{av}\omega(\omega+i/\tau)] - i\beta N(t) \qquad (1)$$

Here m_{av} is a properly averaged (over Fermi sphere) optical mass which for spherical bands (not PbTe) can be simplified as

$$\frac{1}{m_{av}} = \int \frac{1}{\hbar^2 k} \cdot \frac{dE}{dk} \cdot f(E)d\underline{k} \; / \int f(E)d\underline{k} \qquad (2)$$

to which we have applied the Kane band model. The Fermi energy is connected to the carrier density through

$$N = (1/3\pi^2) \; (2m_{de}k_B T/\hbar^2)^{3/2} \; {}^o L_o^{3/2} \; (E_g, E_F) \qquad (3)$$

where ${}^o L_o^{3/2}$ is a generalized Fermi integral and m_{de} the density of states effective mass. Equations (1), (2) and (3) allow a self-consistent connection between N, m_{av}, and E_F. In Eq. (1) τ is a scattering time and β a phenomenological damping coefficient of the plasma by single particle processes (e.g. interband). Figure 3 shows the calculated reflectivity of PbTe (T = 300K) as a function of excess electron-hole pair density. The highly anisotropic masses have been properly averaged in the degenerate limit considered here. The single particle damping of the plasma edge has been incorporated by including known values of intraband absorption cross section modified by bandfilling (saturation) effects. The usefulness of such a calculation

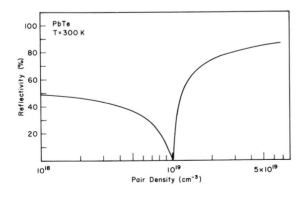

Figure 3 Calculated reflectivity vs. pair density in PbTe

305

is that it identifies the range of excess carrier densities for which the reflectivity is strongly varying. Consequently, the origin of time in Figure 1 has been chosen to correspond to the onset of departure from condition of high reflectivity. The interpretation for the difference in Figures (1a) and (1b) is that for higher excitation the system simply persists longer in the highly reflecting state. The "plasma edge" should thus provide a rather direct indication of the excess carrier concentration (up to $\sim N(\omega_p)$)and its development.

We have analyzed data such as in Figure 1 (while also accounting for the finite infrared pulse widths) by this model to obtain a measure of recombination rates in the narrow-gap materials of interest. It is well known that for semiconductors such as InSb and $Hg_{1-x}Cd_xTe$ Auger recombination is a strong process, dominating other recombination channels at carrier densities already as low as $10^{16}cm^{-3}$ [2]. Recently it has also been shown that for a multivalley "mirror band" semiconductor such as PbTe, scattering of carriers in different valleys of a band can enhance the otherwise weaker Auger rate [3]. Analysis of our data for PbTe at 300K shows that a reasonable agreement with a simple model can be made according to $dN/dt = -WN^3$, where W is an Auger rate coefficient and equals approximately 1×10^{-28} cm^{+6} sec^{-1}. This value is consistent with earlier measurements in the nondegenerate regime, a rather surprising result considering the high degree of carrier degeneracy in our case. This is in strong contrast with our results for InSb and $Hg_{.70}Cd_{.30}Te$ which show very large deviations from extrapolations from a nondegenerate case. In particular, the decay of excess population ≤ 20 psec for $Hg_{.70}Cd_{.30}Te$ with $N\sim4 \times 10^{19}cm^{-3}$) is significantly slower than indicated by such predictions. In a single valley model it has been pointed out by Haug [4] how deviations in a degenerate case may arise to greatly weaken the N^3 dependence through (Fermi) statistical factors together with effects of screening on the appropriate matrix elements. At present our experimental data does not fully permit a unique determination of both the Auger cross-section and the concentration dependence.

The measurement at 5μm of interband contributions by the dense electron-hole gas in PbTe (Fig. 2) shows temporal decays which are consistent with the intraband data. A more serious difficulty concerns the explanation of the magnitude in the observed "gain". Direct calculations of maximum gain from a Kane band model in PbTe yield values appreciably smaller than our observations. We have also made calculations for the renormalization of the energy gap by many-body effects (exchange and correlation energies). While such corrections improve the agreement somewhat, discrepancies remain. Ongoing work is under way to clarify existing questions. In principle, however, the interband spectroscopy can be a valuable tool for the highly excited narrow-gap semiconductors while showing the influence of many-body effects on single particle properties.

In conclusion we have examined in initial experiments how picosecond infrared techniques may be applied to the study of high density electron-hole systems in the narrow-gap materials, notably PbTe. While this work has yielded for the first time a direct measure of carrier relaxation rates in these materials at such densities, our understanding of the basic characteristics of the high-density phases remain far from complete.

We would like to thank Dr. H. Holloway of Ford Research and Engineering Center for the PbTe samples and Dr. S. H. Shin of Rockwell Science Center for the $Hg_{1-x}Cd_xTe$ samples.

This work was supported by AFOSR through Grant #77-3199C.

References

1. S. A. Jamison and A. V. Nurmikko, Appl. Phys. Lett. <u>33</u>, 598 (1978)

2. A. R. Beattie and P. T. Landsberg, Proc. Roy. Soc. <u>A249</u>, 16 (1959)

3. P. R. Emtage, J. Appl. Phys. <u>47</u>, 2565 (1976)

4. A. Haug, Solid State Electr. <u>21</u>, 1281 (1978)

Observation of Picosecond Photon Echoes in GaP:N

P. Hu, S. Chu, and H.M. Gibbs
Bell Laboratories, Murray Hill, NJ 07974, USA

Abstract

Photon echo decays show phase memory times from 97 down to 8 ps as a function of increasing nitrogen concentration in GaP:N.

We report the first observation of photon echo in a semiconductor, in this particular case GaP:N. Phase memory times, measured with the help of the simple angularly resolved technique [1], range from 100 ps at low N concentration down to 8 ps for N concentration up to $2.5 \times 10^{18}/cm^3$. This implies a strongly concentration dependent homogeneous linewidth; in contrast, the total linewidth of about 2 Å is much less sensitive to concentration. Knowledge of dephasing times in semiconductors is sparse but could contribute to the understanding of semiconductor physics and the operation of semiconductor lasers. It is hoped that this investigation can also shed light on the dynamic interaction between the impurity excitons.

A mode-locked Argon laser is used to synchronously pump a dye laser tuned to the A line of the bound exciton in GaP:N. A two-stage amplifier pumped by a 50 kW N_2 laser operating at 50 Hz is used to generate $\sim$ 10 ps pulses with peak intensities of $\sim$ 300 kW at the 535 nm wavelength of the exciton. By splitting the light pulses with a beam splitter and delaying one with an optical delay line (mirror on a translator), the simple setup is complete. The two pulses spatially separated $\sim$ 1 cm transversely, but parallel, are then focused by a 30 cm lens onto the sample. The signal, separated from the second pulse by the same angle θ as the second is from the first, is easily observable either by eye or a simple diode. At zero delay between the two excitation pulses, self-scattering from the grating created by the two pulses can be observed. The scattered light appears as two separate beams, displaced at an angle θ with respect to the first and second beams, respectively. Optimization of these grating signals serves as a convenient way to align the optics for the echo experiment.

In Fig. 1, we show the echo signal as a function of pulse separation for three different samples. For the Monsanto sample, the echo decay could not be clearly resolved with the present pulses.

The fact that we see diffraction from a grating resulting from the interference of our two laser beams shows that there can not be spatial mobility of the excitons over distances on the order of the interference periodicity on a $\sim$ 10 ps time scale. By using counter propagating pulses and a third delayed probe beam, we should be able to measure energy transfer with spatial resolution of a few hundred angstroms in the lifetime of the exciton.

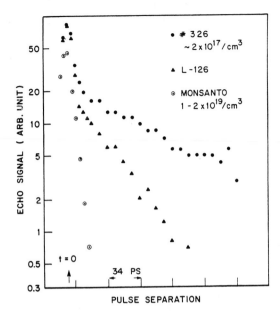

Fig. 1 Photon echo decays from bound excitons in GaP:N with nitrogen concentration as the variable parameter.

At concentrations $\approx 10^{18}$ nitrogen atoms per cm^3, evidence for exciton energy transfer by tunneling has been presented [2]. At these concentrations the average nitrogen spacing is about 100 Å whereas the exciton diameter is almost 50 Å. Thus this technique would allow us to directly probe "single-hop" energy transfers between near-neighbor nitrogen sites. Finally, by examining both photon echo and transient grating decay times, the possibility exists for investigating coherent energy transfer.

References

1. I. D. Abella, N. A. Kurnit, and S. R. Hartmann, Phys. Rev. Lett. 141, 391 (1966). P. Hu and H. M. Gibbs, J. Opt. Soc. Am. 68, 1631 (1978).

2. P. J. Weisner, R. A. Street, H. D. Wolf, Phys. Rev. Lett. 35, 1366 (1975).

Pulsewidth Dependence of Picosecond Laser-Induced Breakdown

E.W. Van~Stryland, A.L. Smirl, and W.E. Williams

Department of Physics, North Texas State University,
Denton, TX 76203, USA

M.J. Soileau

Michelson Laboratory, Physics Division, Naval Weapons Center,
China Lake, CA 93555, USA

The laser-induced breakdown fields for fused SiO_2, single crystal NaCl and air were measured at 1.06 μm for laser pulsewidths ranging from 40 psec to 31 nsec. These experiments represent the first such measurements over a range of 10^3 in pulsewidth while keeping other parameters such as specimen, laser frequency and focal volume constant. The laser-induced breakdown fields of fused SiO_2 and single crystal NaCl were found to be only slightly dependent on pulsewidth, i.e., $E_B \sim t_p^{-x}$ where $x \lesssim 1/8$, whereas the air breakdown field varied as $E_B \sim t_p^{-\frac{1}{4}}$. The NaCl and SiO_2 results are in sharp contrast to the predictions of various models [1,2] and previous published pulsewidth dependence data [2, 3]. The work of BETTIS et al. [1] predicts that the breakdown threshold field will scale as $t_p^{-\frac{1}{4}}$. In contrast, work by SMITH et al. [3] indicated an approximately $t_p^{-\frac{1}{2}}$ dependence of the breakdown field for NaCl using pulses in the picosecond region. The latter workers compared their 30 psec data with previous data from FRADIN et al. [4]. SMITH et al. [3] attempted to correct their data for self-focusing effects by using the technique proposed by ZVEREV and PASHKOV [5] but did not similarly reduce the data from FRADIN et al. [4], even though the latter authors exceeded the critical power for self-focusing by as much as an order of magnitude. In addition, the measurements performed by these two groups were taken with different focal volumes. Recent work has shown that in many cases laser-induced breakdown is dependent on focal volume and varies greatly among specimens of a given material [6,7]. Thus, pulsewidth dependence data is difficult to interpret unless all other parameters are held constant and the same specimen is used at all pulsewidths.

The laser source for the picosecond data presented in this work was a passively mode-locked Quantel, Nd:YAG system operating at 1.06 μm. A single pulse from the mode-locked train was switched out and amplified to produce single pulses of measured Gaussian intensity profile and Gaussian temporal distribution. The temporal pulsewidth was variable between 30 and 200 psec by selecting various etalons as the output coupler. The width of each pulse was determined by monitoring the ratio, R, of the square of the energy in the fundamental (1.06 μm) to the energy in the second harmonic produced in a $LiIO_3$ crystal. This ratio is directly proportional to the pulsewidth and was calibrated by measuring the pulsewidth using second harmonic autocorrelation scans and accepting only a narrow range in ratios, R. Damage data was then selected according to the temporal width of individual pulses. The breakdown threshold was taken to be that intensity at a given pulsewidth which produced damage 50% of the time. Each site was irradiated only once. The energy on target was varied by changing the angle between a calibrated pair of Glan polarizers. The nanosecond data was taken on the same samples using a Q-switched Nd:glass laser. The same focusing lenses were used, and the energy monitor was directly calibrated with the one used for the picosecond measurements.

AIR			SiO$_2$			NaCl		
ω_o (μm)	t_p (psec)	E_B (MV/cm)	ω_o (μm)	t_p (psec)	E_B (MV/cm)	ω_o (μm)	t_p (psec)	E_B (MV/cm)
19.3	31,000		19.3	31,000		19.3	31,000	2.16 ± 0.22
	191 ± 25	36.5 ± 2.3		133 ± 13	8.4 ± 0.4		134 ± 13	2.58 ± 0.22
	104 ± 15	40.5 ± 3.0		75 ± 7	9.2 ± 0.5		92 ± 8	2.58 ± 0.13
	40 ± 3	55.6 ± 5.2		40 ± 2	8.4 ± 0.3		42 ± 3	2.74 ± 0.17
10.3	31,000	14.2 ± 1.4	10.3	31,000	8.36 ± 0.8	10.3	31,000	2.57 ± 0.26
	183 ± 17	42.2 ± 1.8		175 ± 12	12.3 ± 0.5		175 ± 25	3.28 ± 0.39
	117 ± 17	51.8 ± 3.4		75 ± 5	12.6 ± 0.6		117 ± 8	3.84 ± 0.20
	47 ± 5	67.4 ± 3.3		47 ± 5	12.3 ± 1.0		47 ± 5	4.20 ± 0.30
6.1	31,000	20.8 ± 2.0	6.1	31,000	11.4 ± 1.1	6.1	31,000	2.93 ± 0.29
	125 ± 8	64.3 ± 2.1		175 ± 20	16.0 ± 1.2		167 ± 16	4.91 ± 0.30
	100 ± 10	68.0 ± 3.4		108 ± 15	16.8 ± 1.1		100 ± 10	5.50 ± 0.40
	43 ± 2	86.1 ± 2.2		47 ± 6	16.7 ± 1.0		45 ± 3	6.06 ± 0.37
5.0	31,000	28.9 ± 3.0	5.0	31,000	12.6 ± 1.3	5.0	31,000	
	134 ± 8	86.1 ± 2.6		134 ± 8	25.9 ± 1.5		125 ± 16	9.10 ± 0.8
	100 ± 8	99.5 ± 3.4		109 ± 5	25.9 ± 1.5		104 ± 7	9.30 ± 0.4
	44 ± 2	135 ± 4.0		43 ± 3	30.1 ± 1.2		43 ± 3	11.60 ± 1.4

Fig.1 A tabulation of the RMS breakdown electric fields, E_B, of SiO$_2$, NaCl and air for focused spot sizes (1/e^2 half-width in intensity) of 5.0, 6.1, 10.3, and 19.3 μm and for various optical pulsewidths (FWHM) in psec

The results of our measurements over a range of 10^3 in pulsewidth for focal spot radii from 5.0 to 19.3 μm are summarized in Fig.1. Self-focusing effects were neglected in calculating the breakdown fields. Some workers claim that observed dependences of breakdown fields on focal spot radii are due to self-focusing [3,5], and they scale their results in accordance with the technique suggested by ZVEREV and PASHKOV [5]. ZVEREV and PASHKOV [5] predict that a plot of P_B^{-1} (P_B is the power at which breakdown occurs) versus ω_o^{-2} yields a straight line given by

$$P_B^{-1} = 2(I_B \, \pi\omega_o^2)^{-1} + P_{CR}^{-1} \qquad (1)$$

where P_{CR} is the critical power for self-focusing and I_B is the breakdown intensity. The basic assumption of this procedure is that I_B is the intrinsic breakdown intensity and is independent of the focal spot radius. It is clearly seen from Fig.2 that this scaling technique cannot be used for our experiments. All of the data for the different pulsewidths follows a similar pattern and cannot be fit to a straight line. Deviations from the straight line fit for large spot sizes have been noted previously [3]. Such data have been disregarded by arguing that for large focal radii and powers near P_{CR}, the constant shape solution to the nonlinear wave equation (on which the ZVEREV and PASHKOV [5] procedure is based) is no longer valid. However, the argument cannot explain the small focal radii data shown in Fig.2.

It has been clearly established that with the possible exception of a small number of specimens tested by MANENKOV [6], the laser-induced breakdown fields are not intrinsic, and vary greatly even for specimens of a given material from the same supplier [7,8,9]. This violation of the basic assumption that the damage is intrinsic casts doubt on previously published data where the ZVEREV and PASHKOV [5] scaling was used to interpret breakdown thresholds.

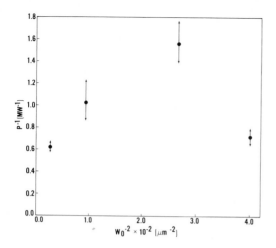

Fig.2 A representative plot of the inverse of the power, P_B, needed to induce breakdown versus ω_0^{-2}, where ω_0 is the $1/e^2$ radius in intensity at the laser focus calculated using linear Gaussian optics. Here data for SiO_2 using 44 psec (FWHM) pulses is presented

We conclude that the role of self-focusing in these experiments cannot be determined, and therefore, the data presented in Fig.1 do not include any self-focusing corrections (even though P_B exceeds many estimates of P_{CR}). This apparent discrepancy with self-focusing theory can be explained by including the effect of plasma defocusing. It has been shown that the negative nonlinear index of refraction, n_2, due to free electrons produced by the intense optical fields counters the positive n_2 resulting from bound electronic effects [7,8].

The pulsewidth dependence as shown in Fig.3 is important for understanding the basic mechanisms of laser-induced failure. Only part of the data of Fig.1 is shown in Fig.3 but that shown is typical of all the data. Air breakdown was observed to scale as $t_p^{-1/4}$ as predicted by the simple avalanche theory; however, bulk damage in both SiO_2 and NaCl show a much slower dependence on t_p than predictions of various models and previous [1,2] pulsewidth dependence data [2,3].

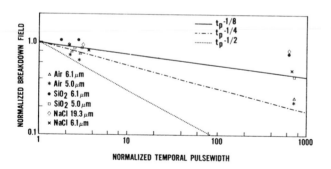

Fig.3 The normalized RMS breakdown field versus normalized temporal pulsewidth at 1.06 μm. The breakdown field was normalized by dividing the field by the breakdown field at the shortest pulsewidth employed for that material and spot size. The pulsewidth was then scaled by the shortest pulsewidth. For example, the breakdown field for SiO_2 at a spot size of 5.0 μm was determined by normalizing the breakdown field by 30.1 MV/cm and the pulsewidth by 43 psec (see Fig.1). This normalization assures that the intercept goes through 1.0

For example, the fit of data from [3] and [4] to predictions based on d.c. breakdown theory has been interpreted as evidence for the avalanche breakdown model. The $t_p^{-1/2}$ dependence of the breakdown field determined in [3] was used to scale 30 psec data at 1.06 μm to 21 psec so that frequency dependence between 1.06 and 0.53 μm could be examined [10]. As a result of this scaling, the breakdown fields at 0.53 μm are found to be greater than those at 1.06 μm further supporting the avalanche model. However, if the scaling results for NaCl and SiO$_2$ shown in Fig.3 are used instead, the breakdown field at 0.53 μm would be lower than at 1.06 μm and would indicate that avalanche breakdown was not the damage mechanism.

There are laser-induced breakdown theories which predict very weak pulsewidth dependence similar to the bulk damage results presented here. For example, SPARKS [11] predicts a $t_p^{-1/6}$ scaling of the laser-induced breakdown fields. BRAUNLICH, et al. [12] have pointed out that multiphoton processes should become increasingly more important as laser pulsewidth decreases. They predict that in the absence of an avalanche, $t_p^{-1} \propto (E_B)^{n+1}$ where n is the order of the multiphoton process. The band gap of SiO$_2$ and NaCl is approximately 7.8 eV. Direct excitation of electrons across the gap would be a 7 photon process for those materials at 1.06 μm for which case the pulsewidth dependence would be $E_B \propto t_p^{-1/8}$. Although it is unlikely that direct excitation of electrons across the gap is the mechanism for laser induced failure of NaCl and SiO$_2$, this and previous [7,8] pulsewidth dependence studies indicate that multiphoton processes may play a more important role in laser-induced breakdown than was previously thought. The data of Fig.1 are qualitatively consistent with the multiphoton-initiated avalanche breakdown model presented in Refs. 7 and 8 [13].

This work was performed as part of a joint NTSU/NWC research program sponsored by the Office of Naval Research (ONR) and the North Texas State Faculty Research Fund. A. L. Smirl also wishes to thank the ONR for support.

[1] J. R. Bettis, R. A. House, and A. H. Guenther, NBS Special Publication 462, p. 338, 1976.
[2] W. Lee Smith, Optical Engineering 17, 489 (1978).
[3] W. Lee Smith, J. H. Bechtel, and N. Bloembergen, Phys. Rev. B 12, 796 (1975).
[4] D. W. Fradin, N. Bloembergen, and J. P. Letelier, Appl. Phys. Lett. 22, 635 (1965).
[5] G. M. Zverev and V. A. Pashkov, Soviet, Phys. JETP 30, 616 (1960).
[6] A. A. Manenkov in NBS Special Publication 509, p. 445, 1977.
[7] M. J. Soileau, Ph.D. Thesis, University of Southern California, 1979.
[8] M. J. Soileau, M. Bass, and P. H. Klein, "Frequency and Focal Volume Dependence of Laser-Induced Breakdown in Wide Band Gap Insulators," in the NBS proceedings of the 1979 symposium on Laser Induced Damage to Optical Materials.
[9] D. Olness, Appl. Phys. Lett. 8, 283, 1966.
[10] W. Lee Smith, J. H. Bechtel, and N. Bloembergen, Phys. Rev. B 15, 4039 (1977).
[11] M. S. Sparks, NBS Special Publication 435, p. 331, 1975.
[12] P. Braunlich, A. Schmid, and P. Kelly, Appl. Phys. Lett. 26, 150 (1975).
[13] E. W. Van Stryland, M. J. Soileau, Arthur L. Smirl, William E. Williams, to be published.

Part VII

Ultrashort Processes/Biology

Electron Transfer from Photoexcited Bacteriopheophytin to p-Benzoquinone in Cationic Micelles

D. Holten[1], S. Sadiq Shah, and M.W. Windsor

Department of Chemistry, Washington State University,
Pullman, WA 99164, USA

Introduction

The primary energy storage reactions in photosynthesis are light-driven charge separation processes that take place within special pigment-protein complexes called reaction centers. In reaction centers isolated from photosynthetic bacteria, it appears that electron transfer occurs from the lowest excited singlet state (P*) of a bacteriochlorophyll dimer to a quinone by way of a molecule of bacteriopheophytin (BPh). Although the initial electron transfer from P* to BPh occurs within 4 ps of excitation by a short pulse of light (1), the reverse electron transfer step that returns the molecules to the ground state is more than 1000 times slower (2). This gives ample time for the electron to move from BPh⁻ to the quinone, which takes about 200 ps (3,4). Both of the forward electron transfer steps have quantum yields of 100%, even at cryogenic temperatures (2,5). Thus, within the bacterial reaction center light is converted to useful chemical potential via a highly efficient multistep charge separation process, initiated by an excited singlet state electron donor, in a time appreciably less than a nanosecond.

In homogeneous solution the situation is quite different. Although both the excited singlet and triplet states of BPh and chlorophyll (Chl) are effectively quenched by p-benzoquinone (BQ) and other electron acceptors in polar organic solvents, radical ions are detected only from the excited triplet state reaction [7-11]. The reason adduced is that reverse electron transfer within the intermediate triplet radical pairs to give the ground singlet state is spin-forbidden, giving the ions more time to diffuse apart [12]. It is not at present understood why the singlet pathway is so effective in the reaction center, whereas only the triplet route operates effectively in homogeneous solution, although several reasons for this difference in behavior have been considered. These include the effect of Franck-Condon factors on reverse electron transfer rates and the possibility that different orbitals may be involved in the forward and reverse electron transfer steps [12-14]. The possible importance of the microscopic environment in assisting charge separation and hindering reverse electron transfer in reaction centers has not explicitly been considered in these discussions. To explore its role, we have studied reactions of photoexcited BPh in aqueous micellar solutions with picosecond and slower spectroscopic techniques.

---

[1] Present address: Department of Chemistry, Washington University, St. Louis, MO 63130, USA

Surfactant (detergent) molecules possess a polar headgroup and a nonpolar hydrocarbon tail. These self-aggregate in aqueous solution to form colloidal complexes of high molecular weight, called micelles [15]. These are roughly spherical structures with a nonpolar inner core comprising the hydrocarbon tails and a charged surface region made up of the headgroups, called the Stern layer, surrounded by the counterions and the bulk aqueous phase. The micellar surface is positively charged (cationic) in the cetyltrimethylammonium bromide (CTAB) micelles. The hydrophobic interactions responsible for micellar stability are similar to those which stabilize biological membranes and globular proteins [15]. Micelles provide a means for imposing structure on the microscopic environment and for controlling the distance between the reactants.

Our purpose in undertaking the studies presented here was to find out if micelles could effect the rate of electron transfer between the reactants, the rate of reverse electron transfer within the radical pair state, and the recombination rate of separated ions. Previous studies in micelles of electron transfer from photosynthetic pigments and other porphyrins to various electron acceptors have had some success in this regard [16-18]. We chose to study BPh + BQ because the excited state photophysics and electron transfer behavior of BPh with this electron acceptor in homogeneous solution are well characterized [8,12-14]. By comparing results obtained in micelles, homogeneous solutions, and reaction centers, we hope to explore the effect of the environment on charge separation in a systematic way.

Experimental

The picosecond apparatus has been described previously [19]. For the present study, 8 ps 530 nm flashes were used to excite sample solutions of 40 μM BPh in 2 or 5 mm path cells. Microsecond flash photolysis studies made use of 200 ns 530 nm excitation flashes from a flashlamp-pumped coumarin-7 dye laser. The detection system had a response time of about 1 μs.

BPh was prepared as described previously [8]. Micellar solutions were prepared by injecting concentrated aliquots of BPh in ethanol into 0.1 M aqueous CTAB followed by addition of the electron acceptor. These solutions were thoroughly degassed by at least 10 freeze-thaw cycles on a high-vacuum line, and sealed. Fluorescence quenching studies were carried out on a Farrand MKI spectrofluorometer.

Sites of solubilization

Previous studies suggest that Chl-a is solubilized within the nonpolar inner core of micelle, the process being assisted by hydrophobic interactions between the hydrocarbon (phytyl) chain of Chl-a and the hydrocarbon chain of the surfactant molecules [15-18]. Based on the structural similarity between Chl-a and BPh, we believe that BPh is incorporated into the inner core of the micelle. The site of solubilization of BQ in CTAB micelles was determined by proton NMR spectroscopy to be on the average near the inner surface of the Stern layer [21]. As described below, the behavior of photoexcited BPh in the presence of BQ supports our views on the sites of solubilization of both molecules.

Excited singlet state quenching

Difference spectra for the formation of the BPh excited singlet state (BPh*), the excited triplet state (BPh^T), and the cation radical (BPh$^+$)

in CTAB micelles are closely similar to those obtained in a number of organic solvents [8,21].

The lifetime of BPh* was obtained from the decay of excited state absorbance at 610 nm and from recovery of BPh ground state bleaching at 755 nm. Lifetimes measured at the two wavelengths were the same within experimental error. Figure 1 shows the first order decay plots for BPh* in CTAB at 755 nm. These measurements give lifetimes for BPh* of 2.2 + 0.3 ns, 1.1 + 0.2 and 0.6 + 0.1 ns in the absence and in the presence of 30 or 70 mM BQ, respectively. The lifetime in the absence of quenchers is the same within experimental error as the value of 2.0 + 0.2 ns obtained previously in acetone:methanol (7:3) solution [8].

The reduction in BPh* lifetime and the fluoresence quenching measurements (not shown) are in good agreement and are linear in BQ concentration up to about 30 mM BQ. These data give a quenching constant $K_Q = 33 + 2$ M^{-1}. Together with a 2.2 ns lifetime in the absence of quencher, this yields a second order quenching rate of $1.5 + 0.2 \times 10^{10}$ $M^{-1}s^{-1}$ in CTAB micelles. Previously a value of $1.7 + 0.2 \times 10^{10}$ $M^{-1}s^{-1}$ was obtained in acetone:methanol (7:3) [8]. Thus, quenching rates of BPh* by BQ are about the same in CTAB micelles and in this homogeneous solution.

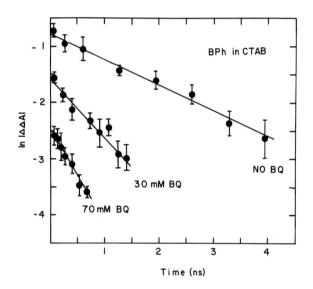

Fig.1 First order decay plots for 20 μM BPh following excitation with 8 ps 530 nm flashes, measured from recovery of ground state bleaching at 755 nm. The ordinate is the absorbance change at time, t, minus that at the asymptote of the decay curve, measured between 6 and 9 ns after excitation.

Above 30 mM the quenching becomes nonlinear in BQ concentration in both CTAB and in organic solvents, but is much more pronounced in CTAB micelles. This does not appear to be due to ground state complexes involving BPh and BQ, because at the highest concentration of BQ used (70 mM) no new bands were observed in the absorption or fluorescence spectra. Such nonlinear quenching behavior has been attributed previously to static quenching and would involve "instantaneous" quenching of those BPh* molecules formed within an active quenching radius of an electron acceptor, and thus becomes increasingly important at higher quencher concentrations [22]. Enhanced static quenching in CTAB micelles at high BQ concentrations suggests that at these concentrations more BQ molecules lie within the effective quenching distance of BPh at the time of excitation than in homogeneous solution. This result is in agreement with the contention that BQ molecules are preferentially solubilized at the inner surface of the Stern layer and that BPh molecules reside within the micellar inner core.

Formation of cation radicals

Irradiation with 200 ns 530 nm flashes of BPh in CTAB micelles in the absence of quenchers produces absorbance changes due to BPh^T, that decay with a time constant of 48 μs. In the presence of BQ, these absorbance changes are replaced by the spectrum of the cation radical, BPh^+. Kinetics for the decay of BPh^+ measured at 420 nm for 10 μM BPh and 10 mM BQ in CTAB were found to be essentially second order. With a value of 1.0 + 0.3 x 10^5 $M^{-1}cm^{-1}$ for the differential extinction coefficient at 420 nm [8,23], we obtain a second order rate constant for the recombination of the separated ions, BPh^+ and BQ^-, of 2.3 + 1.5 x 10^{10} $M^{-1}s^{-1}$ in CTAB micelles. A value of 1.8 $\pm$ 1.0 x 10^{10} $M^{-1}s^{-1}$ was obtained in acetone:methanol (7:3).

With the ps apparatus, no additional absorbance changes near 850 nm, where BPh^+ is known to have weak absorption [8,23], were detectable for 40 μM BPh in CTAB micelles in the presence of 30 or 70 mM BQ. Since the detection limit for absorbance changes is 0.025 with our ps apparatus, the yield of BPh^+ could be as much as 40% at the 40 μM BPh concentrated used and still go unobserved at 850 nm. Thus, in the present study we were not able to determine whether BPh^+ in CTAB micelles forms preferentially via BPh* or BPh^T.

A better region for monitoring the formation of BPh^+ would be at 400 nm where it absorbs strongly. Unfortunately, we could not make ps observations in this region because of strong sample absorbance and declining probe intensity. Efforts are currently underway to obtain enhanced probe intensity in the 400 nm region by using 530 nm light to generate the picosecond continuum and by improvements in the optics.

The fact that BPh is readily seen with 200 ns excitation flashes but goes undetected in the ps experiments is due to multiple recycling of the system during the much longer 200 ns flashes. Thus our observation of BPh^+ in this case does not necessarily imply a high intrinsic quantum yield of formation [8].

Conclusions

Electron transfer from BPh* to BQ is more efficient in CTAB micelles than in acetone:methanol solution at high BQ concentrations. At low BQ concentrations the quenching rate is about the same in the two systems, as is the recombination rate of separated ions. Similar observations have

320

been reported on the chlorophyll sensitized reduction of methyl viologen in nonionic micelles [16], whereas other systems have shown more pronounced effects [17,18].

Additional studies with an expanded series of electron donors and acceptors are underway to elucidate this behavior and to further investigate the effects of various types of micelles on the rate of electron transfer quenching, the yield of ion separation and the rate of charge recombination. We also hope to resolve optically the formation of radical ions from excited singlet and triplet state reactions in micelles, as has been done in homogeneous solution [8], since this is crucial for devising efficient models for photosynthetic electron transfer.

Acknolwedgements

This work was supported by the U.S. Army Research Office under grant DAA629-76-9-0275 and by the NSF under grant PCM 79-02911.

References

1. D. Holten, C. Hoganson, M. W. Windsor, C. C. Schenk, W. W. Parson, A. Migus, R. L. Fork, and C. V. Shank, Biochim. Biophys. Acta
2. W. W. Parson, R. K. Clayton and R. J. Cogdell, Biochim. Biophys. Acta 416, 105 (1975).
3. Kaufmann, K. J., P. L. Dutton, T. L. Netzel, J. S. Leigh and P. M. Rentzepis, Science 188, 1301 (1975).
4. M. G. Rockley, M. W. Windsor, R. J. Cogdell and W. W. Parson, Proc. Natl. Acad. Sci. USA 72, 2251 (1975).
5. R. K. Clayton and T. Yamamoto, Photochem. Photobiol. 24 , 67 (1976).
7. D. Huppert, P. M. Rentzepis and G. Tollin, Biochim. Biophys. Acta 440, 356 (1976).
8. D. Holten, M. Gouterman, W. W. Parson, M. W. Windsor and M. G. Rockley, Photochem. Photobiol. 23, 415 (1976).
9. A. A. Lamola, M. L. Manion, H. D. Roth and G. Tollin, Proc. Natl. Acad. Sci. USA 72, 3265 (1975).
10. N. E. Andreava, G. V. Zakharova, V. V. Shubin and A. K. Chibisov, Chem. Phys. Lett. 53, 317 (1978).
11. G. Tollin, F. Castelli, G. Cheddar and F. Rizzuto, Photochem. Photobiol. 29, 147, 153 (1978).
12. M. Gouterman, and D. Holten, Photochem. Photobiol. 25, 85 (1977).
13. D. Holten, and M. W. Windsor, Ann. Rev. Biophys. Bioeng. 7, 189 (1978).
14. D. Holten, M. Gouterman, W. W. Parson and M. W. Windsor, Photochem. Photobiol. 28, 951 (1978).
15. J. H. Fendler, and E. J. Fendler, Catalysis in Micellar and Macromolecular Systems, Academic Press, N.Y. (1975).
16. K. Kalyanasundaram, and G. Porter, Proc. Roy. Soc. A., (1978).
17. C. Wolf, and M. Gratzel, Chem. Phys. Lett. 52, 542 (1977).
18. G. R. Seely in Topics in Photosynthesis (J. Barber, ed.) pp. 36-37, Elsevier, Heidelberg (1977).
19. D. Magde, and M. W. Windsor, Chem. Phys. Lett. 27, 31 (1974).
20. J. Simplicio, Biochem. 11, 2525 (1972).
21. S. S. Shah, D. Holten and M. W. Windsor, Photobiol. Photobiophys., submitted.
22. J. B. Birks, Photophysics of Aromatic Molecules, Wiley, New York (1970).
23 J. Fajer, D. C. Borg, A. Forman, R. H. Felton, D. Dolphin and L. Vegh, Proc. Natl. Acad. Sci. USA, 71, 994 (1974).

A Report on Picosecond Studies of Electron Transfer in Photosynthetic Models

T.L. Netzel

Department of Chemistry, Brookhaven National Laboratory,
Upton, NY 11973, USA

R.R. Bucks and S.G. Boxer

Department of Chemistry, Stanford University,
Stanford, CA 94305, USA

J. Fujita

Department of Energy and Environment, Brookhaven National Laboratory,
Upton, NY 11973, USA

1. Introduction

Considerable spectroscopic work on reaction centers (RC) from photosynthetic bacteria [1] and on photosystem I (PSI) particles from green plants [2] has established that the initial photochemical step in these systems is the subnanosecond transfer of an electron resulting in the creation of an oxidized donor and a reduced acceptor. For both of these systems the electron donor is a dimer [3]. The acceptor for bacterial RC's is bacteriopheophytin, a metal-free bacteriochlorophyll. The acceptor for PSI is thought to be chlorophyll$_a$. Recently a tetrameric complex was reported [4] to duplicate the $\leq$ 6 ps charge transfer (CT) reaction found in bacterial RC's. This complex was present in an equilibrium mixture of pyropheophorbide$_a$ (PPheo$_a$) ethylene glycol monoester and pyrochlorophyll$_a$ (PChl$_a$) in toluene. The PChl$_a$ dimeric core of the tetramer was reported to donate an electron to one of the attached PPheo$_a$ monoesters. Problems of interpretation arising from excess PPheo$_a$ in solution were not addressed.

To further our understanding of the light driven electron transfer reactions found in natural photosystems, we studied dimeric and trimeric model molecules containing PChl$_a$. However, rather than relying on chemical equilibria to join the potential electron donors and acceptors, we covalently attached all of the subunits to form a single large molecule [5]. We altered the distance between the donor and acceptor subunits by using both 10 atom and 5 atom chains. Also, the effects of altering the relative orientation of the donor and acceptor were probed by contrasting the kinetics observed with added pyridine to those observed with added alcohol.

The addition of pyridine prevents the dimer and trimer models from aggregating. However, the addition of alcohol causes intramolecular bonding of the model's subunits though R-OH bridges. The oxygen of the alcohol is coordinated to the Mg in PChl$_a$, while its proton is hydrogen-bonded to the keto group on ring V of a neighboring subunit. The keto group is present on both PChl$_a$ and PPheo$_a$, but PPheo$_a$ does not contain a Mg atom.

Because the dielectric constant (ε) of the solvent directly affects the kinetics of electron transfer reactions [6], several solvents were used: toluene (ε=2), CH$_2$Cl$_2$ (ε=9) and CH$_3$CN (ε=39). Also, since a goal of this type of research is to correlate electrochemical, spectroscopic and structural information to predict the likelihood of electron transfer reactions, we varied the redox span of the potential photoproducts.

2.   Results

A comparison of the difference spectra produced upon excitation of PChl$_a$ and PPheo$_a$ with those known to occur upon oxidation and reduction of the same molecules shows that identifying CT photoproducts solely on the basis of ps absorbance difference spectra is difficult. This is because these molecules have similar absorbance spectra for their excited singlet states and for their cationic and anionic forms. Fortunately fluorescence quantum yield measurements as well as reference to published streak camera results can provide the necessary additional input to accurately interpret the ps data.

2.1 Dimer Model Results

Of course the occurrence of a rapid $S_1 \rightarrow$ CT reaction implies a greatly reduced S_1 lifetime relative to that of the unreacted PChl$_a$ monomer. It also implies a greatly reduced fluorescence quantum yield. The monomer fluorescence yield is therefore a necessary baseline. Both PChl$_a$ and PPheo$_a$ give essentially the same fluorescence yield in the three solvents mentioned earlier, $\sim$ 100 arbitrary counts. There is no trend with respect to the addition of ethanol or pyridine. Neither is there a trend with respect to increasing solvent dielectric constant.

Table 1 presents data on the fluorescence quantum yields for seven dimeric molecules.

Table 1	Fluorescence Yields* (arbitrary counts)		
	Toluene ($\varepsilon=2$)	CH_2Cl_2 ($\varepsilon=9$)	CH_3CN ($\varepsilon=39$)
1. PChl$_a$ $\sim$ PChl$_a$ + Py	98 $(6.2\pm0.8ns)^a$	78	27
2. PChl$_a$ $\sim$ PPheo$_a$ + Py	100 $(7.5\pm1ns)^a$	33 $(1.8\pm0.2ns)^a$	18 $(0.85\pm0.10ns)^a$
3. PChl$_a$ $\sim$ PPheo$_a$ + EtOH	47 $(3.8\pm0.8ns)^a$	36	18
4. PChl$_a$ $\sim$ Pheo$_a$ + Py	80	38	23
5. PChl$_a$ $\sim$ Pheo$_a$ + EtOH	56	40	27
6. Cl$^-$: PChl$_a$ $\sim$ PPheo$_a$	---	38	20
7. Cl$^-$: PChl$_a$ $\sim$ Pheo$_a$	---	40	24

* Py = 0.5M pyridine; EtOH = 0.5M ethanol; Cl$^-$ = 0.1M[N(C_2H_5)$_4$]Cl

[a] Lifetime obtained from an absorption transient.

The lifetimes obtained from ps absorption spectroscopy are shown in parentheses for five of the entries. While several interesting trends can be seen, the key points are 1) the fluorescence yields are high and 2) the relative absorption lifetimes correlate very well with the relative fluorescence yields. Therefore, only excited singlet states and not CT states are observed in these dimers.

This conclusion is not surprising for the first entry in the table, PChl$_a$ $\sim$ PChl$_a$, because the formation of a CT product from this dimer's S_1 state is only weakly exothermic. However, the exothermicity increases as one goes down the table. It is at least 150 meV for entries 2 and 3; at least 250 meV for entries 4, 5 and 6; and at least 350 meV for entry 7. Apparently, there is a kinetic bottleneck preventing a rapid $S_1 \rightarrow$ CT reaction in these dimers.

This lack of reaction for dimers that are singly linked with a 10 atom chain contrasts vividly with the $\leq$ 6 ps $S_1 \rightarrow$ CT reaction found in other work [7] for doubly linked diporphyrin PSII models. An obvious difference is that the average donor to acceptor distance is much larger in these dimers than the $\sim 4\overset{\circ}{A}$ interplane distance in the diporphyrins. Also, the cofacial arrangement of the doubly linked diporphyrins provides good electronic coupling between the initial S_1 and final CT configurations.

2.2 Trimer Model Results

The above work seems to have bounded the RC modeling problem: chlorophyll type dimers that are single linked with 10 atom chains don't transfer electrons from their S_1 states, but cofacial diporphyrins that are held about $4\overset{\circ}{A}$ apart do. In this section we examine an intermediate case: the donor to acceptor distance is shortened by using a 5 atom chain and a dimer, $(PChl_a)_2$, is used as the electron donor.

2.2.1 Trimer Model in Toluene

An elegant model of the primary electron transfer system of purple bacteria can be constructed by linking two $PChl_a$'s with a 10 atom chain and then linking one of them to a $PPheo_a$ with a 5 atom chain. The construction of the model is completed by adding 0.2M ethanol to the trimer in toluene. This step folds the two $PChl_a$'s with double R-OH bridges. The $PPheo_a$ acceptor is thought to reside over the folded $(PChl_a)_2$ [5]. Electrochemical measurements on another doubly linked $(PChl_a)_2$ molecule showed that the dimer was 70 meV easier to oxidize than the monomer [8]. Therefore an $S_1 \rightarrow$ CT reaction for this $(PChl_a)_2 \sim\!\!\sim PPheo_a$ trimer in its folded conformation should be 150 to 220 meV exothermic.

Unfortunately, this trimer in toluene has a fluorescence quantum yield that is only 50 percent lower than that of a monomer of either $PChl_a$ or $PPheo_a$. Analysis of the ps absorbance kinetics shows a 60 ps energy transfer step in which excited $PPheo_a$ transfers singlet excitation energy to the lower energy S_1 state of the $(PChl_a)_2$ subunit. Parallel with this in the 710 nm region, the trimer amplifies probe light through stimulated emission. The maximum of the gain occurs in the 50 to 100 ps time interval. The recovery of the ground state bleach in the 1 to 5 ns time interval proceeds with a 3.5 ns lifetime. This is half the lifetime of the S_1 state of $PChl_a$. The above evidence indicates that this trimeric RC model does not undergo an $S_1 \rightarrow$ CT reaction.

2.2.2 Trimer Model in CH_2Cl_2

The potential electron donor in the above trimeric RC model is the folded $(PChl_a)_2$ subunit. PELLIN et al. [9] have measured the S_1 state lifetime of this folded dimer to be 110 ps in CH_2Cl_2. FREED [10] has provided a quantum mechanical description of the unusual properties of the excited singlet state of this folded dimer in terms of symmetric and antisymmetric "exciton-like" states. The $\sim$ 60 fold decrease in S_1 lifetime on going from toluene to CH_2Cl_2 is striking enough to warrant asking whether or not the folded $(PChl_a)_2$ subunit will undergo an $S_1 \rightarrow$ CT reaction if the trimer is placed in CH_2Cl_2 rather than in toluene.

Ps absorbance measurements on folded $(PChl_a)_2$ in CH_2Cl_2 show that an initial 110 ± 20 ps recovery of ground state bleach is followed by a long lived state that does not decay in the 1 to 5 ns time region. PERIASAMY et. al. [11] showed that at times > 30 ns after photoexcitation folded $(PChl_a)_2$ yields an unfolded configuration with one subunit a triplet and the other a ground state singlet. We therefore interpret the ps absorbance data as showing a 110 ps S_1 state decaying to a T_1 state. Some repopulation of the ground state may occur also. The fluorescence quantum yield of the folded $(PChl_a)_2$ in CH_2Cl_1 was 13 (arbitrary counts): about 8 times less than that of the $PChl_a$ monomer.

When an electron acceptor, $PPheo_a$, is attached to the folded $(PChl_a)_2$ in CH_2Cl_2, the S_1 lifetime shortens to 64 ± 5 ps and the fluorescence quantum yield decreases to 9. This is followed by a long lived state that does not decay in the 1 to 5 ns time interval. This is essentially the same result as is found when $PPheo_a$ is not attached to the folded $(PChl_a)_2$ electron donor. Therefore, there is no basis for assigning anything other than an $S_1 \rightarrow T_1$ process to this trimer.

Substitution of pheophorbide$_a$ (Pheo$_a$) for pyropheophorbide$_a$ (PPheo$_a$) as the electron acceptor makes the $S_1 \rightarrow CT$ reaction in the folded trimer model more exothermic by 110 meV. The total exothermicity is then 260 to 330 meV. The S_1 lifetime found in this case is 110 ± 20 ps and the fluorescence quantum yield is 13. However, the state that is subsequently formed decays with a lifetime of 3.0 ± 0.7 ns. This may be a trimer triplet state with a 3 ns lifetime. However, this seems unlikely. Another possibility is that the S_1 state decays into a CT state. CT states in diporphyrin PSII reaction center models have been observed to have lifetimes as long as 2 ns [12]. However, the nearly 50 percent decay of the initial ground state bleaching by 1 ns suggests that the yield of this CT photoproduct is not high.

3. Conclusions

The attempts to produce $S_1 \rightarrow CT$ reactions whose rates were $> 10^{11}$ s^{-1} in dimeric and trimeric RC models containing $PChl_a$ were unsuccessful. This occurred in spite of 1) using a 5 atom linking chain between the donor and acceptor, 2) using a dimeric as well as a monomeric electron donor, 3) decreasing the average donor to acceptor distance by adding alcohol as a folding agent, 4) increasing the solvent dielectric constant, and 5) increasing the exothermicity of the possible $S_1 \rightarrow CT$ reaction to over 300 meV.

In one system, $(PChl_a)_2$ ∿∿ $Pheo_a$ with 0.5M ethanol in CH_2Cl_2, there is evidence that an $S_1 \rightarrow CT$ reaction occurs with a formation time of 110 ps. The CT photoproduct decays with a lifetime of 3 ns. However, the yield of the CT product is probably less than 50 percent.

These results suggest that to achieve high yield, rapid $S_1 \rightarrow CT$ reactions in Chl$_a$ type RC models the donor and acceptor probably will have to be held less than 7Å apart and the relative orientation of the donor and acceptor may be crucial. A close to cofacial arrangement may be necessary. To define more accurately the microscopic molecular requirements for rapid $S_1 \rightarrow CT$ reactions we have recently synthesized [13] a cofacial dimer of $PChl_a$ and $PPheo_a$. The subunits are tied at alternating, rather than at adjacent, corners with 5 atom chains. Preliminary CD measurements show significant exciton interactions between the two subunits. Future work will explore the ps photophysics of this compound.

References

1. M. G. Rockley, M. W. Windsor, R. J. Cogdell and W. W. Parson, Proc. Natl. Acad. Sci. US 72, 2251 (1975); K. J. Kaufmann, P. L. Dutton, T. L. Netzel, J. S. Leigh and P. M. Rentzepis, Science 188, 1301 (1975); V. A. Shuvalov, A. V. Klevanik, A. V. Sharkov, J. A. Matveetz and P. G. Krukov, FEBS Lett. 91, 135 (1978).

2. V. A. Shuvalov, E. Dolan and B. Ke, Proc. Natl. Acad. Sci. US 76, 770 (1979); V. A. Shuvalov, A. E. Klevanik, A. V. Sharkov, P. G. Kryukov and B. Ke, FEBS Lett. 107, 313 (1979).

3. J. R. Norris, R. A. Uphaus, R. A. Crespi, H. L. Katz and J. J. Katz, Proc. Natl. Acad. Sci. US 68, 625 (1971); G. Feher, A. J. Hoff, R. A. Isaacson and J. D. McElroy, Biophys. Soc. Abstracts 13, 61 (1973).

4. M. J. Pellin, K. J. Kaufmann and M. R. Wasielewski, Nature 278, 54 (1979).

5. S. G. Boxer and G. L. Closs, J. Amer. Chem. Soc. 98, 5406 (1976); S. G. Boxer and R. R. Bucks, J. Amer. Chem. Soc. 101, 1883 (1979).

6. N. Sutin in Inorganic Biochemistry , Vol. II, (G. L. Eichorn, ed.) pp 611-653, Elsevier Scientific Publishing Company, New York, 1977.

7. T. L. Netzel, C. K. Chang, I. Fujita and J. Fajer, Chem. Phys. Lett. 67, 223 (1979).

8. M. R. Wasielewski, W. A. Svec and B. T. Cope, J. Amer. Chem. Soc. 100, 1961 (1978).

9. M. J. Pellin, M. R. Wasielewski and K. J. Kaufmann, J. Amer. Chem. Soc., 102, 1868 (1980).

10. K. F. Freed, J. Amer. Chem. Soc. 102, 3130 (1980).

11. N. Periasamy, H. Linschitz, G. L. Closs and S. G. Boxer, Proc. Natl. Acad. Sci. US 75, 2563 (1978).

12. I. Fujita, J. Fajer, C.-B. Wang and T. L. Netzel, manuscript in preparation.

13. R. R. Bucks and S. G. Boxer, work in progress.

Research carried out at Brookhaven National Laboratory under contract with the U. S. Department of Energy and supported by its Office of Basic Energy Sciences.

Picosecond Studies on Bile Pigments

M.E. Lippitsch, A. Leitner, M. Riegler, and F.R. Aussenegg

Institut für Experimentalphysik, Universität Graz,
Universitätsplatz 5, A-8010 Graz, Austria

1. Introduction

Bile pigments are biological dyes built up essentially of four
pyrrol rings. Thus they show a certain chemical relationship to
haem and chlorophyll, but the pyrrol rings do not form a closed
porphyrine ring like in these molecules. Bile pigments are found
in various biological systems (e.g. mammalian liver and bile,
bird egg-shells, insect wings and cuticula pigments, red and blue
algae, higher plants). They play a key role in a number of important photochemical processes, namely
1) photo therapy of hepatitis in new-born babies
2) photosynthesis in red and blue algae
3) higher plant morphogenesis.
In the first two processes, the presence of oxygene is assumed
to be essential, while the third is presumably an anaerobic one.
The picosecond investigations presented here center on the primary
events of anaerobic photochemistry in bile pigments.

2. The Phytochrome Problem

Morphogenesis in higher plants is governed by a chromoprotein
named phytochrome. Phytochrome consists of a large protein (molecular weight 60 - 120 000) and a chromophore of the biliverdine
type. Excitation with light of λ = 660 nm transforms the pigment
to a biological active modification, while λ = 730 nm has the
reversal effect. There exist essentially two hypotheses concerning the transformation of the chromophore after light absorption.
The first one [1] assumes steric isomerisation around an exocyclic double bond, while the second one [2] proposes chemical as
well as structural changes. To decide between these hypotheses,
an investigation of the fluorescence and absorption properties
on a picosecond time scale could be of considerable help. The
chemical structure of the chromophore is not yet clear to the
last details, but biliverdines are assumed to be useful model
substances. In this paper, results on aetiobiliverdine and one
of its partial structures, pyromethenone (containing only two
pyrrol rings) are presented.

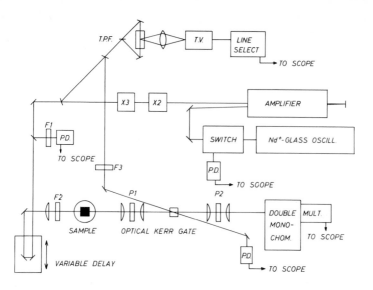

<u>Fig.1</u> Experimental arrangement. F 1,2,3 filters, PD photo diode, P 1,2 polarizers

3.   Experimental

The experimental arrangement to measure fluorescence life times in these compounds was the following (Fig.1): A single pulse extracted from a train produced by a Nd$^+$-glass laser (passively mode-locked by Kodak 9860 dye) was amplified twice to ∿10 mJ and hereafter frequency doubled and tripled. The different wave lengths were isolated by selective mirrors and appropriate filters. The green and UV pulses were used to excite the sample, respectively, while the IR pulse produced birefringence in a CS_2 cuvette placed between crossed polarizers, forming an optical Kerr-switch. The time function of the whole arrangement is shown in Fig.2 (dashed curve). During the experiment, the IR pulses were monitored by two-photon fluorescence with TV-camera registration.

Fluorescence from the sample, after passing the Kerr-switch, was focussed to the slit of a double monochromator and registrated by a photo multiplier.

4. Results and Discussion

Pyromethenone was investigated in a polar (chloroform) as well as a non-polar solvent (n-hexane). Excitation was performed with $\lambda = 354$ nm, which is well within the singlet absorption band ($\lambda_{max} \sim 370$ nm). The fluorescence was detected at $\lambda = 440$ nm. The results are shown in Fig.2. In the polar solvent chloroform (10^{-6} M) the decay time was found to be ∿6 ps. Since the inherent life time of the excited singlet state is calculated to

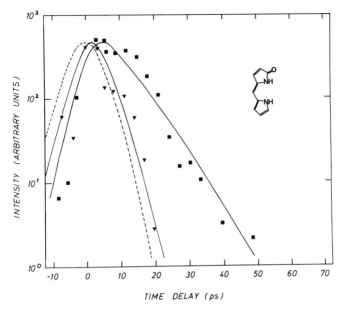

Fig.2 Fluorescence decay of pyrromethenone in chloroform (▪)
and n-hexane (▼) at λ = 430 nm and time function of Kerr-switch
(dashed line)

be $\tau_0 = 4.4 \times 10^{-9}$ s [3] a fast competing process has to be assumed. Since in monomeric pyrromethenone steric isomerisation
around the exocyclic double bond has been shown to occur with
high quantum yield (Φ = 0.25 in chloroform [3]), it seems reasonable to assume this isomerisation to be responsible for the short
life time of the excited singlet state. From chemical investigations on stilbene it is known, that there is a certain thermic
barrier for isomerisation in the excited state, which should
account for the quantum yield as well as for the life time. Since
in this case the life time should be temperature dependent low-
temperature measurements will elucidate this question.In nonpolar
solvents, an intermolecular proton transfer should dissipate the
excitation very efficiently [4]. The result (Fig.2) for n-hexane
solution (10^{-5} M) corroborates this hypothesis (decay time $\lesssim$ 2 ps).

The integral pigment aetiobiliverdine in freshly prepared methanol solution (5×10^{-6} M) gives a fluorescence at 700 nm. In
contrast to a predicted life-time of $\sim$ 1 ps [4], our experiment
yielded 68 $\pm$ 3 ps (Fig.3). From the fluorescence yield it can be
concluded, that a very fast radiationless process ($\lesssim$ 1 ps) occurs,
before the slower fluorescence becomes observable. A steric isomerisation as responsible for this radiationless energy dissipation seems to be unlikely with regard to the very short time
available. This finding may be helpful in the discussion about
the transformation of the phytochrome chromophore.

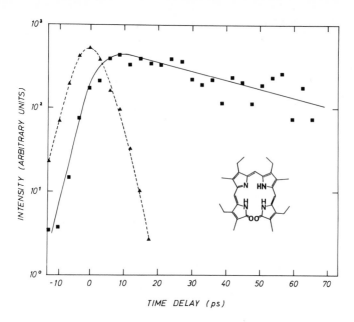

<u>Fig.3</u> Fluorescence decay of aetiobiliverdine in methanol (■)
at λ = 700 nm and time function of Kerr-switch (▲)

<u>References</u>

1. M.J. Burke, D.C. Pratt and A. Moscowitz, Biochemistry <u>11</u>,
4025 (1972)
2. S. Grombein, W. Rüdiger and H. Zimmermann, Z. Physiol. Chem.
<u>356</u>, 1709 (1975)
3. H. Falk and F. Neufingerl, Mh. Chem. <u>110</u>, 1243 (1979)
4. H. Falk, K. Grubmayr and F. Neufingerl, Mh. Chem. <u>110</u>, 1127
(1979)

Financial support of this work by the Fonds zur Förderung der
wissenschaftlichen Forschung (Project Nr. 4031) is gratefully
acknowledged.

Picosecond Torsional Dynamics of DNA

R.J. Robbins, D.P. Millar, and A.H. Zewail

Arthur Amos Noyes Laboratory of Chemical Physics,
California Institute of Technology,
Pasadena, CA 91125, USA

Studies of the structure and conformational dynamics of DNA are central to an understanding of its biological function. Although the long-range segmental motions of the DNA helix have been well characterized by a variety of physical techniques [1], very little is known about the more rapid internal motions in DNA [2-4]. Our objective here is to investigate the torsional (twisting) dynamics of DNA using the techniques of picosecond time-dependent fluorescence depolarization, and to compare the results with the predictions of the elastic model presented by BARKLEY and ZIMM [5].

DNA has a duplex structure, consisting of two (antiparallel) helical polydeoxyribonucleotide chains coiled around a common axis. The chains are held together by hydrogen bonds between complementary purine-pyrimidine base pairs, which are stacked almost perpendicular to the helix axis. Dyes such as ethidium bromide bind rigidly to DNA by intercalation between adjacent base pairs such that the plane of the phenanthridinium ring lies parallel to that of the bases. It is the reorientation of the intercalated dye that is monitored in our fluorescence depolarization experiments.

Several processes can reorient an intercalated ethidium: tumbling and rotation of the entire DNA molecule; long-range internal motions corresponding to end-over-end rotation and translation of persistence-length segments; local torsion and bending of the DNA helix; and, wobbling of the ethidium within its intercalation site. The calf thymus DNA used in this work is very long ($\sim 10,000$ base pairs), while the duplex structure confers stiffness on the chain, as measured by its persistence length of $\sim 600\text{Å}$ [6]. Therefore, rotation of the entire molecule and long-range segmental motions are slow, at least of the order of microseconds, and cannot contribute to the fluorescence depolarization since the decay time of the ethidium-DNA complex is only 22.6 ns. Bending of the DNA helix is also unimportant on the fluorescence time scale due to its stiffness [5]. What we expect to see in the fluorescence depolarization experiments then, are rapid torsional motions of the DNA chain and, possibly, wobbling of the ethidium within its intercalation site.

331

Time-resolved fluorescence depolarization measurements were performed using the mode-locked argon ion laser (Spectra Physics Model 171/342) and single photon counting apparatus shown in Fig. 1. The response of this instrument had a half-width of 250 ps. Samples contained 230 μg/ml calf thymus DNA (Sigma) and 3 μg/ml ethidium bromide dissolved in 0.1 M tris buffer at pH 7.7 with 0.15 M NaCl. Under these conditions essentially all of the ethidium is bound to DNA and the average distance between ethidium molecules is > 200Å so that dye-dye energy transfer is negligible. All experiments were performed at room temperature (22° C).

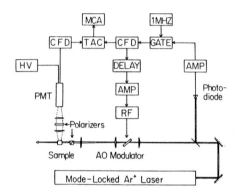

Fig. 1   Block diagram of the single photon counting apparatus. PMT, photomultiplier (Philips XP2020Q); CFD, constant fraction discriminator; TAC, time-to-amplitude converter; MCA, multichannel pulse-height analyzer; AMP, wide-band amplifier; RF, rf driver for the acousto-optic modulator; GATE, fast linear gate; 1 MHZ, pulse generator; HV, high-voltage power supply

Figure 2a shows the experimental fluorescence components polarized parallel $I_{\parallel}(t)$ and perpendicular $I_{\perp}(t)$ to the polarization of the exciting light for ethidium bromide intercalated in DNA. The time-dependent fluorescence polarization anisotropy r(t), which is related to these intensities by

$$r(t) = [I_{\parallel}(t) - I_{\perp}(t)] / [I_{\parallel}(t) + 2I_{\perp}(t)], \tag{1}$$

is shown in Fig. 2b; the anisotropy decay is highly nonexponential, showing an initial rapid decay and tending to level off at longer times. The fluorescence lifetime of the ethidium-DNA complex, obtained by setting the emission polarizer at the 'magic angle' (54.7°), was $\tau = 22.6 \pm 0.2$ ns.

The connection between the time dependent fluorescence polarization anisotropy r(t) and the conformational dynamics of DNA has been analyzed in detail both by BARKLEY and ZIMM [5] and by ALLISON and SCHURR

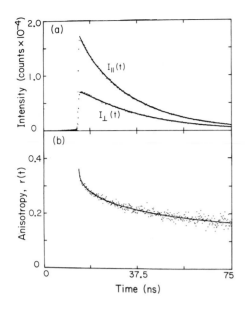

Fig. 2 Experimental data for the ethidium-DNA complex: (a) $I_\parallel(t)$ and $I_\perp(t)$; (b) the fluorescence polarization anisotropy $r(t)$ obtained from the data in (a). The solid lines are the best fit ($\chi^2 = 1.16$) of the data to the model described in the text, with $\tau = 22.6$ ns, $b^2\eta C = 2.4 \times 10^{-35}$ erg^{2}s and $r_0 = 0.36$.

[1]. We will discuss our results in terms of the elastic model of BARKLEY and ZIMM [5]. According to this model, the DNA chain is treated as a thin, flexible rod of length 2L, circular cross section of radius b, and uniform elasticity, immersed in a viscous fluid at thermal equilibrium. The dynamical equations describing the twisting of the chain are derived using results from classical elasticity and hydrodynamic theories. In the case where the fluorescence depolarization is due solely to torsional motions of the DNA chain, $r(t)$ is given by

$$r(t) = r_0 \left\{ \tfrac{1}{4} + \tfrac{3}{4} e^{-\Gamma(t)} \right\} \qquad (2)$$

where r_0 is the anisotropy at time $t = 0$ (the limiting anisotropy) and $\Gamma(t)$ is the torsion decay function. In general, $\Gamma(t)$ is a complicated sum describing the superposition of relaxations due to all normal torsional modes of the chain. However, at sufficiently short times it can be approximated by a simple function [5]:

$$\Gamma(t) = (2k_BT/\pi) \sqrt{t/b^2\eta C} \qquad t < 4\pi\eta b^2 L^2/C \qquad (3)$$

where C is the torsional rigidity of the helix and η is the viscosity of the medium.

The fluorescence polarization anisotropy curve $\Gamma(t)$ shown in Fig. 2b is of the general form expected on the basis of the preceding model. We can, however, verify the exponential-$t^{1/2}$ torsional relaxation more directly by plotting the experimental data in the form log $[r(t) - 0.1]$ vs $t^{1/2}$; a straight line should result. The plot (Fig. 3) is indeed linear within the experimental uncertainty. Quantitative values of the torsional rigidity C

333

are more easily obtained by fitting the experimental data to the theoretical model using the method of nonlinear least squares. Analysis of the data in Fig. 2 gives $b^2 \eta C = 2.4 \pm 0.5 \times 10^{-35}$ erg^{2}s and $r_0 = 0.36 \pm 0.01$; the calculated curve is an excellent fit to the data ($x_r^2 = 1.16$) and provides further support for the exponential-$t^{1/2}$ decay law. Extraction of the torsional rigidity C from this value for $b^2 \eta C$ requires an estimate of the radius b of the helix. By using the translational hydrodynamic radius b = 13.5Å [7], we calculate $C = 1.3 \pm 0.2 \times 10^{-9}$ erg cm for our calf thymus DNA sample.

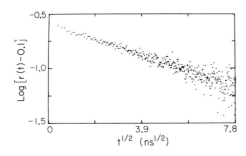

Fig. 3   Plot of the experimental data of Fig. 3 in the form log $[r(t) - 0.1]$ vs $t^{1/2}$.

The torsional rigidity of DNA can also be calculated using classical elasticity theory, wherein the torsional rigidity C is related through Poisson's ratio σ to the flexural or bending rigidity, which is in turn related to the persistence length P [5]. The result is

$$C = k_B T P/(1 + \sigma). \tag{4}$$

Taking a persistence length P = 600Å [6], and a Poisson ratio $\sigma = 0.5$ characteristic of bulk polymeric materials [5], we obtain $C = 1.6 \times 10^{-9}$ erg cm.

The results of this work demonstrate the power of picosecond fluorescence depolarization measurements for probing internal motions in biological macromolecules. In particular, we have verified the applicability of classical elasticity and hydrodynamic theories to the torsional dynamics of a linear duplex DNA both qualitatively, in the prediction of the exponential-$t^{1/2}$ decay of the fluorescence polarization anisotropy of an intercalated dye, and quantitatively, in the ability to calculate the torsional rigidity from known persistence length data. We hope to extend these investigations to related systems in order to explore the effects of molecular structure (primary, secondary and tertiary) on the internal dynamics. A more complete account of the DNA work will be published elsewhere [8].

Acknowledgments

This work was supported in part by the National Science Foundation under grant CHE79-05683. AHZ is the recipient of an Alfred P. Sloan Foundation Fellowship and a Camille and Henry Dreyfus Foundation teacher-scholar award. This is Contribution No. 6263 from the Division of Chemistry and Chemical Engineering of the California Institute of Technology.

References

1. S. A. Allison, J. M. Schurr: Chem. Phys. , 41, 35-39 (1979), and references therein.
2. Ph. Wahl, J. Paoletti, J. -B. Le Pecq: Proc. Natl. Acad. Sci USA, 65, 417-421 (1970)
3. M. E. Hogan, O. Jardetsky: Proc. Natl. Acad. Sci. USA, 76, 6341-6345 (1979)
4. P. H. Bolton, T. L. James: J. Amer. Chem. Soc. , 102, 25-31 (1980)
5. M. D. Barkley, B. H. Zimm: J. Chem. Phys. , 70, 2991-3007 (1979)
6. D. Jolly, H. Eisenberg: Biopolymers 13, 61-95 (1976)
7. R. T. Kovacic, K. E. Van Holde: Biochemistry 16, 1490-1498 (1977)
8. D. P. Millar, R. J. Robbins, A. H. Zewail: Proc. Natl Acad Sci USA, submitted.

335

High-Power UV Ultrashort Laser Action on DNA and Its Components

D.A. Angelov, G.G. Gruzadyan, P.G. Kryukov, V.S. Letokhov,
D.N. Nikogosyan, and A.A. Oraevsky

Institute of Spectroscopy, USSR Academy of Sciences,
Troitzk, Moscow Region, 142092, USSR

Our studies in high-power laser UV action on DNA and its components were stimulated by interest shown in realization of selective action on complex biomolecules. In the very first experiments [1,2] on irradiating the aqueous solutions of nucleic
acid bases by high-power laser UV radiation ($I = 10^9$ w/cm^2,
$\lambda = 266$ nm, $\tau_p = 30$ ps) we revealed that such an action was
characterized by irreversible decrease of absorption in the
first electronic band, the decrease of absorption observed
being proportional to the squared intensity. This points to a
two - step character of the excitation process, and both the
first singlet S_1 state and the first triplet T_1 state can be
served as its intermediate state. When absorbing successively
two UV quanta the molecule under irradiation acquires energy
~ 9.3 ev that exceeds the ionization limit. As a result, some
photoproducts are formed which qualitatively differ from those
formed by ordinary low-intensity UV irradiation [2].

The present report contains the results of our recent experiments on studies into the mechanism of two-step photodecomposition of DNA components in aqueous solutions under powerful
laser UV radiation.

In our research on photoconductivity of aqueous solutions we
revealed [3] a strong photocurrent arising in pure water irradiated by picosecond laser UV pulses ($I = 10^8 \div 10^{10}$ w/cm^2,
$\lambda = 266$ nm, $\tau_p = 30$ ps). The photocurrent amplitude is quadratic in UV radiation intensity. This points to a two-photon
mechanism of water photolysis. The two-photon absorption coefficient of water measured at the wavelength $\lambda = 266$ nm is equal
to $8.1 \cdot 10^{-11}$ cm/w. The kinetic and spectral measurements show,
that under high-power UV laser irradiation water breaks down
to some radicals including a hydrated electron e^-_{aq}. The quantum
yield of hydrated electron formation $\gamma(e^-_{aq}) \approx 10^4\%$.

To clear up the role of water radicals in the process of two-
step photodecomposition we studied the dependence of the photodecomposition efficiency Φ on solution concentration, the
optical density being constant, for thymine - one of the DNA
bases [4]. In the experiment (Fig.1a) an increase of photodecomposition efficiency was observed with a decrease in the
solution concentration which corresponded to an increase of
water radical concentration. This points to the fact that the
products of two-photon water photolysis participate in the

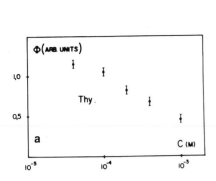 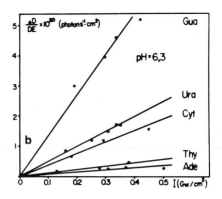

Fig.1 (a) Dependence of the photodecomposition efficiency on concentration of thymine aqueous solution. The excitation wavelength is 266 nm. (b) Dependences of the photodecomposition efficiency on radiation intensity for all five nucleic acid bases. The excitation wavelength is 289 nm.

process of two-step photodecomposition. In this case they interact with the highly excited molecules of the bases. This is confirmed by the experiment on guanine photodecomposition in the mixture of DNA and RNA bases by UV radiation at λ = 289 nm [5] (Fig.1b). It is known that under powerful laser UV irradiation with λ = 266 nm near the maximum of the first electronic band the chemically less stable pyrimidine bases (thymine, cytosine and uracil) decompose more than the purine ones (adenine and guanine). At λ = 289 nm it is mainly guanine that absorbs. Therefore, if the water photolysis products interacted with unexcited molecules, the excitation at λ = 289 nm would also decompose principally the pyrimidine bases.

At high-power laser UV irradiation of concentrated (10^{-3} M) aqueous solutions of DNA bases we could also observe charged adducts [4]. It was revealed that such adducts were formed at the highly excited electron states of the molecules probably through the photoionization stage.

The facts stated result in the following model of two-step photodecomposition of DNA components in aqueous solutions (Fig.2). Under high-power laser UV action two-step ionization of the molecules under irradiation and two-photon water photolysis take place simultaneously. The resulting free radicals of water interact with highly excited molecules of the DNA components which leads to the formation of photoproducts.

It should be said that photoproducts can be formed at both highly excited singlet and triplet states. One can show that the former case may be realized with picosecond laser UV pulses, while under nanosecond laser UV irradiation the latter is realized. We managed to measure the two-step photodecomposition

337

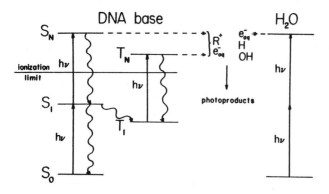

Fig.2   Model of two-step photodecomposition

efficiency in both cases using the fourth harmonic of picosecond Nd:YAG laser radiation (λ = 266 nm, τ_p = 30 ps) and
KrF laser radiation   (λ = 248 nm, τ_p = 20 ns). It turned out
[6], that photodecomposition efficiency under picosecond UV
irradiation is more than ten times higher than under nanosecond
one. This fact is explained by a small (~1%) yield of intersystem crossing $S_1 \rightarrow T_1$. The evaluations performed show that
in both cases the two-step photodecomposition of DNA bases
within intensity range $10^8 \div 10^{10}$ w/cm^2 occure in terms of
water radical excess. This makes it possible to get the information about excited-states properties of nucleic acid components (electronic-state lifetimes, absorption cross sections,
transition rate constants, etc.).

Since the process of two-step excitation of molecules under
irradiation depends on a lot of laser radiation parameters,
such as the radiation wavelength at the first and second steps,
the time delay between pulses, the intensity, etc., it is
possible to make the process of two-step photodecomposition
more effective for a desired type of nucleic acid bases by
choosing laser radiation parameters [5,7]. Fig.1b demonstrates
the selective action on guanine in the mixture of all nucleic
acid bases.

As in the case of high-power laser UV irradiation of DNA components the qualitatively novel photochemistry takes place, so
it is certainly interesting to observe the results of high-power
laser UV irradiation of biological objects. We have studied the
mechanism of inactivation of the viruses λ and φX174 and the
bacterial plasmid pBR322 under high-power laser UV irradiation.

It is shown in [8,9] that at UV irradiation intensities from
10^7 to 10^9 w/cm^2 the inactivation of the viruses and plasmids
is mainly due to single-strand breaks in the DNA, while in the
case of less-power UV irradiation the inactivation occurs
because of the formation of pyrimidine dimers of cyclobutane
type.

Thus under high-power laser UV irradiation of nucleic acids and its components two-quantum photochemical processes take place. Estimations of excited electronic states properties can be made using the dependences of photodecomposition efficiency on the radiation intensity. There is a similarity between the action of powerful UV laser radiation and ionizing one. However, with difference to the ionizing radiation one can, in principle, realize the selective photodecomposition of nucleic acids or their components of desired type varying laser radiation parameters.

References

1. P.G.Kryukov, V.S.Letokhov, Yu.A.Matveetz, D.N.Nikogosyan, A.V.Sharkov: in Picosecond Phenomena, ed. by C.V.Shank, E.P. Ippen,S.L. Shapiro,Springer Series in Chemical Physics, Vol.4 (Springer,Berlin,Heidelberg,New York 1978) p. 158
2. P.G.Kryukov, V.S.Letokhov, D.N.Nikogosyan, A.V.Borodavkin, E.I.Budowsky, N.A.Simukova: Chem.Phys.Lett.61,375(1979)
3. D.N.Nikogosyan, D.A.Angelov: Dokl.Akad.Nauk SSSR (Russian) 1980, in press
4. D.A.Angelov, D.N.Nikogosyan, A.A.Oraevsky: Kvantovaya Elektronika (Russian) 7,N 11(1980)
5. D.A.Angelov, P.G.Kryukov, V.S.Letokhov, D.N.Nikogosyan, A.A.Oraevsky: Kvantovaya Elektronika (Russian), 7,1304(1980).
6. D.A.Angelov, P.G.Kryukov, V.S.Letokhov, D.N.Nikogosyan, A.A.Oraevsky: Kvantovaya Elektronika (Russian), to be published
7. D.A.Angelov, P.G.Kryukov, V.S.Letokhov, D.N.Nikogosyan, A.A.Oraevsky: Appl.Phys.21,391(1980)
8. A.A.Belogurov, G.B.Zavilgelskij, D.A.Angelov, P.G.Kryukov, V.S.Letokhov, D.N.Nikogosyan: Studia Biophysica 78,N 2(1980)
9. G.G.Gurzadyan, D.N.Nikogosyan, P.G.Kryukov, V.S.Letokhov, T.S.Balmukhanov, A.A.Belogurov, G.B.Zavilgelskij: Biophysika (Russian), to be published

Part VIII

Spectroscopic Techniques

Surface Picosecond Raman Gain Spectra of a Molecular Monolayer and Ultrathin Films

J.P. Heritage

Bell Telephone Laboratories, Holmdel, NJ 07733, USA

We employ our recently developed [1] surface picosecond Raman gain spectroscopy (SPRG) to obtain high resolution Raman spectra of ultrathin molecular films and of a monolayer of chemisorbed p-nitrobenzoic acid (PNBA) on aluminum oxide. This is the first measurement of a Raman gain spectrum from a surface monolayer on a low-area dielectric substrate which definitely [2] does not provide anomalous enhancement of the Raman scattering cross section [2]. This technique has broad applications since it does not require special surface materials and morphology. Our spectrum of PNBA are taken in a region near 1610 cm^{-1} where a strong mode has been observed for PNBA monolayers in surface enhanced Raman spectroscopy [3,4].

Vibrational spectroscopy of adsorbed species with optical techniques has been confined until recently to various infrared reflection techniques [5]. Raman spectroscopy of surfaces would be highly desirable, not only for the complementary information available from different selection rules but also from the potential to study new very low frequency vibrations as well as to penetrate substrates or structures opaque to infrared radiation. With the exception of surface enhanced Raman spectroscopy [2] on silver and a few other metal surfaces, ordinary Raman scattering from low area surfaces suffers from extremely weak signals. Furthermore, even weak luminescence can completely obscure such feeble signals.

We introduced the technique of SPRG as a means of overcoming the sensitivity limitation and luminescence background problems of conventional Raman spectroscopy [1]. Raman gain spectra are obtained with synchronized [6], continuous trains of picosecond pulses. The high sensitivity of SPRG is a result of two factors. First, the high intensity of focused picosecond pulses can yield significant gain without use of excessive energy. Second, nearly shot noise limited operation of the continuously mode-locked laser is obtained for modulation of the pump pulse train and synchronous detection of the probe pulse average power at 10 MHz, yielding low-noise performance comparable to single-mode lasers employed in continuous Raman gain spectroscopy [7]. A simple calculation shows that the expected signal to noise ratio for one monolayer can be greater than unity for only a few seconds of integration [1]. When the pulse trains are synchronized and spatially overlapped, a straighforward approach for employing SPRG is realized for molecular monolayers on dielectric surfaces. In this case substrate absorption is not important and a simple transmission geometry may be employed. Ultrathin molecular films and chemisorbed monolayers were prepared on thin oxide films deposited on a sodium fluoride substrate. NaF contributes no first order Raman effect with negligible linear absorption. This is important since the confocal length of even tightly focused laser beams are $\simeq 10^4$

times the thickness of a single monolayer. All material within the confocal volume can contribute unwanted signals. The sandwich structure of NaF - deposited film - monolayer effectively isolates the surface - monolayer interface for picosecond Raman gain spectroscopy.

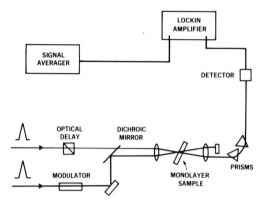

Fig.1 Experimental arrangement. Synchronized trains of picosecond pulses are generated by a pair of modelocked dye lasers. The pump train is inten- sity modulated at 10.1 MHz and the change in the intensity of the Stokes beam is synchronously detected

The experimental arrangement is depicted in Fig.1 Two dye lasers are modelocked by synchronous pumping with a modelocked pulse train from an argon ion laser. In these experiments 3-plate Lyot tuners were employed in both lasers yielding pulses of $\simeq$ 8 psec duration. Synchronism was maintain- ed to better than one pulse width. The lasers were tuned in the vicinity of 0.58 μm (Rhodamine 6G) and 0.64 μm (Rhodamine B). In this configuration the spectral resolution was measured to be $\simeq$ 6 cm^{-1}. The pump (yellow) beam is amplitude modulated at 10.7 MHz with an electrooptic modulator. The beams were suitably delayed and brought colinear with the aid of a dichroic mirror. The beams are focused to a spot diameter of $\simeq$ 8 μm with a microscope objec- tive and collimated after transmission through the monolayer - aluminum oxide - NaF sandwich. An incidence angle of $\simeq$ 30° minimizes back scattered light. The transmitted beams (both p-polarized) are separated with a prism array and color filters. The Stokes beam (red) is incident on a silicon photodiode. Linear detection was obtained with 10 mW of average power incident on the detector. Preamplification and prefiltering are employed before the high frequency lock-in amplifier. The detected signal is sent to a multichannel analyzer for signal averaging. Raman spectra are obtained by tuning the pump laser with a precision stepper motor driven rotator that orients the Lyot filter. Spectral scans of $\simeq$ 90 cm^{-1} are employed while multiple scans are accumulated in the signal averager.

A direct experimental measurement verified the high sensitivity and the utility of the transmission SPRG technique. A NaF-Aluminum oxide-Poly- styrene structure was constructed with a polystyrene film thickness of 900 Å [8]. In Fig.2 we display a SPRG spectrum of this ultrathin film taken in a 70 cm^{-1} band near 1000 cm^{-1}. The features near 1000 cm^{-1} and 1025 cm^{-1} are due to phenyl ring vibrations and agree with spectra taken in the bulk.

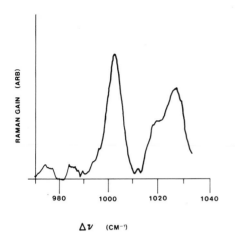

Fig.2 SPRG spectrum of a 900 Å film
of polystyrene. The signal to noise
ratio is ≃ 100:1

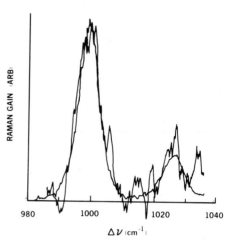

Fig.3 SPRG spectrum of two film thickness of (1 μm and 0.1 μm) polymethylphenylsiloxane. Thin oil films are
used for alignment purposes when
monolayer structures are investigated

The signal to noise ratio is 100:1 for 15 minutes of signal averaging. This
result directly shows that a single monolayer spectrum is obtainable with
this technique and that the thin film structure is reasonably stable.

Similar results were obtained with ultrathin films of various oils. Oil
films are important since they may be readily placed over a monolayer and
used for alignment purposes. Raman active modes of oil films around 1000 Å
thick yield strong signatures that permit perfect alignment of the pump and
Stokes beams at the surface. In Fig.3 we present an example of a Raman
spectrum of two thin films of polymethylphenylsiloxane [9], one film is 1 μm
thick and the second is 0.1 μm thick. The spectra clearly show the two
phenyl ring vibrations at ≃ 1000 cm^{-1} and ≃ 1025 cm^{-1}.

We now turn to a discussion of the measurement of a Raman gain spectrum
of a single monolayer of p-nitrobenzoic acid (PNBA). Monolayer samples of
PNBA were prepared by spinning an ethanol solution of PNBA on an oxidized
aluminum surface and rinsing in chlorobenzene as described by Allara et al.
[4]. The oxide film was prepared by vacuum deposition and oxidation of a
≃ 40 Å layer of aluminum on a polished sodium fluoride crystal. Identical
blank samples were prepared without PNBA for experimental comparison. The
monolayer spectrum was obtained by deliberately applying a thin layer (≃ 1 μm)
of liquid oleic acid on the sample surface [9] as an overlayer on the monolayer of PNBA. The line at 1667 cm^{-1} was used to align the optics as described above. Careful alignment of the focal overlap and placement of the
beam waists at the interface yields a distinct minimum in background contributions together with optimum gain. The monolayer spectrum was then obtained
by spectral shifting into the appropriate wavelength region for the monolayer
and where the overlayer is Raman inactive.

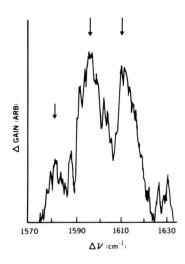

Fig.4 SPRG spectrum of a monolayer of p-nitrobenzoic acid (PNBA) on a thin film of aluminum oxide supported by sodium fluoride. Three principal features are marked.

More detailed discussion of the experimental technique is presented else-where [10]. We obtain the spectrum presented in Fig.4. These peaks at 1610, 1595 and 1580 cm^{-1} are assigned as Raman peaks since their location did not shift when the Stokes laser was shifted by 35 cm^{-1} and repeat spectra were obtained by scanning the pump laser wavelength. It is important to note that great care was exercised in determining that no systematic errors were present that could have given false spectra. We attribute the peak at 1610 cm^{-1} to the PNBA ring quadrant stretch observed in infrared and enhanced Raman spectra reported elsewhere [4]. The features at 1600 and 1580 are unique to this measurement and suggest that the condition of the monolayer has changed during our experiment. We have studied our samples using en-hanced spontaneous Raman scattering by applying a $\simeq$ 50 Å layer of silver over freshly prepared PNBA monolayer. The spectrum is identical to that reported by others [4], showing a single peak at 1610 cm^{-1}. This result confirms that our monolayer is chemically adsorbed PNBA. The evidence suggests that the peaks at 1595 and 1580 are new species formed by photochemical and/or thermal reactions induced by the focused beams unique to our measurement. Support for this hypothesis is provided by the experimental observation that pro-longed exposure of the sample to the laser beams at one site yields spectra with relatively stronger lines at 1595 and 1580 than the PNBA line at 1610.

We conclude from this experiment that vibrational spectra of surface mono-layers may be obtained with good signal to noise ratio using picosecond Raman gain spectroscopy. This technique is ideally suited for smooth transparent dielectric substrates but will be extended to include highly reflecting substrates. SPRG will be a valuable optical tool for obtaining high reso-lution vibrational spectra of surface species, providing complementary in-formation to infrared reflection spectroscopy and the opportunity to study very low frequency surface vibrations.

References

1. J. P. Heritage, J. G. Bergman, A. Pinczuk and J. M. Worlock, Chem. Phys. Letters $\underline{67}$ (1979) 229; J. P. Heritage, Appl. Phys. Letters $\underline{34}$ (1979) 470; B. F. Levine, C. V. Shank and J. P. Heritage, IEEE J. Quantum Electron. QE-15 (1979) 1418.

2. R. P. van Duyne, in: Chemical and Biochemical Applications of Lasers, Vol. 4, ed. C. B. Moore (Academic Press, New York, 1978) Ch. 5.

3. J. C. Tsang, J. R. Kirtley and J. A. Bradley, Phys. Rev. Letters $\underline{43}$ (1979) 772.

4. D. L. Allara, C. A. Murray and M. Rhinewine, to be published.

5. See for example, D. Allara in: Advances in Chemistry, American Chemical Society, in press.

6. R. K. Jain and J. P. Heritage, Appl. Phys. Letters $\underline{32}$ (1978) 41.

7. A. Owyoung, IEEE J. Quantum Electron. QE-14 (1978).

8. The polystyrene films were prepared by D. L. Allara.

9. J. P. Heritage, J. G. Bergman and C. J. Weschler, to be published.

10. J. P. Heritage and D. L. Allara, to be published in Chem. Phys. Lett.

Pulse Sequenced CARS: Background Suppression and Nonlinear Interferences

M.G. Sceats, F. Kamga, and D. Podolski

Department of Chemistry and the Institute of Optics, University of Rochester, Rochester, NY 14627, USA

Coherent Anti-Stokes Raman Scattering (CARS) is a convenient spectroscopic tool for studies of solutions and mixtures in either solid, liquid or gas phase. Unfortunately, the CARS process generates a background signal even when the lasers are tuned off resonance due to the non-linear response of the electrons to the applied field (the electronic Kerr effect). The background interferes with the resonant signal, since they are both coherently generated, to produce a distorted line shape. In the case of a weak Raman mode or in dilute mixtures the non-resonant background signal can be easily comparable to or larger than the resonant contribution of interest, at best producing a distorted bandshape through interference and at worst completely covering the resonance. This severely limits the usefulness of CARS in these situations.

In pulse sequenced CARS (PUSCARS) the background is eliminated by time resolving the various contributions to the polarization [1]. If a time delay t_D is introduced between the sequences of events which lead to the excitation of real vibrational states and the subsequent events leading to anti-Stokes generation, then non-resonant contributions will be eliminated. If t_D is of the order of the dephasing time for level λ, then resonance CARS generation will be obtained with only small loss in signal intensity which depends on the laser bandwidths and the spectral linewidths. The detected signal will be proportional to $(\mathrm{Im}\chi)^2$ and thus a resonant lineshape is restored. We note, however, that the observed lineshape will correspond to the convolution of the resonant response function and the frequency spectrum of the applied fields.

We have demonstrated the PUSCARS technique on the 656 cm^{-1} vibrational Raman mode of CS_2 in a mixture with toluene. A Coherent Radiation mode-locked Argon ion laser (CR 10) synchronously pumps two mode-locked dye lasers. A Rhodamine 6G laser is tuned to a frequency $\omega_S = 17452$ cm^{-1}, and the pulsewidth is set at about 30 psec with etalons to provide a relatively narrow spectral bandwidth. A second laser, a sodium fluorescein dye laser, is tuned to $\omega_L = 18109$ cm^{-1} and provides both pump and probe pulses ω_{L1} and ω_{L2} by use of a beam splitter and a delay line. The pulse width of the laser is tuned to give about 3-5 psec wide pulses. All three beams are focused by a 17 cm achromat lens into a one centimeter cell containing a mixture of 10% CS_2 in toluene. Input average powers are 15 mW approximately for all three beams.

The crossing angles are such that phase matching is only achieved when the condition $\underset{\sim}{k}_{AS} = \underset{\sim}{k}_{L1} + \underset{\sim}{k}_{L2} - \underset{\sim}{k}_S$ is satisfied

For each setting of ω_S, the delay line in the ω_{L2} beam was scanned giving $I_{AS}(2\omega_L - \omega_S)$ at all desired delays t_D, as shown in Figure 1. The peak corresponds to $t_D = 0$ and as the delay increases, the intensity initially falls rapidly and then falls more slowly with an exponential decay. The initial decay corresponds to the removal of the non-resonant background and the polarized vibrational component associated with rapid rotational reorientation, and is the same as that observed for the pure toluene reference.

The second decay is associated with the 20 psec dephasing time of the real vibrational excitations of CS_2.

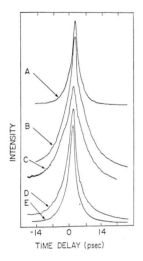

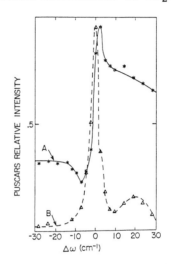

Fig.1 PUSCARS decay at fixed detuning
A: 20 cm^{-1}; B: 3 cm^{-1}; C: 0 cm^{-1}; D: -3 cm^{-1}; E: -20 cm^{-1}

Fig.2 PUSCARS spectra: 10% CS_2 in toluene A: $t_D = 0$; B: $t_D = 30$ psec

From these decay curves at set Stokes frequencies we can synthesize a CARS spectrum at a set delay t_D. This is illustrated in Figure 2, where the PUSCARS spectrum, with $t_D = 0$ and $t_D = 30$ psec, near the 656 cm^{-1} vibration of CS_2 is shown. The signal intensity at resonance was normalized to one. Toluene was chosen because of the high value of its non-resonant contribution χ^{NR}. In Figure 2, far from resonance, the non-resonant contribution is 25% as large as the on-resonance signal for the high frequency shift side of the resonance and larger for the low frequency shift in the spectrum at $t_D = 0$. However, the PUSCARS spectrum at $t_D = 30$ psec is similar to the Raman spectrum illustrating that the non-resonant interferences have been suppressed. The ratio of peak to background at a 100 cm^{-1} shift is between two to three orders of magnitude in the spectrum at $t_D = 30$ psec. The suppression is limited by the pulse overlap and is a strong function of the quality of the mode locking of the

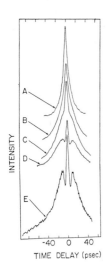

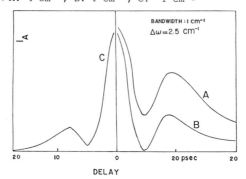

Fig.3 PUSCARS decay for $\Delta\omega \neq 0$. Bandpass B: 20cm^{-1}; C: 6 cm^{-1}; D: 3 cm^{-1}; E: 1 cm^{-1}! A is pump laser autocorrelation

Fig.4 Computed PUSCARS decay for a fixed detuning, showing beats. Monochromator setting from resonance A: 4 cm^{-1}; B: 1 cm^{-1}; C: -1 cm^{-1}

dye lasers. The 629 cm^{-1} mode of toluene is observed at $\Delta\omega = 20$ cm^{-1} in the spectrum at $t_D = 30$ psec but not at $t_D = 0$.

A monochromator was used to filter the anti-Stokes signal from the incident laser fields. When $\omega_L - \omega_S$ is tuned to the high frequency side of the vibrational resonance and decay surves are run for different slit widths, an interference beat develops as the monochromator bandpass is narrowed, as shown in Figure 3. The interference is a beat between the adiabatic non-resonant background and the long lived coherent vibrational response. Theoretical considerations show that the beat develops as the CS_2 oscillators relax from the driving frequency centered at $\omega_L - \omega_S$ towards their natural frequency ω_0 as the excitation pulse is turned off. These relaxed, but coherent, oscillators produce an optical response which can beat with the non-resonant contribution to the susceptibility which remains at the driving frequency. The monochromator is used to select a narrow distribution of frequencies because the incident pulses contain a broad distribution of "driving frequencies" which leads to destructive interference and the disappearance of the beats. Computer simulations of the beats based on semiclassical model are shown in Figure 4.

1. F.M. Kamga and M.G. Sceats, Optics Letters, $\underline{5}$, 126 (1980)

Raman Scattering from Excited States with Picosecond Lifetimes

M. Asano and J.A. Koningstein

Metal Ions Group, Department of Chemistry, Carleton University,
Ottawa, Ontario K1S 5B6, Canada

Results and Discussion

If $\sigma(\lambda)$ is the absorption cross section of an excited* ← ground state
transition and $\sigma^*(\lambda)$ that of an higher excited ← excited state* transition,
both at λ_{laser}, then the latter becomes observable if:

$$\frac{\sigma(\lambda)}{\sigma(\lambda) + \sigma^*(\lambda)} \, N_p^{SN} = N \frac{1}{\tau^*} \qquad (1)$$

where N_p^{SN} is the number of photons/sec required to saturate the excited
state* with a lifetime τ^* and N is the number of absorbing species in the
trace of the laser beam.

If pulsed lasers are used, (1) is only valid if τ^* is shorter than the
duration of the pulse. The results displayed in Fig. 1 are directly applicable to the above and serve as an illustration of the usefulness of (1).
Shown there is the behaviour of the transmission at 488.0 nm of chlorophyll
a in pyridine (1.45 x 10^{-4} molar) as a function of I_{laser}(cw) and I_{laser}
(pulsed). The transmission decreases upon an increase of I_ℓ and reaches
a constant value. Of paramount importance here is the fact that, although
the laser energy is absorbed by vibronic components associated with the
singlet states (see Fig.2), the value of T at effective population
saturation is that in the triplet-triplet absorption spectrum at λ = 488.0 nm
Apparently part of the laser energy at λ = 488.0 nm relaxes to S_1 and reaches
T_0 via an intersystem crossing process. Effective population saturation of
T_0 is achieved because T remains constant at large I_ℓ but (1) cannot be
directly applied because only part of the photon flux reaches T_0($S_1 \to S_0$
fluorescence is excited). Also, for the pulsed laser experiments (3.2 ns
FWHM, repetition rate 25 Hz) it appears that Δt_p = 3.2 ns is much shorter
than τ_{T_0} which suggest that if all photons would reach T_0:

$$\frac{\sigma(\lambda)}{\sigma(\lambda) + \sigma^*(\lambda)} \, N^{SN} = N \qquad (2)$$

where N^{SN} is the number of photons of the pulse used to saturate the
excited state. Energy losses to reach T_0 for cw pulsed laser operating
at 488.0 nm are equal and we find that

$$\tau_{T_0} = \frac{N^{SN}}{N_p^{SN}} = 70 \ \mu s.$$

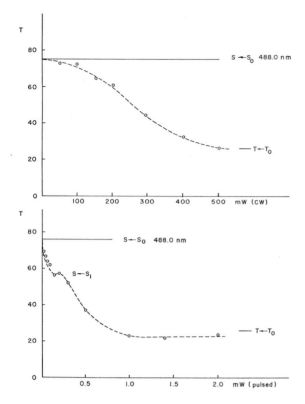

Fig.1 Transmission studies of 488.0 nm (cw and pulsed) of Chl a in
pyridine. For unfocussed laser beam radiation $S \leftarrow S_0$ (see Fig.2)
determines the value of T but at high laser powers (focussed beam) the
value of T indicates that transitions occur in the triplet-triplet
absorption spectrum.

This because $\sim$ 500-500 mW corresponding to 1.2×10^{18} photons/sec are
effectively involved in the saturation of T_0 in the cw case and
$.8 \times 10^{14}$ photon/pulse for the pulsed laser. Fig. 1 shows that with the
pulsed laser radiation, another state gets populated (before T_0) and a
detailed analysis shows that this state is S_1. By direct pumping into the
$S_1 \leftarrow S_0$ band system and measuring T as a function of I_{laser} we obtain from
(1) τ_{S_1} = 45 ns. By direct pumping into the Soret band system in the blue
a value of τ_{S_4} = 110 ps is computed while in addition, the quantum efficiency
(Q_{ISC}) for the $S_1 \rightarrow T_0$ intersystem crossing process is calculated to be
Q_{ISC} = 40%. These and other experimental results show that with the nanosecond pulses excited states having pico (or even subpico) second lifetimes
can easily be populated to say nothing of the first excited (singlet) state
of the chlorophyll a dimer in pyridine which has a lifetime of 145 ns. The
fluorescence spectrum of the dimer shown in Fig. 3 is recorded (making use
of the fact that $\tau_{S_1}^{dimer} \gg \tau_{S_1}^{monomer}$) by moving the aperture duration of a

352

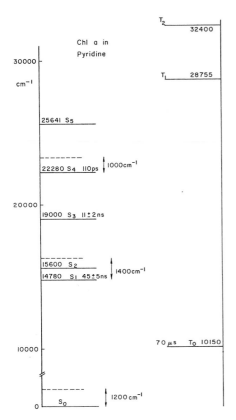

Fig.2 Part of the energy level diagram of monomer Chl a in pyridine.

boxcar some 250 ns away from the position in time which coincides with maximum intensity of the laser pulse itself. We obtain from the band widths an exciton splitting of $\sim$ 105 cm^{-1} for the dimer in pyridine and $\sim$ 260 cm^{-1} in acetone. A weak Raman band in the Rayleigh wing is recorded for the former solution at $\sim$ 100 cm^{-1} using λ_ℓ = 475 nm which radiation i) creates the excited state of the dimer ($\sim$ 0.5 x 10^{-4} molar of the 1.45 x 10^{-4} molar chl a solution) while ii) the remaining part of the laser energy serves as a (resonance) electronic Raman probe. Similar optical pumped excited state Raman transitions were induced with cw laser radiation of Cr^{3+} in ZnAl$_2$O$_4$, αAl$_2$O$_3$, Y$_3$Al$_5$O$_{12}$ and beryl (Fig.4). The transitions occur between the component millisecond lived levels of ^{2}E which are readily populated with cw lasers. At low temperatures we find that the width of these component states are dominated by the inhomogeneous broadening process which suggest that the electronic Raman transition is extremely narrow. On the other hand at high T homogeneous broadening effects dominate and the width of the Raman band decreases rapidly [1].

A rather interesting situation is encountered for the tris(2,2'-bipyridine) Cr(III) complex and the relative part of the energy level diagram is given in Fig. 5. Upon exposing a 10^{-3} molar solution to pulsed laser radiation with λ_ℓ = 458.0 nm one learns from studies similar as those shown in Fig. 1 that population takes place of vibrational states of the ^{4}T$_2$ electronic state.

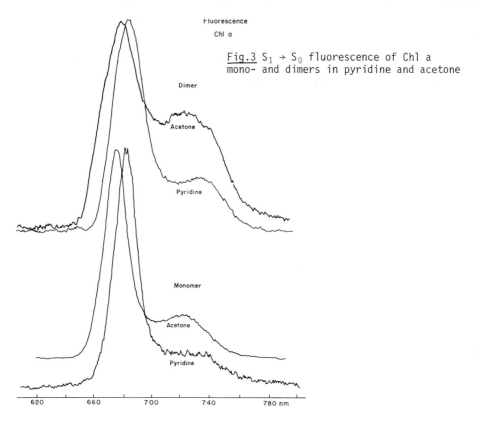

Fluorescence

Chl a

Fig.3 $S_1 \rightarrow S_0$ fluorescence of Chl a mono- and dimers in pyridine and acetone

Dimer

Acetone

Pyridine

Monomer

Acetone

Pyridine

620   660   700   740   780 nm

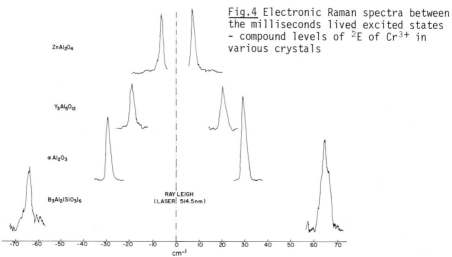

Fig.4 Electronic Raman spectra between the milliseconds lived excited states - compound levels of ^2E of Cr^{3+} in various crystals

$ZnAl_2O_4$

$Y_3Al_5O_{12}$

$\alpha\ Al_2O_3$

$B_3Al_2(SiO_3)_6$

RAYLEIGH
(LASER 514.5nm)

-70 -60 -50 -40 -30 -20 -10   0   10   20   30   40   50   60   70
cm^{-1}

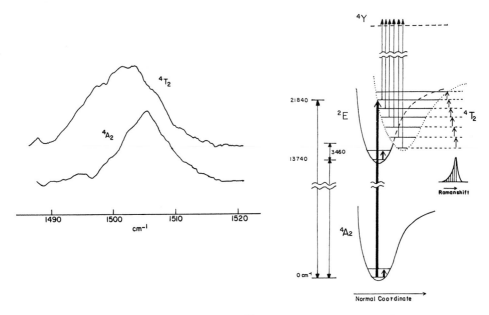

Fig.5 Energy level diagram of $Cr(bpy)_3{}^{3+}$ and a comparison of a Raman band for ground state $^4A_2$ and the 10 ps lived $^4T_2$ excited state.

During the rise time of the pulse one finds that population of vibrational levels in the top of the electronic surface is favoured over that in the bottom and vice versa during the trailing end of the pulse. Realizing that the intensity of a vibrational Raman transition $v'' \leftrightarrow v'$ is related to the intensity of the transition from $v = 1 \leftrightarrow v = 0$ by

$$I_{v'} = (v' + 1)I_0$$

we anticipate that the integrated intensity of the vibrational band is greater during the risetime of the pulse than that during the trailing end of the pulse if the band is recorded with the aperture duration of the boxcar adjusted to isolate respectively the rising and trailing end of the pulse. This is indeed observed and in addition we expect that the Raman band of a normal mode of this chromium complex should show a broadening towards the lower energy side as the result of anharmonicity effects. The results - see Fig.5 - support the assignment of scattering of $^4T_2:Cr(bpy)_3{}^{3+}$, which state has a lifetime of 10 ps.

In order to deliver the large flux of photons to saturate the excited states one must in nearly all cases focus the laser beam down into the sample and of course λ_ℓ should be in resonance with an absorption band of the system. This suggests that a close experimental relationship exists between resonance Raman and Optical pumped excited state Raman spectroscopy. A particular interesting case may be the resonance Raman spectrum which was recorded for Chlorophyll a with a cw laser operating at 441.0 nm [2]. This radiation is absorbed by the $S_4 \leftarrow S_0$ band system and population of T_0 is achieved even for 50 mW. Consequently, Chlorophyll a in solution may be a molecule for which the recorded resonance Raman spectrum is actually that of normal modes associated with T_0 at $\sim$ 10150 cm^{-1} and not with the S_0 ground state.

From a practical point of view it seems worthwhile to point out that, while recording light scattering of short lived excited states with an aperture duration of the boxcar equal to or smaller than the width of the laser pulse, the noise of the photomultiplier tube is not detectable. Instead, preamplifier and boxcar noise itself are the limiting factors and the Raman spectra recorded with 1 mJ of power from pulsed laser sources can have the same signal to noise features as those excited with $\sim$ 100 mW cw power. The critical component of the light scattering apparatus is (as usual) the photomultiplier tube, in particular its rise time and gain. It appears that - at this moment - a risetime of 100 ps is the lower limit which suggests that good quality Raman spectra could be recorded from tens of femto second lived excited states, if picosecond pulses instead of nano second pulses are used in combination with this fast risetime PMT.

References

1. D. Nicollin and J.A. Koningstein, Chem. Phys. 42, 277 (1979).
2. M. Lutz, J. Raman Spectrosc. 2, 497 (1974).

Picosecond Backward Echo in Sodium Vapor — Relaxation and Quantum Beat Modulation

M. Matsuoka

Department of Physics, Faculty of Science, Kyoto University,
Kyoto 606, Japan, and
Institute for Molecular Science, Okazaki 444, Japan

H. Nakatsuka[1] and M. Fujita

Department of Physics, Faculty of Science, Kyoto University,
Kyoto 606, Japan

1. Introduction

The backward echo [1-3] is a three-pulse stimulated photon echo produced backward to either one or two of the excitation pulses keeping the phase matching conditions among the three excitation pulses and the echo. It has been pointed out [1] that this echo is readily separated from the excitation pulses without the use of optical shutters. This feature facilitates its use in picosecond experiments [1]. A similar principle leads to the phase conjugated free induction decay (FID) [3], in which the signal propagates opposite to the initial excitation pulse. The use of this backward FID in picosecond experiments has been proposed [3].

Elimination of the excitation pulses without shutters may be accomplished by beam angling and polarization. Diagrams of the usual two-pulse echo and

Fig.1 Diagrams showing wavevectors and polarizations of (a) the two-pulse echo and (b) the forward three-pulse echo. The dotted arrow in (a) indicates the echo-polarization wave having phase mismatch with the echo-light wave at large θ.

(a) (b)

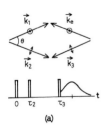

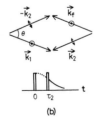

Fig.2 Diagrams showing wavevectors, polarizations and pulse sequences of (a) the backward echo and (b) the backward FID *in gaseous media*. The echo-pulse shape in (a) illustrate the case T_2^* is long compared with τ_2. The pulse shape of FID in (b) can be determined by varying τ_2 and using a slow detector.

(a) (b)

[1]Present address: IBM Thomas J. Watson Reseach Center, Yorktown Heights, New York 10598 USA.

the (forward) three-pulse echo are shown in Fig.1, where k_p (p = 1,2 and 3) and k_e are the wavevector of the p-th excitation pulse and the echo, respectively. The polarizations were chosen to facilitate the rejection of the excitation pulses. In the case of the two-pulse echo (Fig.1(a)), the second pulse can be eliminated by the combined use of a small aperture centered on the echo beam and an analyzer crossed to the second pulse. The first pulse, on the other hand, must be eliminated with the aperture alone. In practice θ larger than a theoretically estimated value is often needed, and this reduces the coherence length of the phase matching between the echo wave and polarization [4]. This difficulty also takes place in the forward three-pulse echo (Fig.1(b)). (See for the two-pulse echoes [5,6].)

Diagrams of the backward echo and the backward FID are shown in the upper part of Fig.2. In the case of the backward echo (Fig.2(a)) the first and the second pulses point opposite to the echo. We have to be concerned only with the third pulse (θ ≪ 90° is assumed). The third pulse is easily eliminated again by the combination of the aperture and the analyzer; for example, a reduction of 10^{-4}~10^{-6} by an aperture and an additional reduction of 10^{-5} by an analyzer are easy. This allows enough reduction factor for detection of weak echoes. The coherence length here is as long as the interaction length [7].

We have made the first observation of the backward echo [8] and FID in the picosecond regime. The experiments were performed on the D_1 line of atomic sodium vapor, and the echo decay as a function of Ar buffer gas pressure was measured. The echo was also observed on molecular sodium. Some experiments were performed using a streak camera, and echo modulation was directly observed reflecting hyperfine splitting (hfs) of the $3S_{1/2}$ ground state [8].

2. Backward Echo in Na

As a first demonstration of the picosecond backward echo we chose atomic sodium vapor with high—pressure Ar buffer gas. At this pressure the echo intensity decays rapidly, so that the echo can only be observed with pulse separations in picosecond regime.

In order to obtain picosecond pulses of a constant area, high-power pulses inversely proportional to the square of the pulse width are required. In the present experiment the excitation pulse was obtained by amplification of a cavity dumped pulse of a mode-locked dye laser (Spectra Physics) synchronously pumped by a cw mode-locked Ar laser. The amplifying system consisted of three dye cells which were pumped by a frequency-doubled Q-switched YAG laser (Quanta Ray DCR-1A) [9]. The cavity dumped mode-locked dye laser produced about 10 psec long 500 W pulses at 5896Å resonant to Na D_1 line. The three amplifier cells contained rapidly flowing solutions of Kiton Red in ethanol. The gains of the first, second and third stage amplifiers were 100, 30 and 4, respectively. The final output power and the pulse width were ∿6 MW and ∿10 psec, respective-

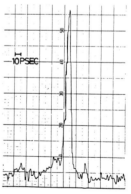

Fig.3 A streak camera trace of an amplified pulse.

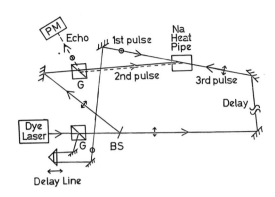

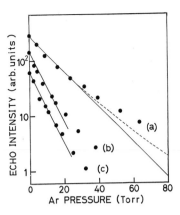

Fig.4 Experimental setup for the
picosecond backward echo

Fig.5 Measured echo intensity
vs Ar pressure for three pairs
of τ_2 and τ_3, where (τ_2, τ_3)
are for curve a (300 psec, 1.1
nsec), for curve b (630 psec,
1.2 nsec), and for curve c
(630 psec, 2.1 nsec).

ly. A streak camera (HTV C979) trace of the output is shown in Fig.3,
where the resolution time of the streak camera was ∿10 psec.

The setup for the echo experiment is shown in Fig.4. The angle between
the first and the second pulses was ∿10 mrad. The separations of the successive excitation pulses were varied from 300 psec to 1.5 nsec, which were
about ten times shorter than those of Ref.1. The atomic sodium density was
9×10^{10} cm^{-3} (420 K).

The echo intensity I(P) was measured by changing Ar buffer gas pressure
P in the heat pipe oven. The result is shown in Fig.5 for three pairs of
the excitation pulse separations. The echo pulse was detected with a photomultiplier. Each point in the figure represents the average of about 200
echo pulses. The intensity ratios of the echo to the excitation pulses and
to the background at the detector were $10^{-5} \sim 10^{-7}$ and 10^3, respectively.
It is found that the curves (b) and (c) in Fig.5 are not different although
the separation between the second and third pulses was varied more than
twice. Since only the velocity changing collisions (VCC) affect the echo
intensity between the second and third pulses, we can conclude that the
effect of VCC on the echo is negligible in the present conditions of Ar
pressure and pulse separations. It is also found from the curves (a) and
(b) that I(P) can be expressed as a function of $P\tau_2$, where τ_2 is the separation of the first and second pulses. These two results are in agreement
with nanosecond experiment of Ref.1, and confirm a reliability of the
present picosecond one. The theoretical values of

$$I(P) = I_0 \exp(-P\tau_2/\gamma) \tag{1}$$

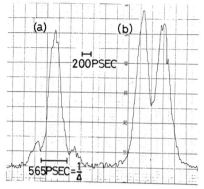

Fig.6 Typical streak camera traces of the quantum beat of the echo pulse; (a) and (b) represent the cases for $\phi=0$ and $\phi=\pi$, respectively.

are shown in Fig.5 by solid lines, where the value of γ is taken from Ref.1, and is 4.1 nsec·Torr. If the excitation pulses are assumed to be much shorter than the Doppler dephasing time T_2^*, the echo pulse width is $\sqrt{2\ln 2}\cdot T_2^*$ (FWHM), and hence Eq.(1) is valid only when $\tau_2 \gg T_2^*$. But in this experiment the latter condition is not valid, therefore Eq.(1) must be modified as

$$I(P) \propto \int_{-\tau_2}^{\infty} \exp(-2t^2/T_2^{*2})\exp\{-P(2\tau_2+t)/2\gamma\}dt. \qquad (2)$$

Equation (2) for the case (a) is shown in Fig.5 by a broken curve which explains well the slowing down of the echo decay at high Ar pressure.

Next, we detected the echo pulse with the streak camera at virtually zero Ar pressure, and observed the quantum beat of the echo pulse arising from the $3S_{1/2}$ hfs (1.77 GHz). If this ground state hfs is considered (hfs of $3P_{1/2}$ is negligible), the echo intensity can be expressed as

$$I(t-R/c) = I_0\exp\{-2(t-R/c-\tau_2-\tau_3)^2/T_2^{*2}\}[A+B\cos\{\Delta(t-R/c-\tau_2-\tau_3)+\phi\}]^2, \qquad (3)$$

where R is the coordinate of the observing position, τ_3 is the separation between the first and third pulses, Δ is the hyperfine splitting, A, B and ϕ are some functions of τ_2, τ_3 and Δ, and are non-negative. Therefore, if τ_2 or τ_3 is changed, the echo pulse changes through A, B and ϕ. Typical example of the echo pulse shape are shown in Fig.6(a) and 6(b), in which the delay line of the first pulse (Fig.4) was varied by 280 psec. Figures 6(a) and 6(b) represent the cases for $\phi=0$ and $\phi=\pi$, respectively, where $A>B$. The separation between the two dips in Fig.6(a) was confirmed to be equal to the inverse of the hfs frequency.

3. Backward Echo in Na$_2$

The picosecond backward echo has been also performed in molecular sodium Na$_2$. The echo was observed at any wavelengths from 565 to 620 nm corresponding to the $X\Sigma-A\Sigma$ transition of Na$_2$. Rhodamine 6G and Rhodamine 640 (Exciton Co.) were used for the amplifying dye at short and long wavelengths, respectively. Temperature of the sample was raised to 650 K. The decay of the echo intensity as a function of Ar pressure followed the linear line of Fig.5 of the atomic sodium, but without the bend at high Ar pressure. This is in contrast with the atomic sodium, and is indicative of an effective inhomogeneous line broadening due to highly dense distribution of the rotational-vibrational lines.

4. Backward FID in Na

The proposed backward FID was tested also in atomic sodium. The most useful diagram of it for gaseous media [3] is shown in Fig.2(b), where the second pulse is given by two counter-propagating waves $\pm k_2$ and the FID signal k_f appears backward to the first pulse k_1. The excitation pulses are again eliminated by the polarizations indicated. The FID signal which appears after the second pulse is a portion of the FID which started at the time of the first pulse.

We have observed this FID signal in a similar setup as in Fig.4, but the third pulse was made antiparallel to and coincident with the second pulse. The signal detected by a photomultiplier was the integrated FID after the second pulse. A tentative experiment showed that the decay of this integrated signal as a function of the pulse separation τ_2 from about zero to 670 psec had a half maximum point at about 140 psec. This should be explained using $T_2^* = 340$ psec and the hyperfine splitting. It is found, however, that an improvement of the experiment is needed, in particular, the zero point of τ_2 has to be determined with care.

5. Concluding Remark

We have demonstrated the backward-wave methods for the first time in picosecond regime. If we go to experiments with shorter pulse separation of the order of a few picoseconds or less, the sample length must be made shorter in order to define the pulse separation well. This length may be the same order of the magnitude as the coherence length of the two-pulse echo in a certain condition. However, our estimate shows that the backward echo and FID are still advantageous and convenient, since they can eliminate the excitations well. We can then find immediate applications of these methods in various molecular systems and ions in solids.

We would like to acknowledge Professor S. R. Hartmann for valuable discussion during his stay in Kyoto University. We thank S. Asaka for his help in the experiment.

References

1. M. Fujita, H. Nakatsuka, H. Nakanishi, and M. Matsuoka, Phys. Rev. Lett. 42, 974 (1979).
2. T. Mossberg, A. Flusberg, R. Kachru, and S. R. Hartmann, Phys. Rev. Lett. 42, 1665 (1979); T. W. Mossberg, R. Kachru, and S. R. Hartmann, Phys. Rev. A20, 1976 (1979); R. Kachru, T. W. Mossberg, and S. R. Hartmann, Opt. Commun. 30, 57 (1979).
3. M. Matsuoka, H. Nakatsuka, and M. Fujita, Appl. Phys. (to be published).
4. For example, a Gaussian beam of the waist diameter 0.2 (0.02) cm, 10^{-6} times reduction of the first pulse is theoretically obtained at $\theta = 6\times10^{-4}$ (6×10^{-3}) rad, and the coherence length $l_c = 86$ (0.86) cm. However, in practice, θ as large as 5×10^{-2} rad is often needed to avoid scatterings, and then l_c becomes as short as 0.012 cm.
5. P. Hu and H. M. Gibbs, J. Opt. Soc Am. 68, 1630 (1978).
6. T. Yajima and Y. Taira, J. Phys. Soc. Japan 47, 1620 (1979).
7. The reduction factor which applies to gaseous media [1, 3] is of the order one in picosecond experiments, and will not be considered.
8. H. Nakatsuka, M. Fujita, and M. Matsuoka, (to be published).
9. E. P. Ippen and C. V. Shank, in Picosecond Phenomena, edited by C. V. Shank, E.P. Ippen, and S.L. Shapiro, Springer Series in Chemical Physics, Vol. 4 (Springer, Berlin, Heidelberg, New York 1978)

The Spatial Growth Rate of Stimulated Raman Scattering: A New Technique for Measuring Picosecond Laser Pulse Intensities

J.R. Valentin, M.A. Lewis, and J.T. Knudtson

Chemistry Department, Northern Illinois University,
DeKalb, IL 60115, USA

1. Introduction

We report a new technique for measurement of the spatial growth rate of a Stokes pulse in liquids generated by a single picosecond laser pulse. Solutions to the coupled differential equations describing the Stokes growth, taken in the region where the depletion of the laser pulse is small, show that the Stokes pulse grows exponentially with a rate equal to $g_S I_\ell$. Experiments performed in ethanol where g_S, the Stokes gain, is known allows us to calculate the initial laser intensity, I_ℓ. This technique shows promise as a convenient method for measuring the absolute intensity of picosecond laser pulses.

2. Theory of Stokes Growth

Ignoring scattering losses and second Stokes generation the growth of the Stokes wave is described by the following coupled equations in which I_ℓ and I_S are laser and Stokes intensities, z is the direction of laser propagation and g_S is the Raman Stokes gain and g_ℓ is $g_S k$ where $k = w\ell/ws$ [1,2].

$$\frac{dI_\ell}{dz} = -g_\ell I_\ell(z) I_S(z)$$

$$\frac{dI_S}{dz} = g_S I_\ell(z) I_S(z)$$

These equations can be solved for $I_S(z)$, the spatial growth of the Stokes plane wave. Integrating $I_S(z)$ over a Gaussian laser spatial profile

$$I_\ell(z,r) = I_\ell(z,0) \exp - (r/r_0)^2$$

yields the equation describing the growth of the Stokes power, $P_S(z)$.

$$P_S(z) = \frac{\pi r_0^2}{g_S kz} \left[\ell n\left\{ \frac{C + k \exp g_S I_\ell(0,0)z}{C + k} \right\} \right] \tag{1}$$

$$C = I_\ell(0)/I_S(0)$$

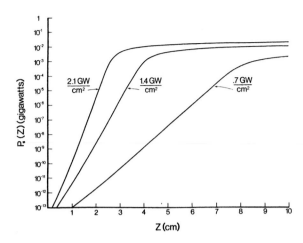

Fig.1 The calculated power of the Stokes pulse versus
distance along the optical axis for three different
initial laser intensities.

Figure 1 shows a plot of the Stokes power versus z from (1) for three
different initial laser intensities. Over much of its growth the Stokes
pulse increases exponentially. In the limit that the laser pulse depletion due to the Stokes growth is small and the region over which the
growth is observed, Δz, is small compared to the total distance that the
Stokes wave has traveled, the Stokes pulse grows exponentially.

$$\ell n P_s(z_2)/P_s(z_1) = g_s I_\ell(0,0)\Delta z \tag{2}$$

$$\Delta z = z_2 - z_1$$

We have measured the Stokes growth at several points along the z axis and
used (2) to determine the $g_s I_\ell(0,0)$ product. Knowledge of the Stokes
gain of the medium allows us to calculate the initial laser intensity
(GW/cm^2).

3. Experimental

A ten element beamsplitter using microscope cover glasses as optical elements was placed in the rectangular glass cell, as shown in Fig.2.
Because the difference in refractive index between glass and ethanol
(the sample) is small and because the beamsplitters are near Brewsters
angle, each plate reflects only 5×10^{-5} of an incident beam. The total
attenuation of the ten plates is not a significant loss to the incident
beams. A Reticon RL-1024S, 1024 element self-scanning diode array serves
as the detector in conjunction with a set of appropriate filters to
observe either the incident 532 nm, frequency-doubled, single mode-locked
Nd:YAG pulse, or the 630 nm Stokes pulse. The output of the diode array
was digitized and stored with a Biomation 805 transient recorder.

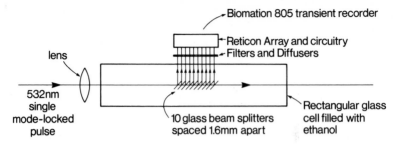

Fig.2 Apparatus for observing the spatial growth of Stokes pulse.

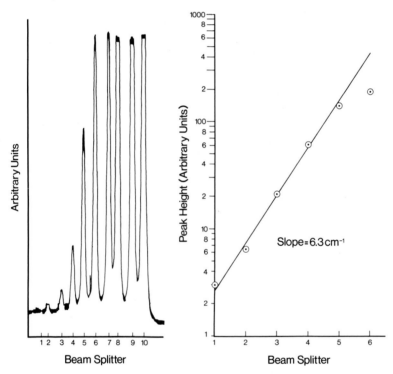

Slope=6.3cm⁻¹

Fig.3 Output of diode array showing Stokes growth.

Fig.4 Log of plots of Stokes power versus beamsplitter.

Figure 3 shows a typical Stokes growth pattern obtained with the multi-element beamsplitter (1.6 mm spacing). The flat-topped peaks from elements 7 through 10 indicate that the array has been exposed past its saturation limit. A log plot of the peak heights, corrected for small (10%) variations in beamsplitter efficiency, versus position along the optical axis is shown in Fig.4. The slope of the line is 6.3 cm⁻¹. Using the Stokes gain in ethanol, the initial laser intensity is 1.2 GW/cm². This is in reasonable agreement with a single pulse energy of a few tenths of a millijoule at 1.06 μ, 10% conversion to 0.532 μ, an estimated 30 picosecond pulsewidth and the approximate factor of 100 decrease in beam diameter due to focusing.

4. Discussion

Previous measurements of picosecond laser intensity require separate measurement of pulse energy, beam area and temporal width [4,5]. The technique described here represents a considerable simplification in laser intensity measurements. It is limited to pulses in the 1GW/cm^2 range and to pulse widths less than a few nanoseconds due to stimulated Brillouin scattering from longer pulses. Also the laser pulsewidth must be large compared to the molecular dephasing time to avoid transient effects [3].

The accuracy could be considerably improved with the use of an optical multichannel analyzer which is about 10^3 more sensitive than our Reticon array. This would permit measurement of the growth rate over a longer portion of the z axis and where it has less power but laser depletion is less important.

Self-focusing of the laser beam in the liquid was considered as a potential source of error. It could be detected by observing the diameter of the laser beam as it propagates through the liquid. The interference pattern produced by the cover glass beamsplitters required the use of a diffuser in front of the array (see Fig.2). This provided reproducible peak heights, but also prevented direct measurement of the focused beam diameter. A beamsplitter made from a thin sheet of mylar was used in place of the 10 element beamsplitter. Only a single mylar beamsplitter could be used because its reflectivity was about one hundred times higher than that of the glass beamsplitter and more than one beamsplitter would significantly attenuate the laser beam.

In order to calculate the focused beam diameter, the diameter of the 532 nm beam was measured with a diode array. When the Pockel cell in the single pulse selector was carefully aligned, the radial profile was found to be almost perfectly Gaussian ($r_0 = 0.075$ cm) with only small, less than 10% shot-to-shot variations. Slight misalignment of the single pulse selector produced irregular beam profiles with strong spatial modulation. Based on r_0 and a calculated divergence of 0.4 mrad the diffraction limited diameter of the focused laser beam should be 160 μ [6]. The focused laser beam diameter was measured throughout the confocal region and found to be almost perfectly Gaussian with a diameter of 180 ± 25 μ, in good agreement with the calculated value. The diameter of the Stokes beam could only be measured near the end of the confocal region and was found to be 200 μ. If self-focusing of the laser beam were present, much smaller beam diameters would be expected [7,8,9]. From the good agreement between the calculated and observed beam diameter we conclude that self-focusing is not important. The high-quality, reproducible TEM$_{00}$ pulses from the Nd:YAG laser are probably responsible for minimizing self-focusing effects.

It may be possible to use this technique to measure laser intensity, energy, beam diameter and pulsewidth (calculated from the first three parameters) simultaneously. The detector records energy, and with proper pellicle beamsplitters both Stokes growth rate (giving intensity) and beam diameter can be measured simultaneously. A single apparatus could therefore provide all the important parameters of a single mode-locked pulse.

This work is supported by the National Science Foundation.

References

1. D. von der Linde, M. Maier, W. Kaiser, Phys. Rev. _178_, 11 (1969).
2. W. Kaiser, M. Maier: In _Laser Handbook_, vol. II. ed. by F.T. Arecchi
    and E.O. Schulz-Dubois (North Holland, Amsterdam 1972).
3. A. Lauberau and W. Kaiser, Rev. Mod. Phys. _50_, 607 (1978).
4. D.J. Bradley, J. Phys. Chem. _82_, 2259 (1978).
5. E.P. Ippen and C.V. Shank: In _Ultrashort Light Pulses_, ed. by
    S.L. Shapiro, Topics in Applied Physics, Vol. 18 (Springer, Berlin,
    Heidelberg, New York 1977)
6. A.E. Siegman, _An Introduction to Lasers and Masers_ (McGraw-Hill, NY
    1971) Chapter 8
7. Y.R. Shen: In _Progress in Quantum Electronics_, Vol. 4, Part 1, ed.
    by J.H. Sanders and S. Stenholm (Pergamon Press, Oxford)
8. J.H. Marburger: In _Progress in Quantum Electronics_, Vol. 4, Part 1,
    ed. by J.H. Sanders and S. Stenholm (Pergamon Press, Oxford)
9. S.A. Akhmanov, R.V. Khokhlov, A.P. Sukhorukov: in _Laser Handbook_,
    Vol. II, ed. by F.T. Arecchi and E.O. Schulz-Dubois (North Holland,
    Amsterdam 1972)

Correction of Excite-and-Probe Spectra with Respect to Variations of Pump Pulse Energy

S. Belke, I. Kapp, W. Triebel, and B. Wilhelmi

Physics Department, University of Jena,
DDR-6900 Jena, German Democratic Republic

Excite-and-probe beam spectra can be applied to investigate the nature of excited molecular states as well as to determine kinetic parameters of these molecules [1] . Often many kinetic processes occur simultaneously and bring about a complex behaviour of the spectra. Furthermore the separation of the contributions of different transitions to the probe beam absorption can be obstructed by a spectral overlap of bands with different kinetics. In most of the experimental arrangements the probe beam transmission is not measured simultaneously for all wavelengths and delay times, and, besides, it is often necessary to repeat the transmission measurements to enlarge accuracy. Because the probe beam transmission is dependent on the pump beam parameters it is advantageous to correct the results with respect to pump pulse variations, if a sufficiently high stabilization of these parameters is not possible. Variations of the primary laser pulses are often enlarged by nonlinear optical processes applied to change the excitation frequency.

If the excitation pulse is short compared with all the molecular life times under consideration, it can be characterized by its energy only. To correct the probe beam spectra with respect to energy variations of the pump pulse one has to take into account the nonlinearity of the corresponding dependence. In the case of time independent rate constants the kinetic of the molecular system can be characterized by the following system of rate equations:

$$\frac{\partial}{\partial t} n_i(t) = \left[K_{ij} + I(t) \Sigma_{ij} \right] n_j(t) \quad , \tag{1}$$

n_i population density of level i,

K_{ij} relaxation matrix $\quad K_{ij} = \begin{cases} k_{ij} & \text{if} \quad i \neq j \\ -\sum_m k_{im} & \text{if} \quad i = j \, , \quad k_{ii} = 0 \, , \end{cases}$

Σ_{ij} matrix of single photon cross sections,

$I(t)$ excitation intensity (photon flux density).

In the case of δ-pulse excitation ($I(t) = E \cdot \delta (t-t_0)$)the solution of (1) found by Laplace transformation can be written in the form

$$n_i(t-t_0) = \Phi_{ij}(t-t_0) A_{jk}(E) n_k(t_0) \quad , \quad (t > t_0) . \tag{2}$$

(Dropping the condition of δ-pulse excitation the general solution of (1) can be obtained in form of a Volterra-development.) The probe beam transmission in dependence on the time coordinate $\eta = t - z/v$, delay time t_D and wavelength λ is given by

$$\tilde{T}(\eta, t_D, \lambda, E)/\tilde{T}_0(\lambda) = exp\left\{-\sigma_i(\lambda)\int_{z_0}^{z_1} dz\, n_i(\eta, z, E) + \sigma_i(\lambda)\, n_i(\eta_0)(z-z_0)\right\}, \tag{3}$$

where $\tilde{T}_0(\lambda)$ is the small signal transmission without excitation and $\sigma_i(\lambda) = \sum_m \sigma_{im}(\lambda)$. By inserting (2) into (3) the dependence of the probe beam transmission on molecular parameters and on the number of photons per area E is obvious. Procedures to correct changes in E can be obtained by solving (3) on appropriate conditions. Simple examples with analytical solutions are given where the energy transmission of the probe pulse T at excitation E_x is transformed into an energy transmission at normal excitation E_0. On the condition of weak excitation

$$T(t_D, \lambda, E_0)/\tilde{T}_0(\lambda) = \left[T(t_D, \lambda, E_x)/\tilde{T}_0(\lambda)\right]^{E_0/E_x} \tag{4}$$

holds. On the condition of a two-level system under arbitrary excitation

$$T(t_D, \lambda, E_0)/\tilde{T}_0(\lambda) = \left[T(t_D, \lambda, E_x)/\tilde{T}_0(\lambda)\right]^{\frac{E_0(z_0) - E_0(z_1)}{E_x(z_0) - E_x(z_1)}} \tag{5}$$

is found. Neglecting the z-dependence of the pump pulse energy

$$T(t_D, \lambda, E_0)/\tilde{T}_0(\lambda) = \left[T(t_D, \lambda, E_x)/T_0(\lambda)\right]^{\frac{1 - exp[-2\sigma_{12} E_0]}{1 - exp[-2\sigma_{12} E_x]}} \tag{6}$$

holds.

By using the normalized probe beam spectrum the spectral and kinetic parameters $\sigma_i(\lambda)$ and $\Phi_{ij}(t_D)$ can be evaluated now. On the condition of high saturation of the pump transition the dependence of T on E becomes small and vanishes finally as can be seen, e.g., from (6). This means, pump energy variations are of minor importance. Therefore it is often of advantage to work, if possible, at E-values high above the saturation value $E_s = 1/\sigma_{12}$.

As an example we give the probe beam spectra of excited copperphthalocyanine-sulfonate (CuPc) solved in water/acetonitrile mixture (Fig. 1a, b, c) and water (Fig. 1d), which stabilize the monomeric and the dimeric form, respectively, as well as the results of evaluation. In both the cases the absorption is caused by two transitions characterized by different kinetic behaviour (Fig. 1b, c, d). In these investigations $E/E_s \approx 10^2$ was used. The influence of energy variation was proved to be neglegible. While in the monomer these two absorption bands are clearly separated (Fig. 1c) a strong spectral overlap is obvious for the dimeric species (Fig. 1d). The dependence of the probe beam spectra on wavelength and time becomes more complex if a higher singlet level is primarily excited instead of the S_1 level [2] .

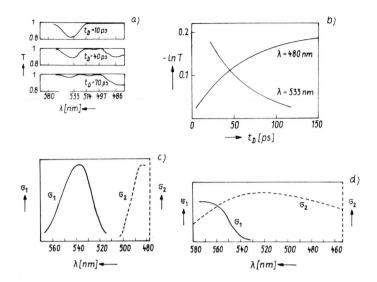

Fig. 1   Excited-state absorption of CuPc: a, b energy transmission of monomeric CuPc in dependence on wavelength
and delay time, respectively; c, d cross sections of
excited-state absorption of CuPc in the monomeric and
dimeric form, respectively

If the conditions concerning the molecular system are not known before, the needed knowledge can be obtained by comparison of calculated and measured dependences on E at one delay time. As an example we measured the energy transmission of probe pulses as a function of pump pulse energy for 2-(2',4'-Dinitro-α-brom-benzyl)-pyridine (Br-DNPB). This molecule is one of the ortho-Nitrobenzylarenes which undergo intramolecular proton-transfer [3] . The probe beam experiments are limited to regions of weak excitation where (4) could be applied. Fig. 2 shows the dependence of energy transmission of probe beam for Br-DNPB at two different delay times between excitation and probe pulses. Changes in energy of the excitation source (higher-harmonics) up to $\pm$ 30 percent causes changes in the population density of the primary excited state of the same order. By application of (4) it is possible to reduce the errors of energy transmission of the delayed probe beam. (In this example by a factor 4.) Fig. 2b shows the time dependence of probe beam transmission determined by application of (4) at each delay time. The errors caused by energy variation in the pump beam are decreased to a level below the errors of other experimental parameters.

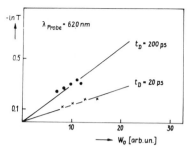

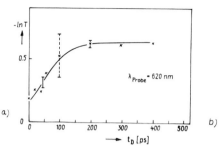

a)

b)

Fig. 2 Transmission of delayed probe beam in BR-DNPB in dependence on excitation energy (a) and delay time (b), respectively. The errors are given by I ; for comparison errors applying (4) are given by I

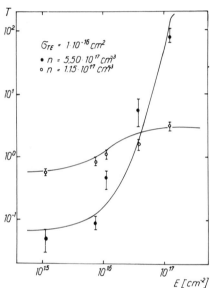

Fig. 3 Probe beam amplification T of BMC in dependence on the number of photons per area E of the pump pulse

Variations of the pump pulse energy can also disturb kinetic measurements by their influence on the probe beam transmission in spectral regions where amplification occurs. This is demonstrated in Fig. 3 for the polymethine dye bis-dimethyl-amino-heptamethine-perchlorate (BMC) solved in ethanol, pumped by the second harmonic radiation of a mode-locked Nd-glass laser and probed at the fluorescence maximum (λ = 542 nm). The maximal energy transmission of the probe beam is depicted in dependence on pump energy for two different dye concentrations. A system of rate equations for all population densities and for the photon number of pump and probe pulses has been solved numerically [4] and the solutions have been fitted to experimental results. The probe beam amplification increases strongly with pump energy at first and attains then a saturation level because of the finite number of excitable molecules (clearly to be seen for the smaller concentration in Fig. 3). The error arrows of the experimental points are the standard deviations of repeated measurements. The emission cross section σ_{TE} has been determined by a fitting procedure.

In connection with amplification experiments it should be noted, that variations of the probe beam energy can also affect the probe beam transmission. For high-emission cross sections of the dye the amplification will be reduced with increasing probe pulse energy, because of the depletion of the population inversion during the amplification process [4]. For this reason the probe energy has to be held below a critical value.

References

1. S.L. Shapiro (ed.): *Ultrashort Light Pulses*, Topics in Applied Physics, Vol. 18 (Springer, Berlin, Heidelberg, New York 1977)
2. B. Wilhelmi: in *Picosecond methods in spectroscopy of molecules, crystals and biological systems*, Tallinn 1978, p. 36 - 62
3. D. Klemm, F. Klemm, A. Graneß, J. Kleinschmidt, Chem. Phys. Lett. 55 (1978), 113; 55 (1978), 503
4. S. Belke, M. Fritsche, J. Herrmann and B. Wilhelmi, in: *Nelinenaja Optika*, Proc. of the Vavilov-Conference, Novosibirsk 1979, p. 99 - 115

Picosecond Phase-Conjugation Reflection and Gain in Saturable Absorbers by Degenerate Four-Wave Mixing

J.O. Tocho, W. Sibbett, and D.J. Bradley

Blackett Laboratory, Imperial College,
London, SW7 2BZ, England

More reliable mode-locking and bandwidth-limited pulses of shorter durations are obtained when the high-reflectivity mirror is immersed in the saturable absorber solutions used to passively mode-lock pulsed and CW dye lasers, Neodymium and ruby lasers [1]. This result had been previously explained [2,3] by preferential saturation of the absorber at the antinodes of the standing waves in the neighbourhood of the mirror, particularly when the dye cell length is comparable to the laser pulse length. With the recent demonstration of phase conjugation reflection in many materials by degenerate non-linear mixing [4] involving local saturation of absorption [5] it seemed timely to investigate picosecond phase-conjugation in DODCI and other saturable absorbers commonly used for mode-locking dye lasers. Our results indicate that, in addition to preferential saturation, phase conjugation effects contribute to the improved performance of the immersed mirror mode-locking dye cell, even when the laser beam is focussed on to the mirror as in the case of CW dye lasers [6]. Photoisomer effects [7] are also shown to be important in the phase-conjugation process, and play a dominating role as the laser is tuned to longer wavelengths.

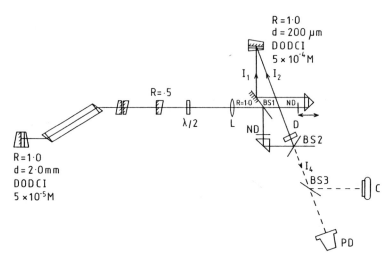

Fig.1 Experimental arrangement for extra-cavity phase conjugation

Experiment

A 1.2μs train of 50 μJ, 5ps pulses from a flashlamp pumped mode-locked dye
laser [1] was divided by a beamsplitter to generate pumping pulses of high-
intensity I_1 and low-intensity I_2 probe pulses, arranged to arrive simultaneously at the extra-cavity retroflecting cell by appropriate adjustment of
the prism optical delay lines. The pulses were focussed into the 200 μm
path length dye cell to give peak intensity of ∿500 MW cm^{-2} in an area of
∿1mm diameter. DODCI, DQTCI and Oxazine I dissolved in a range of solvents
(ethanol, methanol, and glycerol) were employed. The results were independent of the solvent used and phase conjugation reflectivity was produced by
all three dyes. Most of the work was carried out with DODCI.

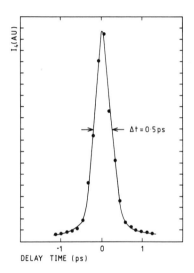

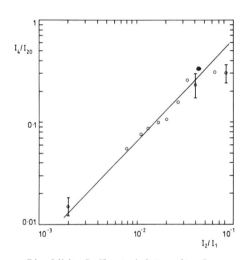

Fig.2(a) Reflected intensity
dependence on delay between
pump and probe pulses

Fig.2(b) Reflected intensity I_4
dependence on probe intensity I_2
(I_{20} = 36 MW cm^{-2}; I_1 = 440 MW cm^{-2})

Figure 2(a) shows the variation of reflected intensity for DODCI as a
function of the delay between the object and pump pulses. The half-width of
0.5ps, averaged over a complete pulse train, compares with a coherence time
of 0.22ps for the 2.3 nm total bandwidth. Self-phase modulation spectral
broadening along the dye lasers pulse train [7] accounts for the difference.
As expected there was a quadratic dependence of reflectivity upon pump intensity [8,9]. The output intensity increased linearly with the object wave
intensity up to ∿17 MW cm^{-2} when saturation set in at a power reflectivity
of ∿50% (Fig.2(b)). The dependence of reflectivity upon DODCI concentration
(Fig.3) is a good fit to the theoretical relation (exp - 2αd) (1 - exp (-2αd))^2
[9] with an optimum concentration of 5.5 x 10^{-4}M, the typical concentration
giving optimum mode-locking performance. Confirmation of wavefront phase
conjugation was obtained by correcting for the effect of a cylindrical lens
distorter. Phase-conjugate reflection was also obtained when the laser was
not mode-locked.

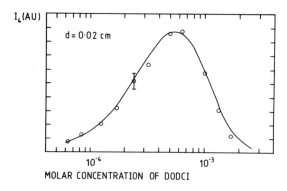

Fig.3 Reflected intensity dependence on DODCI concentration. Smooth curve
I_4 $\propto$(exp - 2αd) (1 - exp (- 2αd))^2, α = absorption coefficient, d = 0.2mm

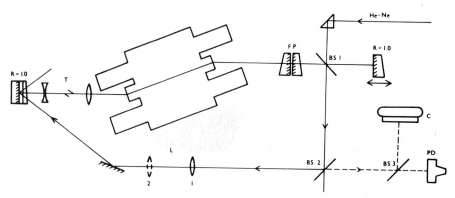

Fig.4 Intra-cavity phase conjugation experimental arrangement.

The experimental arrangement for intracavity phase conjugation is shown in
Fig.4. At a DODCI concentration of 2.5 x 10^{-4}M in a cell of 0.2 mm thickness
a reflectivity of ∿1% was obtained. Fig.5 shows the return beam profile when
(a) the cell mirror was at the focus of lens L; (b) the lens was displaced by
5 cm; the fluorescence patterns (c and d) for these two positions respectively
and (e) the probe beam reflected by a mirror. Inside the mode-locked laser
cavity, phase-conjugation reflection should be produced with greater fidelity
[10] because the pump beams and the effective probe beams experience the same
distortion in passing through the laser medium. In the arrangement of Fig.4
this was not necessarily the case since the pump beams will have made an extra
transit through the laser dye solution. Also, the temporal structure of the
pump beam will have been changed [1] by this extra transit through the
amplifying medium. As with the extra-cavity arrangement coincidence of the
pump and probe pulses to within the coherence time was needed to achieve
phase conjugation reflectivity. Thus intra-cavity phase-conjugation will
both correct for thermal distortion and tend to produce bandwidth-limited
pulses, since bandwidth-limited structures from the initial intensity
fluctuations [1] will be preferentially reflected in multiple passes through
the saturable absorber.

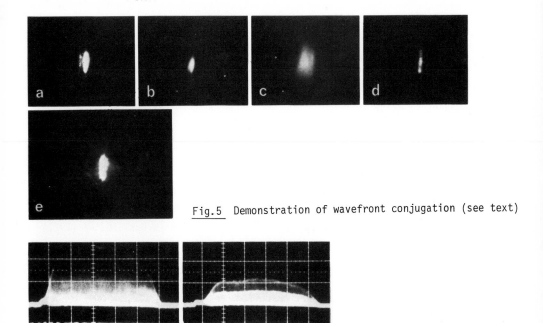

Fig.5 Demonstration of wavefront conjugation (see text)

Fig.6(a) Upper Pump mode-locked
pulse train (605 nm) Lower
Reflected pulse train.

Fig.6(b) As for Fig.6(a) but laser tuned
to operate at 617 nm. Note restructuring
of mode-locked pulse train envelope due
to generation of DODCI photoisomer and
subsequent phase-conjugation reflection.

The effect of photoisomer generation in DODCI [7] was clearly seen by a
reflected pulse delay of ∿500 ns, compared with the pump pulse, for a DODCI
(5 x 10^{-4}M) cell pumped by a 617 nm unmode-locked laser pulse. At 586 nm the
efficiency reached 150% with zero delay. That the delay at longer wavelengths
arises from photoisomer generation is confirmed by Fig.6. Fig.6(a) shows the
pump and reflected pulses for a 605 nm mode-locked pulse train, while Fig.6(b)
shows a drastic restructuring and delay (time scale 200 ns major division)
when the laser was tuned to 617 nm. At this longer wavelength phase conjugation would be more efficient for the DODCI photoisomer created at the beginning of the train, than for the normal form [1]. The delay in the build-up
of the photoisomer concentration could be manipulated for the production of
wavelength—dependent multiplexing for applications in real—time holography,
data and picture processing and wavelength filtering, all with picosecond
response times.

Thus with a tunable dye laser it is possible to produce phase conjugation with gain and with picosecond time response. By exploiting photoisomer effects wavelength-dependent variable delays and pulse envelope restructuring is obtained. Since phase conjugation requires coincidence within the coherence time, bandwidth-limited pulses will be preferentially produced by a retro-reflecting absorber dye cell. Amplification and phase conjugation in BDN saturable absorber by 13 ns pulses of a ND:YAG Q-switched laser and simultaneous intra-cavity Q-switching and phase-conjugation reflection has also recently been reported [11].

References

1. D.J. Bradley: Topics in Applied Physics 18, "Ultra-short light pulses" ed: S.L. Shapiro (Springer-Verlag, Heidelberg) 17-81 and refs.therein

2. D.J. Bradley, G.H.C. New and S.J. Caughey: Opt.Commun. 2, 41 (1970)

3. G.H.C. New: IEEE J.Quant.Electron.QE-10,115 (1974)

4. B.I.Stepanov, E.V.Ivakhin and A.S.Rubanov: Sov.Phys.Dokl.16, 46 (1971)

5. A.D. Kudriaytzeva et.al.: Opt.Commun. 26, 446 (1978)

6. I.S. Ruddock and D.J. Bradley: Appl.Phys.Lett. 29, 296 (1976)

7. E.G. Arthurs, D.J. Bradley and A.G. Roddie: Appl.Phys.Lett. 19, 480 (1971) 20, 125 (1972)

8. P.F. Liao and D.M. Bloom: Opt.Lett. 3,4 (1978)

9. R.L. Abrams and R.C. Lind: Opt.Lett.2, 94 (1978)

10. W.M. Grossman and D.M. Shemwell: J.Appl.Phys.51, 914 (1980)

11. E.I. Moses and F.Y.Wu: Opt.Lett. 5, 64 (1980)

Picosecond Resonance Raman Spectroscopy:
The Initial Photolytic Event in Rhodopsin and Isorhodopsin

G. Hayward, W. Carlsen, A. Siegman, and L. Stryer

Department of Applied Physics and Edward L. Ginzton Laboratory,
Stanford University, and
Department of Structural Biology and Sherman Fairchild Center,
Stanford University School of Medicine

1. Introduction

We report here the first picosecond resonance Raman studies of visual pigments. Flowing samples of rhodopsin and isorhodopsin are photolysed and
simultaneously probed by 30 psec pulses at 532 nm from a low-energy, high-
repetition-rate doubled Nd:YAG laser. At intensities sufficient to cause
moderate photoalteration during each pulse, both molecules exhibit Raman
bands in the 850 to 940 cm^{-1} region that are distinctive to bathorhodopsin
in a twisted all-<u>trans</u> configuration. The presence of these lines in our
resonance Raman spectra indicates that the photoisomerization of retinal
to its all-<u>trans</u> form must be essentially complete within picoseconds after
the absorption of a photon.

2. Background Information

The visual pigment rhodopsin consists of a light-absorbing molecule retinal
(similar to Vitamin A) buried inside a large protein opsin. The retinal
chromophore in rhodopsin has a strong absorption peak centered at ~500 nm
produced by the chromophore in the 11-<u>cis</u> configuration. The synthetic
pigment isorhodopsin, which is associated with the same retinal chromophore
in the 9-<u>cis</u> configuration, has a similar peak at ~490 nm. Absorption of a
photon by either pigment leads to the formation of a photolytic intermediate
bathorhodopsin, which has a strong absorption peak at ~540 nm, and which is
associated with the photoisomerization of the retinal chromophore from the
9-<u>cis</u> or 11-<u>cis</u> form to the all-<u>trans</u> form.

Earlier picosecond absorption studies have shown that following photolysis by strong visible light pulses the absorption lines characteristic
of rhodopsin or isorhodopsin are depleted, and the broad 540 nm absorption
band characteristic of bathorhodopsin appears, within times less than 6
psec. There has been considerable debate, however, as to the exact nature
of this very fast initial photolysis event. There has been doubt in particular as to whether the sizable atomic motion involved in the photoisomerization of retinal could take place on a picosecond time scale. Alternative
hypotheses involving very rapid proton transfer have also been advanced.

Fast-flow cw resonance Raman experiments have recently shown that certain
long-wavelength Raman lines at 853, 874 and 921 cm^{-1} are characteristic only
of fully formed bathorhodopsin, and that these lines appear within at least
a few μsec after photolysis, if not faster. It has also been confirmed that
these distinctive lines arise from a strained all-<u>trans</u> form of the retinal
chromophore (they do not appear, for example, in the protonated Schiff base
form of all-<u>trans</u> retinal).

We have now performed picosecond resonance Raman experiments on rhodopsin and isorhodopsin, using a new picosecond Raman technique, and have shown that these distinctive low—frequency lines characteristic of isomerized all-trans bathorhodopsin appear within less than 30 picoseconds following photolysis of either pigment. The appearance of these lines in our resonance Raman spectra appears to confirm that full isomerization to the all-trans form is complete within picoseconds after the absorption of a photon.

3. Experimental Procedure and Results

The strategy of our experiment is to illuminate the samples with relatively weak picosecond laser pulses of intensity just sufficient to cause small to moderate photolysis in the illuminated volume, at a repetition rate sufficient to accumulate a detectable signal from the very weak Raman scattering process within a reasonable period of time. The pulses were generated with a flash-pumped, electrooptically Q-switched and passively mode-locked Nd:YAG laser operated at 30 pulses per second. A single laser pulse selected from each train of pulses was doubled to 532 nm in an angle-matched CDA crystal. Samples of rhodopsin prepared from bovine retinas, or of isorhodopsin synthesized from opsin and 9-cis retinal, were flowed through the illuminated volume at a flow velocity sufficient to assure a fresh sample volume for each laser shot. The Raman spectra were collected by a Spex 1401 double Raman monochromator with a cooled photomultiplier. The laser, monochromator and data collection were controlled by an LSI-11 computer. Laser reference and Raman scattering signals were accumulated on each shot, and data from unacceptable laser shots were discarded.

The resulting resonance Raman spectra in the distinctive low-frequency range are shown in Fig.1. The spectra as shown ride on top of an essentially unavoidable fluorescent background of intensity approximately 3 times the largest peak shown. The fraction of molecules in the illuminated volume isomerized by a single laser pulse is of order $[1-\exp(-F)]$ where F is the photoalteration parameter. This parameter was limited in our high-flux spectra to a value $F \simeq 1$. This requires a photon flux less than $\sim 10^{16}$ photons/cm^3, or less than ~ 10 μJ into an illuminated volume ~ 350 μm in diameter and 2 mm in length. Low-flux spectra were taken at levels corresponding to $F \simeq 0.1$. Each spectrum is the sum of three scans from 800 to 1000 cm^{-1}, with 256 laser pulses per step in each scan. Spectrometer resolution was 6 cm^{-1} and step size was 3 cm^{-1}. Total bleaching of the recirculated sample during each run was less than 30%.

The spectrum of rhodopsin at very low flux levels (top curve) contains only a single peak which matches the distinctive 971 cm^{-1} line of unphotolysed rhodopsin. At higher flux levels, however, sufficient to cause significant photoalteration, the spectrum of rhodopsin (middle curve) exhibits new bands that correspond closely to the distinctive 853, 874 and 921 cm^{-1} lines known to be associated with all-trans bathorhodopsin. This spectrum also exhibits a shoulder at ~ 960 cm^{-1} which probably arises from isorhodopsin molecules produced (by reverse photolysis of bathorhodopsin) during the same 30 psec pulse. The high-flux spectrum of isorhodopsin (bottom curve) exhibits very similar Raman bands at the bathorhodopsin positions. In addition the 960 cm^{-1} band characteristic of isorhodopsin in this spectrum also has a shoulder near 969 cm^{-1} which is probably due to rhodopsin molecules formed during the pulse.

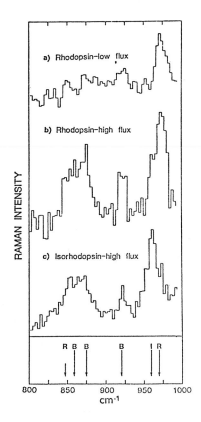

Fig.1 Resonance Raman spectra of flowing samples of rhodopsin and isorhodopsin photolysed and simultaneously probed by low-energy, 30 psec, 532 nm pulses. The arrows indicated positions of known Raman lines characteristic of rhodopsin, isorhodopsin, and strained all-_trans_ bathorhodopsin.

4. Conclusions

The presence of common bathorhodopsin lines in both these 30 psec spectra strongly supports the inference that a common isomerized bathorhodopsin intermediate is formed from either rhodopsin or isorhodopsin within picoseconds of the absorption of a photon by either pigment. This conclusion is reinforced by the finding of an isorhodopsin shoulder in the high-flux rhodopsin spectrum, and vice versa.

Further resonance Raman studies using separate photolysis and delayed probe pulses at different wavelengths are planned, to further elucidate the conformational changes in both the retinal chromophore and the host protein following photolysis. Despite the very much weaker signals associated with Raman scattering as contrasted to conventional pulse-probe absorption experiments, the very much greater structural detail revealed by the Raman signals appears to make this new picosecond technique promising and important.

5. Acknowledgements

The authors thank Sandra Slaughter for excellent technical assistance and Professor Richard Mathies for stimulating discussion. This research was supported by a grant from the National Eye Institute.

Index of Contributors

Lasers in Photomedicine and Photobiology

Proceedings of the European Physical Society, Quantum Electronics Division, Conference, Florence, Italy, September 3-6, 1979
Editors: R. Pratesi, C. A. Sacchi
1980. 108 figures, 20 tables. XIII, 235 pages
(Springer Series in Optical Sciences, Volume 22)
ISBN 3-540-10178-0

Contents: General Introduction to Photomedicine and Photobiology. – Photodynamical Therapy of Tumors. – Photodermatology. – Phototherapy of Hyperbilirubinemia. – Absorption and Flourescence Spectroscopy. – Raman and Picosecond Spectroscopy.

Laser Spectroscopy III

Proceedings of the Third International Conference, Jackson Lake Lodge, Wyoming, USA, July 4-8, 1977
Editors: J. L. Hall, J. L. Carlsten
1977. 296 figures. XII, 468 pages
(Springer Series in Optical Sciences, Volume 7)
ISBN 3-540-08543-2

Contents: Fundamental Physical Applications of Laser Spectroscopy. – Multiple Photon Dissociation. – New Sub-Doppler Interaction Techniques. – Highly Excited States, Ionization, and High Intensity Interactions. – Optical Transients. – High Resolution and Double Resonance. – Laser Spectroscopic Applications. – Laser Sources. – Laser Wavelength Measurements. – Postdeadline Papers.

Laser Spectroscopy IV

Proceedings of the Fourth International Conference Rottach-Egern, Fed. Rep. of Germany, June 11-15, 1979
Editors: H. Walther, K. W. Rothe
1979. 411 figures, 19 tables. XIII, 652 pages
(Springer Series in Optical Sciences, Volume 21)
ISBN 3-540-09766-X

Contents: Introduction. – Fundamental Physical Applications of Laser Spectroscopy. – Two and Three Level Atoms/High Resolution Spectroscopy. – Rydberg States. – Multi-

photon Dissociation,, Multiphoton Excitation. – Nonlinear Processes, Laser Induced Collisions, Multiphoton Ionization. – Coherent Transients, Time Domain Spectroscopy, Optical Bistability, Superradiance. – Laser Spectroscopic Applications. – Laser Sources. – Postdeadline Papers. – Index of Contributors.

V. S. Letokhov, V. P. Chebotayev

Nonlinear Laser Spectroscopy

1977. 193 figures, 22 tables. XVI, 466 pages
(Springer Series in Optical Sciences, Volume 4)
ISBN 3-540-08044-9

Contents: Introduction. – Elements of the Theory of Resonant Interaction of a Laser Field and Gas. – Narrow Saturation Resonances on Doppler-Broadened Transition. – Narrow Resonances of Two-Photon Transitions Without Doppler Broadening. – Nonlinear Resonances on Coupled Doppler-Broadened Transitions. – Narrow Nonlinear Resonances in Spectroscopy. – Nonlinear Atomic Laser Spectroscopy. – Nonlinear Molecular Laser Spectroscopy. – Nonlinear Narrow Resonances in Quantum Electronics. – Narrow Nonlinear Resonances in Experimental Physics.

Tunable Lasers and Applications

Proceedings of the Loen Conference, Norway, 1976
Editors: A. Mooradian, T. Jaeger, P. Stokseth
1976. 238 figures. VIII, 404 pages
(Springer Series in Optical Sciences, Volume 3)
ISBN 3-540-07968-8

Contents: Tunable and High Energy UV-Visible Lasers. – Tunable IR Laser Systems. – Isotope Separation and Laser Driven Chemical Reactions. – Nonlinear Excitation of Molecules. – Laser Photokinetics. – Atmospheric Photochemistry and Diagnostics. – Photobiology. – Spectroscopic Applications of Tunable Lasers.

Springer-Verlag
Berlin Heidelberg New York

Ultrashort Light Pulses

Picosecond Techniques and Applications

Editor: S. L. Shapiro

1977. 173 figures. XI, 389 pages
(Topics in Applied Physics, Volume 18)
ISBN 3-540-08103-8

Contents: S. L. Shapiro: Introduction –
A Historical Overview. – D. J. Bradley:
Methods of Generation. – E. P. Ippen,
C. V. Shank: Techniques for Measurement. –
D. H. Auston: Picosecond Nonlinear Optics. –
D. v. d. Linde: Picosecond Interactions in
Liquids and Solids. – K. B. Eisenthal: Picose-
cond Relaxation Processes in Chemistry. –
A. J. Campillo, S. L. Shapiro: Picosecond
Relaxation Measurements in Biology.

With the advent of picosecond light pulses a
decade ago numerous scientists recognized
that new methods of prime importance for ex-
ploring molecular interactions were feasible
and also that extremely rapid devices based on
new principles were possible. Now that the
basic fundamentals of picosecond technology
are well understood, and many of the early ex-
ploration goals have been realized, it is important to present the first comprehensive treatment by distinguished experts on both pulse
generation and pulse interactions with matter-
for like the spectroscopy field (where phenomena are understood through measurements
in the frequency domain), the picosecond
pulsefield (where phenomena are analyzed in
the time domain) is destined to be enduring
and everlasting. The book aims to summarize
the state of the art now that picosecond technology is rapidly emerging as a general tool in
many different disciplines. Students and
professionals in the sciences and engineering
will be fascinated by the new developments
and will profit from understanding these new
general techniques.

The basic fundamental concepts are introduced in the opening chapters. Broad applications in engineering, chemistry, physics, and
biology are described in later chapters.
Discussions are included of the most rapid
engineering devices developed to date.

Picosecond Phenomena

Proceedings of the First International Conference on Picosecond Phenomena Hilton
Head, South Carolina, USA, May 24–26, 1978

Editors: C. V. Shank, E. P. Ippen, S. L. Shapiro

1978. 22 figures, 10 tables. XII, 359 pages
(Springer Series in Chemical Physics,
Volume 4)
ISBN 3-540-09054-1

Contents: Interactions in Liquids and Molecules. – Poster Session. – Sources and Techniques. – Biological Processes. – Poster
Session. – Coherent Techniques and Molecules. – Solids. – High-Power Lasers and
Plasmas. – Postdeadline Papers. –

After more than a decade of active research, the
sudden availability of more sophisticated and
more powerful techniques has caused an
upsurge of interest in picosecond phenomena.
The first international conference of active research devoted to this subject was held in May
1978, bringing together scientists from such
diverse disciplines as chemistry, physics,
biology and engineering. These proceedings of
the conference provide graduate-level
students and researchers with a comprehensive and up-to-date survey of progress in
understanding picosecond phenomena theory
and experimentation.

Springer-Verlag
Berlin
Heidelberg
New York